D1447720

Chernobyl Record

Chernobyl Record

The Definitive History of the Chernobyl Catastrophe

R F Mould

British Library Cataloguing-in-Publication Data

A catalogue record for this book is available from the British Library.

ISBN 0 7503 0670 X hbk

Library of Congress Cataloging-in-Publication Data are available

Publisher: Nicki Dennis
Production Editor: Simon Laurenson
Production Control: Sarah Plenty
Cover Design: Victoria Le Billon
Marketing Executive: Colin Fenton

Published by Institute of Physics Publishing, wholly owned by The Institute of Physics, London

Institute of Physics Publishing, Dirac House, Temple Back, Bristol BS1 6BE, UK

US Office: Institute of Physics Publishing, The Public Ledger Building, Suite 1035, 150 South Independence Mall West, Philadelphia, PA 19106, USA

Typeset in TEX using the IOP Bookmaker Macros
Printed in the UK by J W Arrowsmith Ltd, Bristol

To
Maureen,
Fiona & Jane,
&
Imogen

Contents

List of Colour Plates xi

Preface xiii

Acknowledgments xv

About the Author xvii

1 Radiation Doses and Effects 1
1.1 Radiation quantities and their units of measurement 1
1.2 Radiation effects 5

2 Nuclear Reactors 12
2.1 Chain reaction 12
2.2 Reactor operation 13
2.3 RBMK-1000 nuclear power units at Chernobyl 15
2.4 Measures to improve the safety of RBMK plants 25

3 Explosion 27
3.1 Eyewitness accounts 27
3.2 Causes of the accident 32
3.3 Countdown by seconds and minutes 33
3.4 Damage to the power plant 39
3.5 Extinguishing the fire 45
3.6 Initial reports of the accident 45

4 Radionuclide Releases 50
4.1 Release sequence and composition 51
4.2 Atmospheric transport 53
4.3 Releases and exposed population groups 53
4.4 Residual activity in the global environment after 70 years 55
4.5 Chernobyl deposition compared to background deposition 55
4.6 Comparison with Hiroshima and Nagasaki 57
4.7 Comparison with Three Mile Island 59
4.8 Comparison with nuclear weapons testing in the atmosphere 60

4.9 Comparison with the Techa river area 64
4.10 Comparison with Tokaimura 69

5 Early Medical Response and Follow-up of Patients with Acute Radiation Syndrome 73

5.1 The first physician to arrive at the power plant after the accident 73
5.2 Medical examinations 75
5.3 Iodine prophylaxis against thyroid cancer 77
5.4 Acute radiation syndrome initial diagnosis 79
5.5 Follow-up of patients with acute radiation syndrome 79
5.6 Transplantation of bone marrow and embryonic liver cells 90
5.7 Case histories 93
5.8 Medical centre in Chernobyl town 94
5.9 Liquidator certificates for benefit entitlements 95

6 Evacuation and Resettlement 103

6.1 Evacuation zones and populations 103
6.2 Pripyat 107
6.3 Evacuation of livestock 108
6.4 Resettlement 110
6.5 Returnees 111
6.6 Radiation phobia 113

7 Sarcophagus 118

7.1 Meteorological, geological and seismic conditions of the site 118
7.2 Cooling slab to prevent contamination of the ground water 119
7.3 Construction of the Sarcophagus 120
7.4 Status of the nuclear fuel 126

8 Nuclear Power Past and Future 134

8.1 Before Chernobyl 134
8.2 After Chernobyl 136
8.3 Future of the Chernobyl nuclear power station including the Sarcophagus 143
8.4 Major radiation accidents 1945–99 147

9 Dose Measurement and Estimation Methods 153

9.1 Military personnel doses 153
9.2 Group dose and itinerary dose 155
9.3 Thyroid and whole-body dosimetry 155
9.4 Electron paramagnetic resonance dosimetry with tooth enamel 158
9.5 Dosimetry based on chromosome aberrations 164

10 Population Doses **170**
10.1 Natural and man-made background radiation 171
10.2 First year dose estimates in European countries other than the USSR 174
10.3 Collective effective doses for populations in Ukraine, Belarus and Russia 175

11 Contamination in Farming, Milk, Wild Animals and Fish **181**
11.1 Exposure pathway 181
11.2 Evaluation of dose from ingestion of food 183
11.3 ^{131}I contamination in milk 183
11.4 Contamination in wild animals and fish 184
11.5 Temporary permissible food contamination levels in USSR 187
11.6 International recommendations for intervention levels 190

12 Decontamination **193**
12.1 Decontamination sequence for the NPP area 194
12.2 Turbine hall roof 194
12.3 Radioactive waste disposal 195
12.4 Forests 196
12.5 Monitoring and decontamination of transport 198
12.6 Work of the chemical forces 198
12.7 Major problems 199
12.8 Post-accident studies on decontamination strategies 199

13 Water Contamination **203**
13.1 Countermeasures 203
13.2 Contamination levels 204

14 Ground Contamination **207**
14.1 ^{137}Cs contamination 209
14.2 ^{90}Sr contamination 212
14.3 ^{239}Pu and ^{240}Pu contamination 213
14.4 Exclusion zone area of the Ukraine 213

15 Psychological Illness **225**
15.1 Liquidators 225
15.2 Residents in contaminated territories 227
15.3 Atomic bomb survivors 232

16 Other Non-Malignant Diseases and Conditions **235**
16.1 Atomic bomb survivors 235
16.2 Brain damage *in utero* and mental retardation 237
16.3 Reproductive health patterns 238
16.4 Haematological diseases 241

16.5 Thyroid diseases 243
16.6 Ocular disease 245

17 Cancer Risk Specification 249
17.1 Absolute risk 250
17.2 Relative risk and excess relative risk 250
17.3 Attributable fraction 251
17.4 Mortality ratio 251
17.5 Confidence interval and confidence limits 252
17.6 Prediction modelling 253
17.7 How does radiation cause cancer? 256

18 Cancer 260
18.1 Incidence and mortality 260
18.2 Thyroid cancer 262
18.3 Leukaemia and other cancers 278

19 The Legasov Testament 287
19.1 My duty is to tell about this... 289
19.2 Defenceless Victor—Margarita Legasov's title of her reminiscences 303

20 Under the Star of Chernobyl 307
20.1 Origin of the names Polissya and Chernobyl 307
20.2 History 308
20.3 Culture 310

Glossary 322

References 332

Index 360

List of Colour Plates

Plate I. Aerial view of the nuclear power plant with the cooling pond in the background and the turbine hall in the foreground. (Courtesy: Chernobylintertinform.)

Plate II. Changes to the skin of a control room operator, four years after the accident. (Courtesy: H Wendhausen and G Kovács.)

Plate III. In Moscow Hospital No. 6, radiation injuries to four of the victims. (Courtesy: A Guskova.)

In Plate II and Plate III, contaminated vehicles, including fire engines, petrol tankers, bulldozers, lorries, military jeeps and helicopters, August 1991, all of which are permanently abandoned in fields within the 30 km zone. (Courtesy: P Pellerin.)

Plate IV. The first photograph of the core of the reactor glowing scarlet: from a military video. (Courtesy: Soviet Embassy, London.) Burning of one of the contaminated cow farms within the 30 km zone. (Courtesy: Chernobylinterinform.)

Plate V. Fuel masses within the Sarcophagus: (*top*) the *Elephant's Foot* and (*bottom*) debris containing the yellow material known as Chernobylite. (Courtesy: Chernobylinterinform.)

Plate VI. Paintings drawn in 1996 by adolescents who were evacuated children in 1986, depicting their perception of the accident and its consequences. *Top left* is from 17-year-old Olena showing a tree superimposed on the international radiation warning sign of a yellow and black trefoil. *Top right* from 16-year-old Katerina shows a black figure sitting on a bench by the river with a yellow thunderbolt coming from the top left. *Bottom left* from 15-year-old Olena shows a wolf howling into the night sky and *bottom right* is from 14-year-old Alla and shows a mother and child in front of a church and three gravestone crosses with a mythical black bird with a skull for a face, hovering over the church in a dark red sky. The children's playground in Pripyat, 1998. (Photographs: R F Mould.)

Plate VII. Wolves and wild boars in the Polissya Ecological Reserve in Ukraine between the rivers Dnieper and Pripyat and some 120 km from Chernobyl, 1999. (Courtesy: L K Sawicki.)

Plate VIII. ^{137}Caesium contamination map centred on the Chernobyl NPP. (Courtesy: International Atomic Energy Agency.)

Preface

The contents of *Chernobyl Record* have taken 14 years to compile and this period of time was necessary to enable information to be released from Soviet sources, measurements to be made in the environment, for estimation of radiation doses and for follow-up of the health of population groups which had been exposed. This time frame also includes the 10th anniversary conferences and the completion of joint projects of the European Commission, Ukraine, Belarus and the Russian Federation. It has also enabled me to visit the power plant site, Chernobyl town and Pripyat relatively soon after the accident and also some 10 years later: December 1987 and June 1998. Without such visits some of the photographs in this *Record* could not have been obtained.

Information is also contained in these pages of comparisons of various aspects of the Chernobyl accident with data from the Three Mile Island accident in the USA in 1979, the Hiroshima and Nagasaki atomic bombs, the highly contaminated Techa river area in the Urals in Russia and the accident in Tokaimura, Japan in 1999.

The first two chapters are introductory in that they describe terminology which is necessary for an understanding of the remaining chapters. Chapters 3–6 describes the early events: including those leading up to the explosion and then what followed in the immediate aftermath. Chapters 7–8 describe the Sarcophagus and the past and future of nuclear power for electricity generation, including the future of the Chernobyl power station.

Chapters 9–11 consider the radiation doses received by various populations, including liquidators, evacuees and those living on contaminated territories: and the contamination of milk by ^{131}I, and the contamination of other parts of the food chain by ^{137}Cs. Chapters 12–14 describe the environmental impact of the accident, as does chapter 11. Chapters 15–18 detail the long-term effects on health, including not only the incidence of cancer, but also of non-malignant diseases and conditions, such as psychosocial illnesses.

Chapter 19 is an English translation from *Pravda* of a short memoir entitled *My duty is to tell about this* by Academician Valery Legasov, the First Deputy Director of the Kurchatov Institute of Atomic Energy,

Moscow, who committed suicide on the 2nd anniversary of the accident, April 1988. Previously he had been one of the leading Soviet proponents of the nuclear power option for electricity generation. Chapter 20 records the local history and culture of Ukranian Polissya, the area which includes most of the 30 km zone.

What I have borne in mind throughout the research for this book, including the eye witness accounts, have been the words of Thomas Gradgrind in the Charles Dickens novel *Hard Times*: 'Now what I want to hear is facts'. This philosophy has, I believe, ensured that what follows is a balanced account of the accident and its aftermath, excluding media hype and biased accounts of self-interest groups, and debunking some of the myths which have surrounded Chernobyl. I have been encouraged in this aim by many people, both within the former USSR and in Europe and the USA, and hope that what I have produced is a valuable historical record.

Richard F Mould
Croydon, United Kingdom
April 2000

Acknowledgments

Any book which has taken 14 years to research and write owes a great debt of gratitude to many people and organizations and in particular I would like to thank Dr Gennadi Souchkevitch of Obninsk, Moscow and WHO, Geneva; and Mr Nikolai Pakhomov and Mrs Lyudmila Pakhomova of Moscow who in 1986 were the TASS Bureau Chief and TASS Photographic Librarian in London. These three friends more than anyone else helped me on my way in the early days after 26 April 1986.

The organizations to which I am most indepted include Chernobylinterinform: formerly AI Kombinat, Kiev and Chernobyl; Chernobyl Museum, Kiev; European Commission, Brussels; IARC, Lyon; IAEA, Vienna; NIST, Gaithersburg; Novosti Press Agency, London and Moscow; RERF, Hiroshima; TASS, London and Moscow; Soviet Television; UNSCEAR ,Vienna; UNESCO, Kiev; USSR State Committee for Atomic Energy, Moscow; and WHO, Geneva; all of whom have been very generous in supplying documents and illustrations.

Considerable assistance was given to me during the August 1986 first post-accident conference at the IAEA, Vienna where the Soviet Delegation was led by Academician Valery Legasov and Academician Leonid Ilyin. It is thanks to the latter for facilitating my December 1987 visit to Chernobyl, which included a meeting with the Director of the NPP, Mr Mikhail Umanets: later to be Minister of Energy in the Ukranian Government. It was also in August 1986 at the post-accident meeting in Vienna that Dr Angelina Guscova first gave me her gracious support for this publication, and this has continued to this day.

For my visit in June 1998 I would like to express my thanks to Professor Angelina Nyagu of the Association of Physicians of Chernobyl, to Dr Souchkevitch again, and to Ms Rimma Kiselitsa of Chernobylinterinform and to Dr Igor Obodovskii of the Research Centre for Radiation Medicine, Kiev.

My interview in 1987 with Colonel Leonid Telyatnikov, Fire Brigade Chief at the NPP at the time of the accident, could not have been organized without the assistance of Mrs Pakhomova; and without the help during my visit to Japan in 1995, of Professor Yutaka Hirokawa, Mr Toshikazu

Hosoda, Professor Itsuzo Shigematsu and Professor Shigenobu Nagataki I would have been unable to obtain such a wealth of follow-up data on the survivors of the Hiroshima and Nagasaki atomic bombs.

My publishers, Institute of Physics Publishing, have been unfailingly helpful, including my commissioning editor Kathryn Cantley, and particularly my production editor, Simon Laurenson, who has been extremely efficient and to whom all credit is due to meeting the deadline for the book's availability at the May 2000 Birmingham IOS Conference: *Imaging, Oncology, Science incorporating Radiology & Med X-Ray*.

The other friends and colleagues who have assisted me in various ways are almost too numerous to mention, and I apologise for any oversight if I have omitted anyone. It would be most unfair to acknowledge them in order of priority and I have therefore resorted to giving their names in alphabetical order. I am though, most grateful to all of them, without whom it would have been impossible to write this *Chernobyl Record*.

Dr Dmitri Afansyev, Dr Bernard Asselain, Dr Akio Awa, Dr Marite Bake, Dr Alexander Baranov, Dr Keith Baverstock, Professor Vladimir Bebeshko, Dr Burton Bennett, Professor Roger Berry, Dr David Bily, Academician Nikolay Bochkov, Dr David Brenner, Dr Werner Burkhart, Mr Dmitro Chukseyev, Mr Paul Colston, Professor Jean-Marc Cosset, Dr Bert Coursey, Dr Marc Desrosiers, Dr Franz Flakus, Ms Frances Fry, Dr Katerina Ganja, Mr Ralph Gibson, Dr John Gittus, Mr Vladimir Gubaryev, Mr Richard Guthrie, Dr Frank Hensley, Professor Yutaka Hirokawa, Dr John Hopewell, Professor Toshihiko Inoue, Dr Viktor Ivanov, Dr Bengt Johansson, Mr Yuri Kanin, Professor Alexander Kaul, Dr Andor Kerekes, Mr Sergei Koshelev, Professor György Kovács, Mr Alexander P. Kovalenko, Professor Alexander N Kovalenko, Mr Bernt Larsson, Dr Michael Lomakin, the late Lord Walter Marshal, Dr Linda Matisane, Dr Fred Mettler, Mr Tom Milne, Colonel Iourii Morozov, Professor Alexander Mrotchek, Dr Aleksey Okeanov, Dr Nataliya Okeanova, Dr Garnets Oksana, Mr Victor Orlik, Professor Pierre Pellerin, Dr Dale Preston, Dr Anatoly Prisyazhniuk, Dr Nicolae Racoveanu, Dr Yuri Riaboukhine, Dr Alexander Romanyuka, Ms Kate Sanders, Mr Leszek Sawicki, Dr Alla Shapiro, Dr Jaak Sinnaeve, Dr Anna Stabrawa, Ir Jan van der Steen, Professor John Surrey, Professor Hans Svensson, Dr Edward Towpik, Professor Anatoli Tsyb, the late Mr Anthony Tucker, Dr Istvan Turái, Academician Evgenyi Velikhov, Dr Vladimir Volodin, Dr Peter Waight, Dr Michael Waligórski, Dr Henning Wendhausen, the late Mr Valery Zufarov, Dr Tija Zvagule.

Last, but not least, I would like to thank my family for being so uncomplaining about having an author in their midst. This book is gratefully dedicated to the girls: my wife Maureen, my two daughters Fiona and Jane, and my grandaughter Imogen.

About the Author

Dr Richard Mould is an internationally known author and speaker and holds an MSc in nuclear physics and a PhD in cancer statistics, has worked as a medical physicist for some 30 years at the Royal Marsden and Westminster Hospitals, London, and has been involved with the Chernobyl accident from the time of the first post-accident meeting, IAEA, Vienna, in August 1986.

He is currently a Scientific Consultant travelling in north America, Europe, South-East Asia and the Far East giving Medical Statistics Courses, editing books and journals, and lecturing on the history of X-rays, radium, Marie Curie and Chernobyl. He also acts as a expert witness in medico-legal matters in the United Kingdom and has for many years been a consultant to WHO and IAEA who sent him to teach in Moscow 1980–83.

His IOPP published books include *Cancer Statistics* (1983), *Radiation Protection in Hospitals* (1985), *Radiotherapy Treatment Planning* 2nd edn (1985), *A Century of X-Rays & Radioactivity in Medicine* (1993), *Mould's Medical Anecdotes Omnibus Edition* (1996), and *Introductory Medical Statistics* 3rd edn (1998).

His first book on the Chernobyl accident was published in 1988 and translated into Japanese, he is a co-editor of the WHO Scientific Report on *Health Consequences of the Chernobyl Accident*(1996) and is on the Organizing Committee of the Chernobyl Conference to be held in Kiev in 2001 sponsored by the WHO and the Association of Physicians of Chernobyl.

Married with a son, two daughters, three grandsons and one grandaughter; he is also an Honorary Member of the Royal College of Radiologists, a Freeman of the City of London and an Honorary Colonel of the Commonwealth of Kentucky, USA.

The name Chernobyl/Chornobyl is from the Russian/Ukranian translation of the name of the herb *wormwood*, in Latin *Artemisia vulgaris*. The woodcut of *A. vulgaris* is from *Zielnik*. *Herbarzem z jezyka lacinskiego zowia et al* which when translated from old Polish is *Herbary that is a Description of Proper Name, Shape, Origin, Effect and Power of all Herbs*. By Szymon Syreniusz it was published in 1613 in Kraków and is now in the library of the Museum of Pharmacy of the Jagiellonian University, Kraków. (Courtesy: Muzeum Farmacji, Collegium Medicum, Uniwersytet Jagielloński.)

Chapter 1

Radiation Doses and Effects

Introduction

This chapter introduces concepts, quantities and terminology which are essential for an understanding of the radiation doses and effects which are detailed in subsequent chapters.

1.1 Radiation quantities and their units of measurement

An interest in radiation doses and effects arose at the end of the 19th century immediately after the discovery of x-rays in 1895 by Wilhelm Röntgen, of radioactivity in 1896 by Henri Becquerel, and of polonium and radium in 1898 by Marie and Pierre Curie. The early investigations concentrated on the measurement of the new rays, x, alpha, beta and gamma, and of proposals for quantities and associated units of dose, intensity, exposure, activity and strength. Such proposals were based on the physical, chemical and biological effects produced by the rays and included fluorescence, photographic film blackening, colour change of platino-barium cyanide, thermoluminescence, ionization and skin erythema[1].

All proposals had been studied by 1905 and some have withstood the test of time, such as ionization-based units, and are relevant to personnel dosimetry of the Chernobyl workers and to radiation surveys of environmental contamination. Thermoluminescence (TLD) and silver bromide film blackening effects have for many years been used as the basis for personnel dosimeters. However, some have been discarded such as the skin erythema unit, but this radiation effect is of particular importance to Chernobyl.

Erythema is reddening of the skin and may come in waves over a few weeks after doses in excess of several Gy (several hundred rad). The firemen and power plant workers who died within three months of the accident suffered severe erythema.

Until 1937, radiation units were separated into two distinct classes, those for x-rays and those for gamma rays, the latter mainly for radium and its daughter product radon, since artificially produced radioactive isotopes were only discovered in 1933 by Irene and Frédéric Joliot-Curie. This led directly to the use of radioactive isotopes other than radium and radon in diagnostic and therapeutic applications in medicine[1] and much later to the atomic bombs at Hiroshima and Nagasaki, and to nuclear power and the Chernobyl accident.

In this 1937 watershed year the International Commission on Radiological Units (ICRU) defined the *roentgen* as 'that amount of x or gamma radiation such that the associated corpuscular emission per 0.001 293 gram of air produces in air ions carrying 1 electrostatic unit of charge of either sign'. Further revisions occurred and in 1953 at the 7th International Congress of Radiology the roentgen was termed the special unit of a quantity called *exposure* and the *rad* was adopted as a unit of absorbed dose of any ionizing radiation and equal to 100 ergs/g for any absorber and not just human tissue.

Now with SI (Système Internationale) units, which have been increasingly used since the mid-1970s, the roentgen has virtually disappeared as a special name for a radiation unit and the SI unit of exposure is 1 coulomb/kg, which is equivalent to 3.876×10^3 roentgen (i.e. 1 roentgen $= 2.58 \times 10^{-4}$ coulomb/kg of air). However, some of the Soviet literature on Chernobyl has still retained the use of the roentgen. The SI unit for absorbed dose is the *Gray* (Gy) where the conversion to the former unit is given by 1 Gy = 100 rad.

1.1.1 Dose

Dose is the general term for a quantity of radiation, but since there are several types of defined dose, e.g. absorbed dose, collective effective dose equivalent, the term *dose* should always be used more precisely.

Exposure (sometimes referred to incorrectly as exposure dose) is a quantity relating to ionization in air and is used to describe a property of x-rays or gamma rays. Exposure does not describe the energy imparted to an irradiated material. That quantity is *absorbed dose* (D) where $D = \Delta E_{\mathrm{d}}/\Delta m$ and its special unit, prior to the introduction of SI units and the Gray, was the rad which, as stated above, equals 100 ergs/g. ΔE_{d} is the energy imparted by ionizing radiation to the matter (e.g. tissue) in a volume element and Δm is that volume element. Absorbed dose and exposure may be related by the formula: Absorbed dose = Exposure $\times f$, where f is a factor dependent on the quality of the radiation beam and the material being irradiated.

In an attempt to measure the maximum amounts of ionizing radiation which persons could safely receive, a quantity called the *relative biological effectiveness* (RBE) was proposed in which selected values of RBE were multiplied by the energy of the ionizing radiation per unit mass. The unit of RBE was the *rem*. However, with the introduction of the rad as a unit of absorbed dose, a quantity called *dose equivalent*, H, replaced the previous RBE dose in 1962, where H is weighted absorbed dose given by H=DQN. Here D is the absorbed dose, Q is one weighting factor which is a quality factor for the ionizing radiation (Q=1 for x-rays, gamma rays and electrons, Q=10 for neutrons and protons, and Q=20 for alpha particles) and usually N=1. The SI unit of dose equivalent is the *sievert* (Sv) where 1 Sv=100 rem. Thus 1 mrem = 10 μSv = 0.01 mSv.

The *collective effective dose equivalent*, which is sometimes termed *collective dose* or *collective dose equivalent*, is the quantity obtained by multiplying the average effective dose equivalent by the number of persons exposed to a given source of radiation. It is expressed in man-Sv units. Its use is illustrated from a commentary in *The Lancet* of 13 September 1986 when it was calculated that the 135 000 evacuees had received a total of 16 000 man-Sv (1.6 million man-rem) from external radiation alone, with some 25 000 of those living 3–15 km from the NPP receiving average doses of 350–500 mSv (35–55 rem). To place these figures in context, the average annual radiation dose to the United Kingdom population from all sources is less than 2 mSv and the annual dose limit for a radiation worker is 50 mSv (5 rem).

The *maximum permissible dose* (MPD) is a concept prior to dose equivalent and is the maximum amount of radiation that may be received by an individual within a specified time period with the expectation of no significantly harmful result. MPD is a regulatory concept and a dose above this level does not mean that harm has been done.

The *lethal dose* is a dose of ionizing radiation sufficient to cause death. The *median lethal dose* (or LD_{50}) is the dose required to kill, within a specified period of time (usually 30 days), half the individuals in a large group of organisms similarly exposed. The $LD_{50/30}$ for humans without medical treatment is about 4–4.5 Gy (400–450 rad).

The *committed tissue* or *organ equivalent dose*, termed *committed dose* or *dose commitment*, is recommended by the ICRP[2] for taking into account the time integral over time τ of the equivalent dose rate in a particular tissue which will be received by an individual following an intake of radioactive material. When the period of integration τ is not known, a period of 50 years is implied for adults and a period of 70 years for children.

1.1.2 ICRP recommended dose limits

The most recent ICRP recommended dose limits for occupational workers and for the public[2] are given in table 1.1.

Table 1.1. ICRP recommended dose limits[2]. These limits apply to the sum of the relevant doses from external exposure in the specified period and the 50 year committed dose (to age 70 years) from intakes in the same period.

Dose	Occupational dose limit	Public dose limit
Effective dose	20 mSv (2 rem) per year averaged over defined periods of 5 years with the further provision that it should not exceed 50 mSv (5 rem) in any single year	1 mSv (100 mrem) in a year but in special circumstances a higher value could be allowed in a single year provided that the average over 5 years does not exceed 1 mSv/yr (100 mrem/yr)
Annual equivalent doses:		
to the lens of the eye	150 mSv (15 rem)	15 mSv (1.5 rem)
to the skin	500 mSv (50 rem)	50 mSv (5.0 rem)
to the hands and feet	500 mSv (50 rem)	—

1.1.3 Activity

The quantity activity is a measure of the amount of radioactivity in a sample of a radioactive isotope (*radionuclide*) and the first proposal for the *curie* (Ci) as a unit of activity was at the 1910 Congress of Radiology when it was defined only for use with radon, as 'the quantity of radon in equilibrium with 1 gram of radium'.

Later it was extended to include all radionuclides as 'a unit of activity which gives 3.700×10^{10} disintegrations per second'. In SI units the basic unit of activity is the *becquerel* (Bq) which equals one disintegration per second and thus

$1 \text{ Ci} = 37 \times 10^9 \text{ Bq}$ or $1 \text{ Bq} = 27.03 \times 10^{-12} \text{ Ci}$
$2 \text{ mCi} = 74 \text{ MBq}$
$100 \text{ mCi} = 3.7 \text{ GBq}$
$1000 \text{ Ci} = 37 \text{ TBq}.$

Contamination densities for the environment around Chernobyl use both units and are generally expressed as either Ci/km^2 or Bq/m^2, although sometimes the unit of *hectare* (ha) or acre is used for the area. The relationships between m, km, ha and acre are:

$1 \text{ ha} = 10\,000 \text{ m}^2 = 2.4711 \text{ acres}$

$1\ \text{km}^2 = 1\,000\,000\ \text{m}^2 = 100\ \text{ha}.$

The radionuclide releases from the accident are given in the literature either as multiples of Bq or multiples of Ci. The internationally accepted prefixes are listed in table 1.2. Thus, for example, the total activity level[3] of the 800 radioactive waste sites within the 30 km zone is approaching 15 PBq, and the total release of radionuclides[4] from the accident is 1–2 EBq which included 630 PBq of ^{131}I and 70 PBq of ^{137}Cs. Prefixes for the smaller factors are used, for example, when stating doses received by a population, thus for an external dose estimate for the populations of Ukraine, Belarus and Russia[3] an appropriate unit is μSv per kBq ^{137}Cs per m^2.

Table 1.2. Prefixes.

Factor	Prefix	Symbol
10^{18}	exa	E
10^{15}	peta	P
10^{12}	tera	T
10^{9}	giga	G
10^{6}	mega	M
10^{3}	kilo	k
10^{2}	hecto	h
10^{1}	deka, deca	da
10^{-1}	deci	d
10^{-2}	centi	c
10^{-3}	milli	m
10^{-6}	micro	μ
10^{-9}	nano	n
10^{-12}	pico	p
10^{-15}	femto	f
10^{-18}	atto	a

1.2 Radiation effects

1.2.1 Acute radiation syndrome

When the whole body, or a major part of it, is exposed to a large acute dose of penetrating radiation (gamma rays, x-rays or neutrons) a pattern of disease develops known as the acute radiation syndrome. The underlying cause of the pattern is the radiosensitivity of three organs, all of which play an essential part in sustaining life. According to the dose delivered, the haematopoietic tissues, the lining of the small intestine and finally the central nervous system are affected. The more of the trunk included in the exposure the worse will be the illness, because of the location there of both the small intestine and a large part of the haematopoietic tissues.

There are four degrees of *radiation syndrome*, also termed *radiation sickness*, which were defined in 1987 for Chernobyl in terms of absorbed dose[5].

1st degree: less than 1 Gy
2nd degree: 1–4 Gy
3rd degree: 4–6 Gy
4th degree: 6–16 Gy

but the dose ranges for 1st to 4th were revised later[6a] to be 0.8–2.1 Gy, 2.2–4.1 Gy, 4.2–6.3 Gy and 6.4–16 Gy, and currently[6b], at least in the Ukraine, they are stated as 1–2 Gy, 2–4 Gy, 4–6 Gy and 6–10 Gy.

Typical characteristics of those with 4th degree of ARS were early primary radiation reactions, in the first 15–30 minutes after exposure and at 7–9 days, vomiting and damage to the digestive tract. The death of 19 Chernobyl cases with 4th degree ARS occurred in the range 14–91 days post-accident, and of seven cases with 3rd degree ARS in the range 14–49 days[7]. Of those who have 1st–3rd degree ARS and survived, some of their symptoms never disappear. These include chronic tiredness.

The severity of radiation burns of the Chernobyl cases have also been defined in four levels[8]. An example of late effects after the accident are shown in figure 1.1 (see also figure 1.2) for a 27 year old man who experienced 3rd degree ARS and 3rd degree radiation burns involving a chronically infected ulcer which required conservative therapy.

1st degree: Painful skin reddening, particularly at the sites of the burns.
2nd degree: Blister formation with sloughing of the skin at the sites of the burns.
3rd degree: Blisters and ulcers at the sites of the burns.
4th degree: Necrosis of affected tissues.

A more recent discussion of the classification and terminology of radiation injuries is to be found in the 1999 volume of the official journal of the Association of Physicians of Chernobyl[6a]. This makes the point that there is still no standardized terminology for radiation injuries developing due to local exposure, and in many cases, no common agreement between terminology used in ICD and in literature devoted to ARS. It is also noted that the dose values for degrees of ARS are often rounded to the nearest integer[9] and instead of being defined as 1st to 4th are classified as follows.

Mild: 1–2 Gy
Moderate: 2–4 Gy

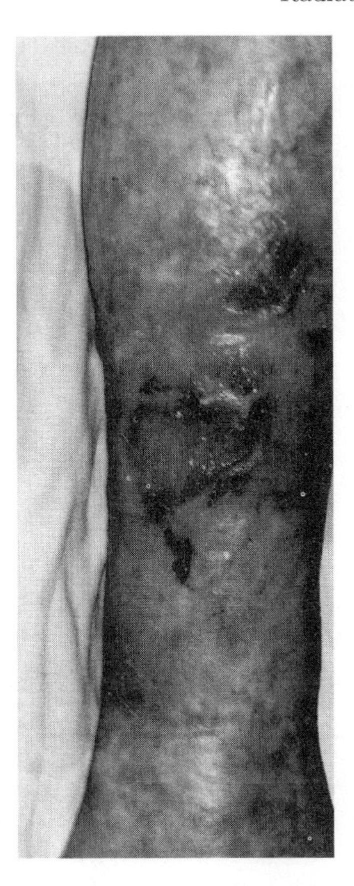

Figure 1.1. Late radiation effects in a 27 year old man with 3rd degree acute radiation syndrome and 3rd degree radiation burns. He was a member of the general public and not a power station worker. The previous evening and in the early morning of 25 April, when the explosion occurred, he was fishing in the power station's cooling pond[8]. (Courtesy: A N Kovalenko and D O Bily.)

Severe: 4–6 Gy
Very severe: 6–8 Gy
Lethal: > 8 Gy

with the onset of symptoms being, respectively, greater than 30 days, 18–28 days, 8–18 days, less than 7 days and less than 3 days and the lethality being respectively, 0%, 0–50% (onset 6–8 weeks), 20–70% (onset 4–8 weeks), 50–100% (onset 1–2 weeks) and 100% (onset 1–2 weeks)[9].

The medical management and treatment for mild ARS is recommended as outpatient observation for a maximum period of one month, but for

Figure 1.2. Satellite picture of the Chernobyl NPP cooling pond. Units No. 1–4 and the site of the unfinished Units No. 5–6 are at top left. (Courtesy: TASS.)

moderate to lethal ARS as hospitalization and isolation as early as possible. Symptomatic therapy only is given for lethal ARS[9].

Bone marrow transplantation (BMT) treatment for the most highly irradiated Chernobyl cases received much reporting in the media and seems to be a logical treatment for victims of accidental whole body irradiation which is sufficiently high to make spontaneous bone marrow recovery impossible. However, BMT has many limitations (table 1.3) and was largely unsuccessful after Chernobyl (see section 5.6). Information now strongly suggests that BMT has only a limited role in treatment, would benefit only a small number of exposed individuals, and should be given only to those receiving doses in the range 8–12 Gy. In the most recent severe accident, on 30 September 1999 at Tokaimura, Japan, see section 4.10, three workers received estimated doses of 10–20 Gy, 6–10 Gy and 1.2–5.5 Gy. Only the most highly irradiated person was initially reported as having a BMT[10] but this was not correct, see section 4.10.

Table 1.3. Limitations of BMT[9].

• Identification of histocompatible donors
• Age constraints
• Human lymphocyte antigen typing in lymphogenic patients
• The need for additional immuno-suppression
• Risk of graft *versus* host disease

1.2.2 Combined radiation injuries

Combined radiation injuries (CRIs) occur whenever radiation effects are combined with mechanical, thermal or chemical injuries, table 1.4. Lethality increases significantly in the presence of CRIs and diagnosis, treatment and prognosis are much more complex than with ARS cases.

Table 1.4. Classification of combined radiation injuries[9].

• Thermal CRI: external and/or internal irradiation with thermal burns
• Mechanical CRI: external and/or internal irradiation with wound or fracture, or with haemorrhage
• Chemical CRI: external and/or internal irradiation with chemical burns or chemical intoxication

1.2.3 Somatic effects

These radiation effects are limited to the exposed individual, and should be distinguished from genetic effects, which also affect subsequent unexposed generations. Large radiation doses can cause somatic effects that are fatal. Lower doses may make the individual noticeably ill, may produce temporary changes in blood cell levels detectable only in the laboratory, or may have no detectable effects. Somatic effects are also sometimes referred to as *non-stochastic* or *deterministic* effects.

Examples of radiation exposure thresholds quoted by UNSCEAR[4] are given in table 1.5 and it is also noted that many organs are damaged by doses in the range 10–20 Gy.

Table 1.5. Radiation exposure thresholds for selected somatic effects[4].

Temporary sterility in males	Single short exposure of 0.15 Gy or prolonged exposure of about 0.4 Gy/year
Permanent sterility in males	3–5 Gy acute exposure (in women this would be 2.5–6 Gy) or 2 Gy/year chronic exposure
Whole bone marrow acute exposure	0.5 Gy produces clinically significant depression of blood formation and 3–5 Gy results in death in 50% of cases
Skin exposure	3–5 Gy produces erythema and dry desquamation with symptoms appearing 3 weeks post-exposure. 20 Gy produces moist desquamation with blistering appearing about 1 month post-exposure
Tissue necrosis	50 Gy
Nephritis	14 Gy

The effects are also classified into two groups, early or *acute effects*, and late or *chronic effects*. Examples of early effects are erythema, blood changes and epilation, table 1.6. Chronic effects include cataract formation, fibrosis, organ atrophy and sterility.

1.2.4 Stochastic effects

These are effects for which the probability of the effect occurring, rather than the severity of the effect, if it occurs, varies with the size of the radiation dose. For such effects, as in the induction of genetic effects or cancer, for many years it has always been assumed that there is no threshold dose below which some effect may not occur. Stochastic effects refer to hereditary or carcinogenic effects of radiation, including the induction of thyroid cancer and leukaemia. These contrast with the effects of high dose radiation exposures, for example, radiation skin burns.

Table 1.6. Latent phase of ARS[9].

	Mild	Moderate	Severe	Very severe	Lethal
Diarrhoea	None	None	Rare	Appears on days 6–9	Appears on days 4–5
Epilation	None	Moderate on day 15 or later	Moderate or complete on days 11–21	Complete earlier than day 11	Complete earlier than day 10
Lymphocytes (g/litre) days 3–6	0.8–1.5	0.5–0.8	0.3–0.5	0.1–0.3	0.0–0.1
Granulocytes (g/litre)	> 2.0	1.5–2.0	1.0–1.5	< 0.05	< 0.01

1.2.5 Low-level radiation exposure and the zero-threshold linear model

This section is included to note that the Main Commission of the ICRP is now considering a simpler approach than previously, based on the concept of what is being called *controllable dose*. This represents a shift in emphasis from societal oriented criteria using *collective dose*[11]. This has been brought about by much discussion as to the applicability of the linear dose response model for low-level doses incorporating a zero-threshold into the model[12–16].

Chapter 2

Nuclear Reactors

Introduction

This chapter provides an introduction to the RBMK reactor units at the Chernobyl nuclear power plant (NPP), which is necessary before chapter 3 which details the events which led up to the explosion and the subsequent radioactive releases into the atmosphere. Major components of a unit are illustrated: turbine hall, central reactor hall, control room and the dosimetry measurement (i.e. monitoring) recording laboratory after the accident which has had to replace the previous system, now destroyed, in the control room. For further reading on the nuclear fuel cycle, reactor operation and types of nuclear reactor, the book by Patterson[1] is recommended.

2.1 Chain reaction

The nuclear chain reaction in ^{235}U is the basis for electricity generation using nuclear reactors, as well as for nuclear weapons such as the atomic bomb at Hiroshima, although a ^{239}Pu chain reaction is also possible such as in the atomic bomb at Nagasaki.

In a mass of uranium there are always a few stray neutrons, produced either by spontaneous fission or by cosmic rays. If one of these stray neutrons produces fission of a ^{235}U nucleus, then, as well as the two fission products, two or three high energy neutrons will also be produced. *Prompt neutrons* are those that emerge at the instant of fission and the probability of this occurring is better than 99:1. However, there is a slight chance that a neutron will not emerge until some seconds later, this is a *delayed neutron*.

The three possibilities open to a high energy fission neutron are that it reaches the surface of the material and escapes; it strikes another nucleus and is absorbed without any breakdown of the nucleus or it strikes another nucleus and causes this nucleus to rupture. This third possibility, of induced fission, depends on the energy of the neutron and on the nucleus it strikes. Occasionally a *fast neutron*, fresh from an earlier fission, will rupture a nucleus, indeed only a fast neutron can rupture a nucleus of ^{238}U. However, if the neutron hits several nuclei, one after the other, giving up some of its energy at each collision, it soon slows down and becomes a *slow neutron*, often termed a *thermal neutron.*

A thermal neutron takes much longer to traverse a nucleus than a fast neutron and is thus much more likely to cause fission in a nucleus of ^{235}U. This radionuclide has three fewer neutrons than ^{238}U and in naturally occurring uranium this nuclide with a mass number of 235 forms only 0.7% of the ore.

If there are enough ^{235}U nuclei close together then the neutrons can induce more and more fissions, releasing more and more neutrons. This is termed a *chain reaction*, which, when out of control as in an atomic bomb, will cause a nuclear explosion, but when controlled can form the basis for nuclear reactor design.

Reactor fuel contains *enriched uranium*, by which it is meant that there is more than 0.7% of ^{235}U but even a small increase, of say 2–3%, can make a marked difference, provided that there are sufficient thermal neutrons. To ensure this, the core of the reactor contains not only the uranium fuel but also a *moderator*. This is a material with light nuclei, such as hydrogen or carbon, which are used in the form of water or graphite.

2.2 Reactor operation

Reactor fuel is sealed in casings termed *cladding*, which confines the fission products which are produced. Assemblies of sealed fuel are termed *fuel elements* and are interspersed with moderators, and also with neutron absorbers such as boron, to control the reaction. These are the *control rods* which in the Chernobyl accident were not inserted quickly enough to stop the explosion.

The region in which the chain reaction occurs is termed the *core* of the reactor. When the reactor becomes *critical* with the establishment of a self-sustaining chain reaction, each neutron lost by causing fission is replaced by exactly one neutron, prompt or thermal, which does likewise. The dependence of the chain reaction on thermal neutrons permits a gradual adjustment of the reaction rate.

Removing an absorber, that is, a control rod, out of a stable chain reaction, is called adding *reactivity*, the neutron density increases and the rate of the chain reaction increases. Inserting an absorber is termed adding *negative reactivity* and produces a reverse effect.

However, before removing the control rods sufficient for the reactor to become critical, precautions must be taken against gamma-radiation and neutrons pouring out of the core, since they can, for example, depending on their energy, travel through metres of concrete. The reactor must therefore be surrounded by enough concrete or other protective material to cut down the radiation level outside. In many types of reactor installations this is achieved by a specially designed *containment building*, such as at Three Mile Island, where the presence of a containment building limited the effects of that disaster. For the RBMK-1000 reactors at Chernobyl there was no such containment building and the *biological shield* of 2000 tonnes on the top of the reactor space was blown out of position and came to rest at an angle of 15° to the vertical.

Normal start-up and shutdown of a reactor are both lengthy processes and may take many hours. If, however, it is necessary to stop the chain reaction, for instance in the event of a malfunction, the emergency shutdown is termed a *scram*. At Chernobyl the emergency scram button was pressed on the orders of the shift foreman at 01:23:40 hours on 26 April 1986 but by then this had no effect and could not stop the explosion which occurred at 01:23:44 hours, due to the rapid increase in power which is estimated to have been 100 times full power.

If an operating reactor is left to itself its reaction rate will gradually fall, in part because of the build-up of fission products which absorb neutrons. One of the most effective fission produced neutron absorbers is ^{135}Xenon and the phenomenon is termed *xenon poisoning*. ^{135}Xe is produced by the decay of ^{135}Tellurium and ^{135}Iodine which are generated for several hours after start-up. ^{135}Xe nuclei which fail to capture a neutron undergo beta decay into ^{135}Caesium. If the chain reaction rate remains constant then the average concentration of ^{135}Xe in the core also remains constant. The half-life of ^{135}Xe is 6.7 hours.

Nuclear fuel eventually has to be replaced because the ^{235}U reduces as fission occurs. The replacement procedure is termed *refuelling* and the *spent fuel*, is then stored. Currently some of the highest radiation dose rates inside the Sarcophagus are in the room where the spent fuel still remains. There are many different fission products which occur, including those of plutonium radionuclides with mass numbers 239, 240 and 241: it was ^{239}Pu which was the fissile material for the atomic bomb dropped on Nagasaki. However, not all fission products are solids, some are gaseous such as xenon and krypton, all of which escaped into the atmosphere at Chernobyl.

Complete fission of all the nucleii in one kilogram of ^{235}U would release energy totalling one million kilowatt-days. The nuclear fuel in the core is so arranged that the heat is given off gradually enough to keep the temperatures manageable. The amount of heat given off per unit volume in a reactor core is termed the *power density*.

Heat is removed from the reactor by pumping a heat-absorbing fluid through the core past the hot fuel elements. This fluid is the *coolant* and may be a gas such as air or carbon dioxide, or a liquid such as water. The cooling system can be open-ended or designed as one or more closed circuits. A closed circuit can be pressurized. For electricity generation the heat released can generate steam to run turbines, as with the Chernobyl power station.

There are several different reactor types, see table 2.1, of which the Chernobyl RBMKs are of a pressure tube design in which the fuel elements lie in vertical pressure tubes filled with light water, as distinct from heavy water, and are surrounded by a graphite moderator.

Table 2.1. Reactor types.

Gas cooled power reactors	e.g. Magnox reactors and advanced gas cooled reactors
Light water reactors	e.g. Pressurized water reactors boiling water reactors
Heavy water reactors	e.g. CANDU (Canadian-deuterium-uranium) reactors
Fast breeder reactors	e.g. Liquid metal FBRs

2.3 RBMK-1000 nuclear power units at Chernobyl

2.3.1 History of RBMK reactors

The history of RBMK type reactors in the Soviet Union had been, until 1986, very successful. After the early development of the system, the USSR went directly to full scale 1000 MW(e) units. The first RBMK-1000 was put into service at Leningrad in 1974. The Leningrad, Kursk and Chernobyl power stations each have four units built in pairs, each unit supplying two 500 MW(e) turbogenerators. The first two (of four) units are operating at Smolensk and two more were being constructed at Chernobyl at the time of the accident.

The first of the two larger, 1500 MW(e), versions of these reactors was put into service at Ignalina, Lithuania, in 1984. Its physical size is similar to that of the RBMK-1000 but it has a fuel power density 50% higher.

In safety terms, there had, before 1986, been practical demonstration that the design can handle significant faults. For example, at Kursk nuclear power station in January 1980, a total loss of station internal load occurred that was sustained satisfactorily, and there have been a number of feedwater system transients. None of these presented severe plant safety problems. Electricity production for the period 1981–85 at the Chernobyl NPP was 106.6×10^9 kilowatt-hours.

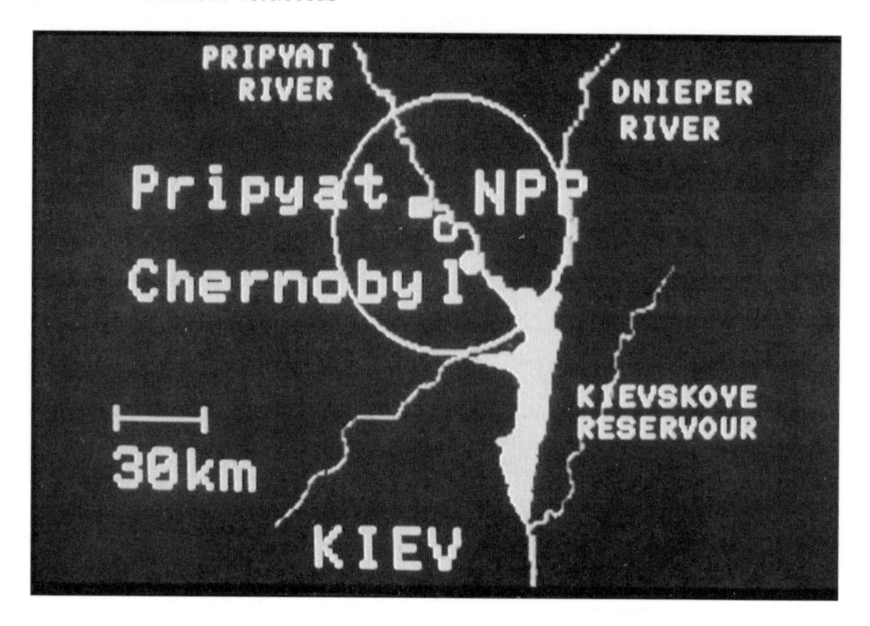

Figure 2.1. Map of the site of the power plant in relation to the city of Kiev and the Kiev reservoir[2]. (Courtesy: USSR KGAE.)

2.3.2 Location of the Chernobyl NPP site

The Chernobyl NPP is situated in the eastern part of a large region known as the Belarussian–Ukranian woodlands, beside the 200–300 m wide river Pripyat which flows into the Dnieper. Figure 2.1 is a map of the immediate area surrounding the NPP including the 30 km exclusion zone. The NPP's cooling pond is linked to the Kiev reservoir. Minsk, the capital of Belarus with a population of 1.3 million, is 320 km from the NPP, and Kiev, the capital of Ukraine with a population of 2.5 million, is 146 km from the NPP. The regional centre is the 12th century town of Chernobyl with a population of 12 500 in 1986, situated 15 km south-east of the NPP. Nearer to the NPP, only 3 km distant, is the town of Pripyat where 45 000 power plant workers and their families lived. The population of all Belarus is 10 million, including 2.3 million children, and of Ukraine is 60 million including 10.8 million children under the age of 15 years.

Figure 2.2 is a map of a wider area and shows the capital cities of Ukraine, Belarus, Poland, Austria, Hungary, Yugoslavia and Romania. One of the earliest concerns was the possibility of contamination of the river Dnieper, all its tributaries, and eventually the Black Sea.

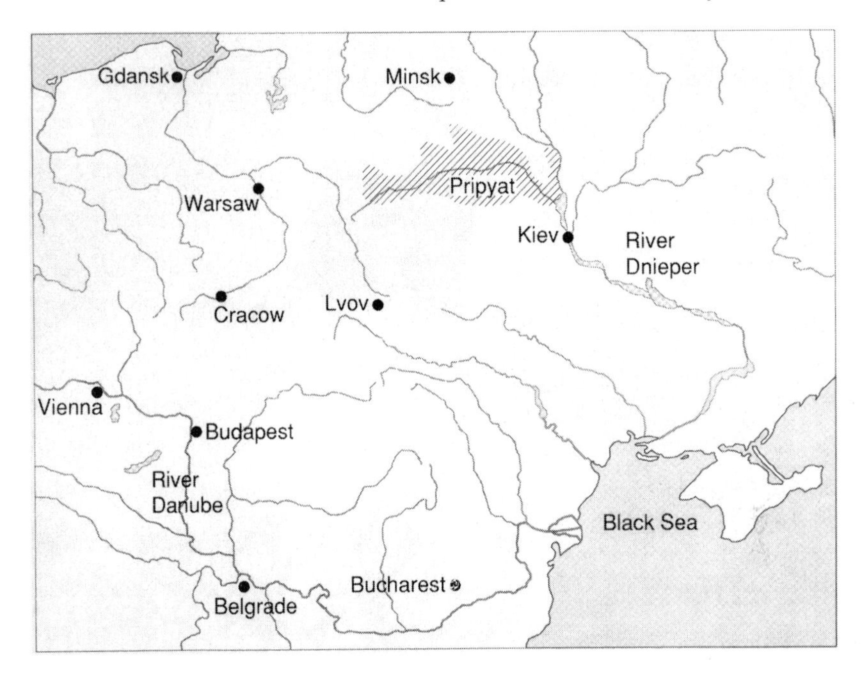

Figure 2.2. Map of Kiev and the Pripyat marshes in relation to Belarus, Poland, Austria, Hungary, Yugoslavia, Romania and the Black Sea.

2.3.3 Construction plans

Construction was planned for three stages with each stage comprising two RBMK-1000 units. The first stage of Units No. 1 and No. 2 was constructed between 1970 and 1977, and the second comprising Units No. 3 and No. 4 was completed in late 1983. It was Unit No. 4 which exploded. In 1981 work was begun on the construction of two more units also using RBMK-1000s, at a site 1.5 km to the south-east of the existing site. They were almost completed for commissioning when the accident occurred but were immediately abandoned and have been left to rust.

2.3.4 Overall view, central reactor hall and turbine hall

Plate I is an aerial view of the NPP taken five months after the accident. The cooling pond is in the background, the tall chimney in the centre is a ventilation stack and was contaminated from top to bottom, the shattered reactor Unit No. 4 is clearly seen and just in front of it, the long white building houses the turbine hall on which damage to the roof is shown. The yellow painted turbines can just be seen through the hole in the roof.

It was on this roof that many of the firemen who died received their high radiation doses. All the forests in this photograph were contaminated and had to be cut down.

Figure 2.3. Turbine hall, November 1982. (Courtesy: TASS.)

Figure 2.3 is a photograph of the turbine hall in 1982 taken at a celebration of the 60th anniversary of the USSR, and figure 2.4 is the central reactor hall of Unit No. 1 in June 1986. The small squares in its centre are the covers to the heads of the fuel rods: these covers were reported to have been blown into the sky at least 1 km when the explosion took place.

2.3.5 Design features

The design features of RBMK nuclear reactors, of which there were four originally operational at the Chernobyl NPP, are well described by INSAG[3] and by UNSCEAR[4].

A cross-section view of a typical unit at Chernobyl NPP is seen in figure 2.5. Each reactor in a pair supplies steam to two 500 MW(e) turbines. The

Figure 2.4. Central reactor hall of Unit No. 1, June 1986. (Courtesy: TASS.)

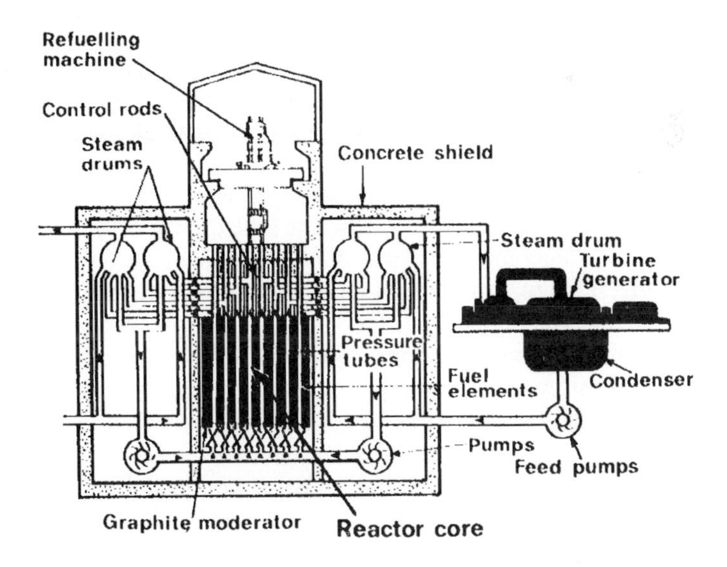

Figure 2.5. Cross-sectional view of the RBMK reactor nuclear power unit[2]. (Courtesy: USSR KGAE.)

two reactors, together with their multiple forced circulation circuits, are located in separate blocks, between which are installed auxiliary systems, and the turbine generator room, figure 2.3, is common to two reactor units. It houses four turbogenerators and associated systems.

An RBMK-1000 reactor is a graphite-moderated light-water cooled system with uranium dioxide (UO_2) fuel in 1661 individual vertical channels. The geometrical arrangement of the core consists of graphite blocks 250 mm $\times$ 250 mm, 600 mm in height, stacked together to form a cylindrical configuration 12 m in diameter and 7 m high. The mass of the graphite moderator is 1700 tonnes. It is located in a leaktight cavity formed by a cylindrical shroud, the bottom support cover and the upper steel cover. In the accident the bottom cover dropped 4 m leaving a gap through which molten fuel could travel. This was not initially realized and the search for the *missing fuel* took a considerable time. Working in the nearest room to the reactor it took 18 months to drill through the adjoining wall. Oil industry engineers were the drillers and the work was completed in October 1988. To their surprise the reactor room was empty. The next approach was to use, because of the high dose rates, a remote controlled device which consisted of a child's toy tank costing 15 roubles, to which a camera was strapped: this was also unsuccessful in locating the fuel. Eventually it was found that this 4 m gap was present and nuclear fuel masses, the lava, were finally located.

Each graphite block has a central hole which provides the space for the fuel channels, thus forming a lattice pitch of 250 mm. Fuel and control rods channels penetrate the lower and upper steel structures and connect to two cooling systems below and above the core. The drives of the control rods are located above the core below the operating floor shield structure.

The fuel, in the form of UO_2 pellets, is sheathed with a zirconium–niobium alloy. A total of 18 fuel pins, approximately 3.5 m in length are arranged in a cylindrical cluster of which two fit on top of each other into each fuel channel. Fuel replacement is done by a refuelling machine located above the core. One to two two fuel channels can be refuelled each day.

The coolant system consists of two loops and the coolant enters the fuel channels from the bottom at a temperature of 270°C, heats up along its upward passage and partially evaporates. The wet steam of each channel is fed to steam drums, see figure 2.5, of which there are two for each cooling loop.

The separated dry steam, with a moisture content of less than 0.1%, is supplied via two steam pipes to two turbines, while the water, after mixing with the turbine condensate, is fed through 12 downcomers to the headers of the main circulation pumps. The condensate from the turbines enters the separators as feedwater, thereby sub-cooling the water at the main circulation pump inlet. The circulation pumps supply the coolant to headers which distribute it to the individual fuel channels of the core.

The coolant flow of each fuel channel can be independently regulated by an individual valve in order to compensate for variations in the power distribution. The flow rate through the core is controlled by circulation pumps. In each loop four pumps are provided, of which one is normally on standby during full power operation.

From the fission reaction approximately 95% of the energy is transferred directly to the coolant. 5% is absorbed within the graphite moderator and mostly transferred to the coolant. The latter part of the fission energy is transferred to the coolant channels by conduction leading to a maximum temperature within the graphite of approximately 700°C. A gas mixture of helium and nitrogen enhances the gap conductance between the graphite blocks and provides chemical control of the graphite and pressure tubes. The control and protection system in the RBMK reactors has the basic functions listed in table 2.2.

Table 2.2. Basic functions of the RBMK control and protection system[3].

- Regulation of the reactor power and reactor period in the range 8×10^{-12} to 1.2 times full power
- Manual regulation of the power distribution to compensate for changes in reactivity due to burnup and other effects
- Automatic stabilization of the radial-azimuthal power distribution
- Controlled power reduction to safe levels when certain plant parameters exceed preset limits
- Emergency shutdown under accident conditions

The system includes 48 measuring devices. These are 24 ionization chambers placed in the reflector region which are used to drive three banks of automatic regulation rods and 24 fission chambers which are in-core detectors located in the central openings of the fuel assemblies which are used to drive the local automatic controllers. There are 211 absorbing rods in the core which are functionally grouped, table 2.3.

When the reactor is started up, the 24 emergency protection rods are the first to be raised to the upper cut-off switches. The speed of the control rods is 0.4 m per second. When a control rod is disconnected from its drive, which is necessary in the case of a power loss, the speed is about 0.4 m per second driven by free fall. Flow resistance precludes a higher velocity. The highest level of emergency is Level 5, which results in the insertion of all the rods (except the 24 shortened absorber rods) into the core up to the lower cut-off switches. The over-power trip set point is set at present power plus 10% of nominal power. The system includes the measurements and subsystems listed in table 2.4. For a full description of the safety systems including the emergency core cooling system, see INSAG[3], and for

Table 2.3. Functional grouping of the 211 absorbing rods[3].

• 24 shortened absorbing rods	
• 24 auto-control rods	12 local auto-control (LAC): regulation rods in 12 zones. 12 average power control: 3 banks of 4 rods per bank.
• 139 manual rods and 24 emergency rods	24 emergency control: uniformly selected. 24 local emergency protection (LEP): 2 rods per zone. 115 manual control.

a summary of the IAEA's co-operative programme for consolidating the technical basis for further upgrading the safety of RBMKs see the 1996 paper by Lederman[5].

Table 2.4. Measurements and subsystems for the RBMK reactor[3].

- Flow rates in all the fuel channels and the control channels: 1661 plus 223 points.
- The temperatures of the graphite core and metal structures: 46 plus 381 points.
- A system of monitoring the main components of the forced circulation system, such as the drum separators, the main circulation pumps and the suction and pressure headers.
- A system for monitoring the power distribution: 130 radial plus 84 axial.

2.3.6 Control room

Major areas of a nuclear power unit, besides the reactor, include the reactor central hall, the turbine hall, and the control room of the unit, which were similar for all four units. The control room is shown before and after the accident in figures 2.6 and 2.7.

2.3.7 Principal specifications

The principal specifications[4] are given in table 2.5.

Figure 2.6. Control room of Unit No. 1, December 1987. (Photograph: R F Mould.)

Figure 2.7. Damaged control room of Unit No. 4, June 1998. (Photograph: R F Mould.)

Table 2.5. Principal specifications of the Chernobyl Unit No. 4 reactor.

Thermal power	3200 MW
Fuel enrichment	2.0%
Mass of uranium in fuel assembly	114.7 kg
Fuel burn-up	20 MW d/kg
Maximum design channel power	3250 kW
Isotopic composition of unloaded fuel	
^{235}U	4.5 kg/t
^{236}U	2.4 kg/t
^{239}Pu	2.6 kg/t
^{240}Pu	1.8 kg/t
^{241}Pu	0.5 kg/t

Figure 2.8. Radiation monitoring inside the Sarcophagus, October 1986. (Courtesy: V Zufarov.)

Figure 2.9. Measurement recording laboratory, June 1998. (Photograph: R F Mould.)

2.3.8 Measurement recording laboratory

Before the accident all the dosimetric and temperature monitoring information required by the power plant operators of Unit No. 4 would have been available in the control room. However, after the accident all the measuring systems were destroyed and had to be replaced.

The initial monitoring was performed manually, as there was no alternative, figure 2.8, but eventually bore holes were made for three types of sensors: those for the measurement of neutron dose rates, gamma dose rates and temperature. The laboratory in which these measurements are recorded is shown in figure 2.9 and seen to be rather basic. In June 1998 the maximum gamma dose rate was 4000 roentgen/hour: recorded in the spent fuel storage pond. The maximum temperature in June 1998 was recorded as 40°C. In 1986 the temperatures were some 300–400°C.

2.4 Measures to improve the safety of RBMK plants

Since the accident, several organizational and technical measures have been developed and implemented to improve the safety of operating RBMK plants. These have been reported to INSAG and table 2.6 summarizes the aims of these measures[6].

Table 2.6. Aims of improvement measures for RBMK plants[6].

• Reducing the positive steam (void) coefficient of reactivity and the effect on reactivity of complete voiding of the core. This has been provided by the installation of additional fixed absorbers, up to 90, into the core, and through the introduction of the use of fuel with 2.4% ^{235}U enrichment.
• Improving the speed of the scram system. The speed of insertion of control and safety rods has been increased with the time for full insertion into the core reduced from 18 s to 12 s.
• Introducing new computational codes for the operational reactivity margin (ORM) with numeric indication of the ORM in the control room. This has been increased to between 43 and 48 control rods, depending on the reactor.
• Precluding the possibility of bypassing the emergency protection system while the reactor is at power, through an operating limit requirement and the introduction of a two key system for the bypass action.
• Avoiding modes of operation leading to reduction of the departure from nuclear boiling margin for the coolant at reactor inlet. This addresses the question of adequate subcooling at the core inlet. Operating instructions have been updated to take into account lessons learned from the accident and among the new provisions is one which now sets a lower limit of 700 MW (th) for steady operation of an RBMK reactor.

Figure 2.10. The Chernobyl NPP was named after V I Lenin and his bust remains in front of the administration building in 1998. One of Lenin's quotes is 'Russia is communism and electrification'. (Photograph: R F Mould.)

Chapter 3

Explosion

Introduction

Reliable eyewitness accounts of a catastrophe supplement the later objective data and can provide a real sense to the reader of *being there* which cannot be captured in any other way, even by photographs. In the case of Chernobyl not many eyewitnesses were in the area outside the NPP at the time of the explosion in the very early hours of 26 April 1986 and there are relatively few records from the power plant workers who survived and the firemen. Other eyewitness accounts are given in chapter 5 from some of the physicians who worked as *liquidators* in the early days of May 1986.

The term *liquidator* is a special one, used to describe the emergency accident workers, EAWs, who were involved in the cleanup operations and the healthcare delivery during the period 1986–89. The estimated number varies according to the definition of a liquidator/EAW and the source used, but drawn from all parts of the USSR they may have numbered as many as 650 000[1]. This chapter also describes the work of the liquidators who were helicopter pilots and firemen who fought under extremely high radiation dose conditions to extinguish the fires in the central reactor hall and on the roof of the turbine hall.

3.1 Eyewitness accounts

3.1.1 A schoolgirl

Only two eyewitness accounts at the time of the explosion have been located apart from the general comments on page 38 relating to time *01:23:48* on 26 April. The first was from a 16-year-old schoolgirl, Natasha Timofeyeva, who, with her relatives, was returning home late from a visit to friends. Her small village of only 55 houses was Chamkov in the Gomel region of Belarus, and was only 6 km from the NPP, on the opposite bank of the river Pripyat. It was quite dark, and she recorded[2] that she saw 'a bright flash over the most distant chimney of the power plant'.

3.1.2 A Polish surgeon and radiation specialist

A more informative account is from a Polish surgeon, Edward Towpik of the Institute of Oncology in Warsaw and the editor of a memorial book for Marie Curie and her work with the discovery of radium[3]. On the night of 25 April he had been hunting black grouse on the Polish–Belorussian border, not too far from Chernobyl, where one has to lurk in the middle of the night, in a special ditch covered with branches, to be prepared for the shoot at daybreak.

'The dawn was surprisingly purple-red and not a single black grouse appeared at the shooting ground.' He began to 'feel sick with fever and an intensive sore throat, just like very acute laryngopharyngitis' and was so ill that he had to return to Warsaw. After his return, everything ceased quickly without any medication. Soon afterwards he learned what happened, from Western radio stations, of course, as the communist media remained silent. Immediately the most sought after medicine in Poland was iodine solution.

3.1.3 Firemen

The initial six firemen, who were on duty at the NPP and who fought the blaze right from the start, all died. The following two eyewitness accounts are from their colleagues Private Andrei Polovinkin and Sergeant Ivan Shavrei who were on backup duty and who were interviewed for a special memorial issue of *Izvestia*[4a].

From Polovinkin:

> We arrived at the scene of the accident in 3–5 minutes and started to turn the fire engine and to prepare for extinguishing ... I went onto the roof of the turbine generator twice to pass on the brigade leader's order: how to deal with it. I would personally like to place on record a favourable mention of Lieutenant Pravik who knowing that he had received severe radiation burns still went and found out everything down to the last detail.

From Shavrei:

> Alexsandr Petrovskii and I went up onto the roof of the machine room; on the way we met the kids from the Specialized Military Fire Brigade No. 6; they were in a bad way. We helped them to the fire ladder, then made our way towards the centre of the fire where we were to the end, until we had extinguished the fire on

> the roof. After finishing the job we went back down, where the ambulance picked us up. We too, were in a bad way.

In a 1990 interview[4b] in Moscow by Dr Fred Mettler of the University of New Mexico, with a 24-year-old fireman, Mr Irmolenko, it was quite clear when asked 'What were the circumstances of your exposure?' that the firemen were given no proper advice whatsoever on radiation protection procedures when they arrived at the NPP some two and a half hours after the explosion.

> My fire brigade of about 12–13 firemen was called to the nuclear station at 4am. We were positioned about 100 metres from the wall of the main reactor building and were told to be ready in case we were needed. We remained in that spot most of the morning without being asked to do anything. At noon we ate lunch and again waited. At about 3pm several of us began to experience nausea and vomiting but suspected this was due to food poisoning. At about 4pm we were allowed to go home and told to return the next morning.
>
> Our brigade returned next day to the same spot and again waited, but before noon several again developed nausea and vomiting. A physician came by (assumed by Dr Mettler to have probably been Academician Leonid Ilyin) and immediately evacuated them for medical care. I was subsequently diagnosed with first degree acute radiation syndrome and recovered with supportive care.

On being asked 'If you could tell firemen in other countries one thing about your experience, what would it be?'. The answer was 'If there was a radiation accident and they have nausea and vomiting they must leave immediately. Nobody even told me that in any training I ever had'.

3.1.4 A radiation monitoring technician

The number of deaths in the first three months were 31 but of these, one was a reactor operator Valery Khodemchuk whose body was never recovered and is entombed in the debris, and one was a NPP worker Vladimir Sashenok who died in Chernobyl Hospital from thermal burns within 12 hours of the accident. Sashenok's death has been described[5] to Yuri Scherbak (author[6], Ukranian Ambassador to the USA in 1996, and in 1988, founder of the Ukranian Green Movement) by Nikolai Gorbachenko, a radiation monitoring technician at the NPP whose shift began at midnight on 25 April.

Gorbachenko was in the duty room drinking tea at this time and not in Unit No. 4 as it was in the process of being shut down as part of the experiment which was being carried out. He heard

> two flat and powerful thuds, the lights went out including the lights on the control panel and it was just as in a horror film. The blast blew open the double doors and black-red dust starting coming out of the ventilation vent. The emergency lights then went on and we put on our gas masks. My boss sent me to Unit No. 4 to find out what was happening.
>
> Two workers entered and asked us to help find one of their comrades: Vladimir Sashenok who had been missing for 30 minutes and was supposed to be in the upper landing across from the turbine room. Everything was a shambles on this landing, steam was coming out in bursts and we were up to our ankles in water. Suddenly we saw him lying unconscious on his side, with bloody foam coming out of his mouth making bubbling sounds. We picked him up by the armpits and carried him down. At the spot on my back where his hand rested I received a radiation burn. Sashenok died without regaining consciousness at 6am.

3.1.5 A control room operator

Oleg Heinrich, a control room operator, related his experiences[7,8] when in Germany in 1990 visiting relatives and taking the opportunity to ask for a hospital check-up for cancer at the Kiel University Hospital, as cancer was a great worry to him. This is his story. Born in April 1960, and therefore aged 26 when the accident happened, he was working in the control room on a second eight-hour shift (because he needed the money) with another operator, an older man.

He was sleeping in a room next to the control room, which was a room with no windows, when the explosion occurred. His older colleague was crying, the window in the control room had broken, he had received a heat burn, the lights had gone out and he was looking for the stairs. Those on the right-hand side of the room were destroyed but on the left they were still usable. Oleg had recently attended a lecture on radiation protection and because of this he took a shower and a change of clothing. His colleague did not, and instead, went to see what had happened, and subsequently died.

Oleg ended up some days later in Moscow Hospital No. 6 where he received skin transplants for his burns, but no bone marrow transplant. Plate II includes a series of four previously unpublished photographs from the Kiel University Hospital case notes[8] showing the post-irradiation skin changes of Oleg Heinrich some four years after the accident.

3.1.6 An operating shift chief

This account[5] is from someone knowledgeable about nuclear physics.

> It seemed as if the world was coming to an end. I could not believe my eyes. I saw the reactor ruined by the explosion. I was the first man in the world to see this. As a nuclear engineer I realized all the consequences of what had happened. It was a nuclear hell. I was gripped by fear.

3.1.7 An air force colonel

Colonel Anatoli Kushnin, when interviewed by a journalist from *Literaturnaya Gazetta*[9], stated that there were 80 helicopters and airplanes of various types deployed in Chernobyl and that he was responsible for the radiation safety of the staff. One of his orders to the helicopter pilots was that they should cover the floors of their machines with lead.

> By 4 May the pilots had buried the reactor core in sand despite conditions that were difficult and dangerous. The dosimetric devices on these helicopters measured radiation levels up to 500 roentgens per hour. In the first days after the accident these dosimeters went off scale. The crews were exposed to enormous radiation doses during their flights over the reactor. The military test pilot Anatoly Grischenko died in 1995 in the United States. He was the one who tried to lift a huge dome over the exploded reactor with the biggest helicopter in the world, the MI-26. He didn't succeed, but he was exposed many times to huge doses of radiation. He wasn't even told about that for a while.

3.1.8 The scientific advisor to President Gorbachev

Evgenii Velikhov[10] (now Director of the Kurchatov Institute of Atomic Energy, Moscow, who in 1986 was one of the Deputy Directors) was told by Nikolai Ryzhkov, the Prime Minister of the USSR to go to Chernobyl to try and find out what had happened. He left the next day expecting to stay for three days but remained for one and a half months. On 6 May from a helicopter he had his first view of the damaged reactor through the holes in the shield and by the light of the burning parachutes which contained the materials (silicates, dolomite and lead) intended to put out the fire. 'I could see no reactor in sight, this was very embarrassing for me as nobody believed me. The problems were not only scientific and technical, but also political and psychological'.

Velikhov also related how he could not initially understand why, as the helicopter lost height flying from the top of the ventilation stack towards the bottom, the radiation dose remained constant. Surely, he thought, the inverse square law of radiation should apply. It was only later that he realized how highly contaminated this stack was and that the source of radiation was not limited to the area at the base of the stack.

3.2 Causes of the accident

The fatal accident sequence was initiated by the power station's management and specialists when they sought to conduct an overnight experiment to test the ability of the turbine generator to power certain of the cooling pumps whilst the generator was free-wheeling to a standstill after its steam supply had been cut off. The purpose of the experiment was to see if the power requirement of Unit No. 4 could be sustained for a short time during a power failure.

It has been admitted[11a] that these tests were not properly planned, had not received the required approval and that the written rules on safety measures said merely that

> All switching operations carried out during the experiments were to have the permission of the plant shift foreman, that in the event of an emergency the staff were to act in accordance with plant instructions and that before the experiments were started the officer in charge would advise the security officer on duty accordingly.

With regard to the *officer in charge*, the principal managers were electrical engineers from Moscow and the person in charge was an electrical engineer who was not a specialist in reactor plants[12] and as *Pravda* reported[13] there was noticeable confusion even in minor matters.

It was also admitted[11a] that

> Apart from the fact that the programme made essentially no provision for additional safety measures, it called for shutting off the reactor's emergency core cooling system. This meant that during the entire rest period, which was about four hours, the safety of the reactor would be substantially reduced.

In addition[11a], 'the question of safety in these experiments had not received the necessary attention, the staff were not adequately prepared for the tests and were not aware of possible dangers'.

The NPP staff conducting the experiment, incredible as it might seem, knowingly departed from the experimental programme which was already of a poor quality. This was in part due to the fact that the experiment was behind schedule and if not completed, could affect the bonuses of the power workers. This created the conditions for the emergency situation which finally led to the accident which no one believed could ever happen.

In summary, therefore, the main causes of the accident, which were technological followed by human activity, are given in table 3.1, and have been drawn from several sources and reviewed by Meshkati[14]. Table 3.2, which to a certain extent overlaps with table 3.1, is a summary from the second INSAG report[11c] and summarizes a number of broader problems, rather than specific problems, which also contributed to the accident.

Table 3.1. Main causes of the accident.

- Faults in the concept of the RBMK: inherent safety not built-in.
- Faults in the engineering implementation of that concept: insufficient safeguard systems.
- Failure to understand the man–machine interface¶.
- The shutdown system was, in the event of the accident, inadequate and might in fact have exacerbated the accident, rather than terminated it.
- There were no physical controls to prevent the staff from operating the reactor in its unstable regime or with safeguard systems seriously disabled or degraded.
- There were no fire drills, no adequate instrumentation and alarms to warn and alert the operators of the danger.
- Lack of proper training as well as deficiencies in the qualifications of the operating personnel.
- Management and organization errors: as distinct from operator's errors.

¶ The man–machine interface was of concern to Valery Legasov, the First Deputy Director of the Kurchatov Atomic Energy Institute in 1986 and also the leader of the Soviet delegation to the post-accident meeting in August 1986 at the IAEA in Vienna[11a]. He has been quoted[15] as saying

> I advocate respect for human engineering and sound man–machine interaction. This is a lesson that Chernobyl taught us. One of the defects of the system was that the designers did not foresee the awkward and silly actions of the operators. The cause was due to human error and problems with the man–machine interface: this was a colossal psychological mistake.

Legasov was one of the casualties of Chernobyl in that in spite of glasnost and perestroika he became too outspoken about the political, managerial and scientific organizational faults which led to the accident. He became increasingly sidelined in Soviet nuclear energy politics and in April 1988 he committed suicide, see Appendix, which the authorities blamed on a diagnosis of leukaemia: this was untrue.

3.3 Countdown by seconds and minutes

The events of the 24 hours leading up to the explosions at 01:24 hours on 26 April 1986 are given in chronological order[11a–b] in terms of the current state of knowledge in August 1986 from Soviet documents[11a] and from the INSAG 1986 report[11b] and commentary is made in the light of furthur studies reported in the INSAG 1992 report[11c].

Table 3.2. Broader problems which contributed to the accident[11c].

• A plant which fell well short of safety standards when it was designed, and even incorporated unsafe features.
• Inadequate safety analysis.
• Insufficient attention to independent safety review.
• Operating procedures not founded satisfactorily in safety analysis.
• Inadequate and ineffective exchange of important safety information both between operators and between operators and designers.
• Inadequate understanding by operators of the safety aspects of their plant.
• Insufficient respect on the part of the operators for the formal requirements of operational and test procedures.
• An insufficient effective regulatory regime that was unable to counter pressures for production.
• A general lack of safety culture in nuclear matters, at the national level as well as locally.

25 April 1986

01:06

Start of reactor power reduction in preparation for the experiments and the planned shutdown of Unit No. 4.

03:47

Reactor power reduced to 1600 MW of *thermal* power, which was 50% of the maximum thermal power of the reactor. (The 1000 in RBMK refers to the maximum *electrical* power of 1000 MW.)

13:05

Unit No. 4 has two turbine generators, numbers 7 and 8, and turbine generator number 7 was *tripped* (terminology for shutdown) from the electricity grid and all its working load, including four of the main circulating pumps, transferred to turbine generator number 8.

14:00

As part of the experimental programme, the reactor's emergency core cooling system was disconnected. However, at this point in time the experiment was subjected to an unplanned delay because of a request by the electricity grid controller in Kiev to continue supplying the grid till 23:10 hours. This was agreed to by the Chernobyl NPP staff, but the reactor's emergency core cooling system was not switched back on.

This as far as it was known in 1986[11a–b] represented a violation of written operating rules and was maintained for just over nine hours. However, recent Soviet information confirmed that isolation of the emergency core cooling system was in fact permissible at Chernobyl if authorized by the Chief Engineer. Although INSAG now believes that this point did not affect the initiation and development of the accident, it is of the opinion that operating the reactor for a prolonged period of 11 hours with a vital safety system unavailable was indicative of an absence of safety culture[11c].

23:10

The reduction of the reactor's thermal power was resumed, since in accordance with experimental procedure the test was to be performed at between 700 MW and 1000 MW thermal power[11a–b]. It became clear after the accident that sustained operation of the reactor at a power level below 700 MW(th) should have been proscribed[11c].

26 April 1986

00:05

Thermal reactor power 720 MW; steady unit power reduction continues.

00:28

Thermal reactor power at around 500 MW. On going to low power, the set of control rods used to control reactor power at high powers, and called local automatic control rods (LACs), were switched off and a set of control rods called the automatic control rods (ACs) were switched on. However, the operators had failed to reset the set point for the ACs and because of this they were unable to prevent the reactor's thermal power falling to only 30 MW, a power level far below the 700–1000 MW intended for the experiment. However, later investigations suggest that the system was not working properly, the cause was unknown and hence there was inability to control the power, and therefore, as such, there was no operator error[11c].

01:00

The operators succeeded in stabilizing the reactor at 200 MW thermal power, although this was made difficult due to xenon poisoning of the reactor. The 200 MW level was only achieved by removing control rods from the core of the nuclear reactor. Nevertheless, 200 MW was still well below the required power level and the experiment should not have proceeded, but it did[11a–b]. Later reports[11c] confirm that the minimum operating reactivity margin (ORM) was indeed violated by 01:00 and was also violated for several hours on 25 April. Also the safety significance of the ORM is much greater than was indicated in the INSAG-1 report[11b].

01:03 to 01:07

The two standby main circulating pumps were switched respectively into the left and right loops of the coolant circuit. Eight main pumps were now working and this procedure was adopted so that when, at the end of the experiment in which four pumps were linked to turbine generator number 8, four pumps would also remain to provide reliable cooling of the reactor core. However, due to the low power of 200 MW and the very high (115–120% of normal) coolant flow rate through the core due to all eight pumps functioning, some pumps were operating beyond their permitted regimes. The effect was a reduction in steam formation and a fall in pressure in the steam drums.

01:19:00

The operators tried to increase the pressure and water level by using the feedwater pumps. The reactor should have tripped because of the low water level in the steam drums, but they had overridden the trip signals and kept the reactor running. The water in the cooling circuit was now nearly at boiling point.

01:19:30

The water level required in the steam drums is reached, but the operator continues to feed water to the drum. The cold water passes into the reactor core and the steam generation falls further, leading to a small steam pressure decrease. To compensate for this, all 12 automatic control rods (ACs) are fully withdrawn from the core. In order to maintain 200 MW thermal power, the operators also withdrew from the core some manual control rods.

01:19:58

A turbine generator bypass valve was closed to slow down the rate of decrease of steam pressure. Steam is not dumped into the condenser. Steam pressure continues to fall.

01:21:50

The operator reduces the feed water flow rate to stop a further rise in the water level. This results in an increase in the temperature of the water passing to the reactor.

01:22:10

Automatic control rods (ACs) start to lower into the core to compensate for an increase in steam quality.

01:22:30

The operator looks at the printout of the parameters of the reactor system. These are such that the operator is required in the written rules to immediately shut down the reactor, since there is no automatic shutdown linked to this forbidden situation. The operator continues with the experiment.

Computer modelling has shown that the number of control rods in the reactor core were now only six, seven or eight, which represents less than one-half the design safety minimum of 15, and less than one-quarter the minimum number of 30 control rods in the operator's instruction manual.

01:23:04

The experiment is started with the reactor power at 200 MW, and the main line valves to the turbine generator number 8 were closed. The automatic safety protection system which trips the reactor when both turbine generators are tripped was deliberately disengaged by the operators, although this instruction was not included in the experimental schedule. After all, operation of the reactor was not required after the start of the experiment. What seemed to be going through the mind of the operator was that if the experiment at first failed, then a second attempt could be made if the reactor was still running. It is difficult to avoid the conclusion that the major priority of the Unit No. 4 operators was to ensure that they completed the experiment during the 1986 rundown to the annual maintenance in 1987. It is hard to imagine a situation where the pressure and stress exerted on experimentalists is such that they would ignore many vital safety procedures. Nevertheless this is just what happened[11a–b].

However, later analysis[11c] shows that although the second turbogenerator was tripped at 01:23:04 the first turbogenerator was tripped at 00:43:27. This trip was in accordance with operational procedures and therefore the operators were not at fault and the original INSAG-1[11b] statement that 'This trip would have saved the reactor' seems not to be valid[11c].

01:23:05

The reactor power begins to rise slowly from 200 MW.

01:23:10

The automatic control rods (ACs) are withdrawn.

01:23:31

The main coolant flow and the feedwater flow are reduced, causing an increase in the temperature of the water entering the reactor, and an increase in steam generation. The operators noted an increase in reactor power.

01:23:40

A reactor power steep rise (sometimes termed a *prompt critical excursion*) was experienced, and the Unit No. 4 shift foreman ordered a full emergency shutdown (an *emergency scram*). Unfortunately the order came too late. Not all the automatically operated control rods reached their lower depth limits in the core and an operator unlatched them in order to allow them to fall to their positional limits under gravity. However, since the rods had been nearly withdrawn, a delay of up to 20 s would have had to occur before the reactor power could have been reduced. This would have been at 01:24:.00.

01:23:43

Emergency alarms operate, but unfortunately the emergency protection is not sufficient to stop reactor runaway. The sharp growth of the fuel temperature produces a heat transfer crisis. Reactor power reaches 530 MW in 3 s and continues to increase exponentially, figure 3.1.

01:23:46

Intensive generation of steam.

01:23:47

Onset of fuel channel rupture.

01:23:48

According to observers outside Unit No. 4, two explosions (these were thermal) occurred about 01:24 one after the other. Burning debris and sparks shot into the air above the reactor, and outbreaks of fire occurred in over 30 places due to high temperature nuclear reactor core fragments falling onto the roofs of buildings adjacent to the now destroyed reactor hall. Diesel

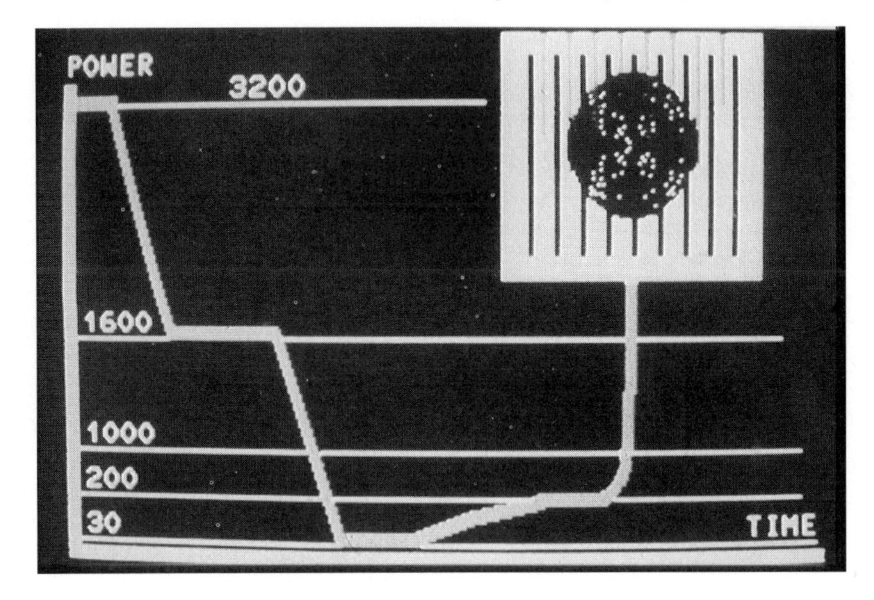

Figure 3.1. Variation of thermal power in MW with time, showing the final exponential rise of power[11a]. (Courtesy: USSR KGAE.)

fuel and hydrogen stores were also threatened and firefighting took precedence over radiation protection, since an even bigger disaster would have occurred if the fires had gone out of control.

There has been considerable further analysis[11c] of the events, including computer modelling, since the presentation[11a–b] by Soviet scientists at the August 1986 Post-Accident Review Meeting and these have led to new insights into the physical characteristics of the RBMK reactor. Most analyses now associate the severity of the accident with defects in the design of control and safety rods in conjunction with the physics design characteristics, which permitted the inadvertent setting up of large positive void coefficients. The scram just before the sharp rise in power that destroyed the reactor may well have been the decisive contributory factor[11c].

The features of the RBMK reactor have also resulted in other pitfalls for the operating staff and any of these, table 3.3, could just as well have caused the initiating event for this or an almost identical accident.

3.4 Damage to the power plant

One of the best descriptions of the damage to Unit No. 4 is given in part of the tender documentation[16] for the building of a second Sarcophagus to

Table 3.3. RBMK pitfalls other than defects in the design of control and safety rods and questions arising from the accident[11c].

Pitfalls
• Pump failure, disturbance of the function of coolant pumping or pump cavitation, combined with the effect of the positive void coefficient. Any of these causes could have led to sudden augmentation of the effect of the positive void coefficient.
• Failure of zirconium alloy fuel channels or of the welds between these and the stainless steel piping, most probably near the core inlet at the bottom of the reactor. Failure of a fuel channel would have been the cause of a sudden local increase in void fraction as the coolant flashed to steam. This would have led to a local reactivity increase which could have triggered a propagating reactivity effect.
Questions
• Which weakness ultimately caused the accident?
• Does it really matter which shortcoming was the actual cause, if any of them could potentially have been the determining factor?

protect the first one which is now crumbling with a likelihood of at least a partial collapse in the not too distant future. Part of this description is reproduced below.

After the explosion, part of the construction in the reactor unit, the ventilation stack, the turbine hall and other structures turned out to be destroyed, figures 3.2 and 3.3[11]. The reactor core was completely destroyed, walls and ceiling in the central reactor hall were demolished, Plate IV, figure 3.4, ceilings in the water separation drum premises were displaced and walls were destroyed. Premises housing the main circulation pumps (MCP) oriented to the north were destroyed completely and premises for the MCP lying to the south partially. Two upper stories of the ventilation stack were demolished and the columns of the building frame were shifted to the side of the turbine room.

The ceiling in the turbine room was destroyed in many places by fire and falling debris, several building girders were deformed and building frame columns were displaced along one axis by the explosion wave. The reactor emergency cooling system was completely destroyed from the north side of the reactor building and buried by debris.

The upper plate of the reactor's biological shield which weighed 2000 tonnes, was with the steam–water pipeline system and various ferroconcrete constructions were displaced so that the shield was inclined at 15° to the vertical and rested against the metal tank edge, figures 3.5 and 3.6[17]. The central reactor hall is filled with debris including materials thrown from

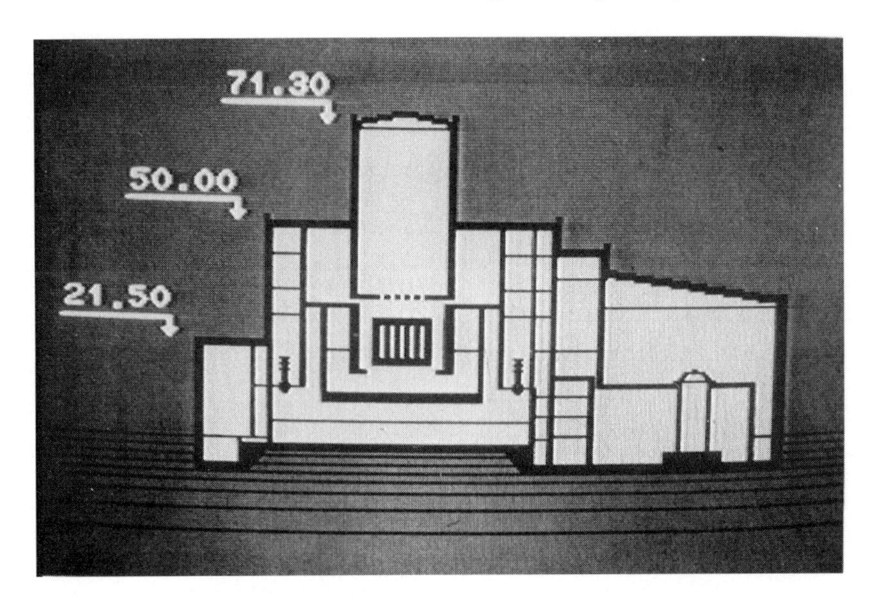

Figure 3.2. Cross-section through Unit No. 4 before the accident[11a]. (Courtesy: USSR KGAE.)

Figure 3.3. Cross-section through Unit No. 4 after the accident[11a]. (Courtesy: USSR KGAE.)

Figure 3.4. Close-up view of the damage to Unit No. 4. (Courtesy: Chernobylinterinform.)

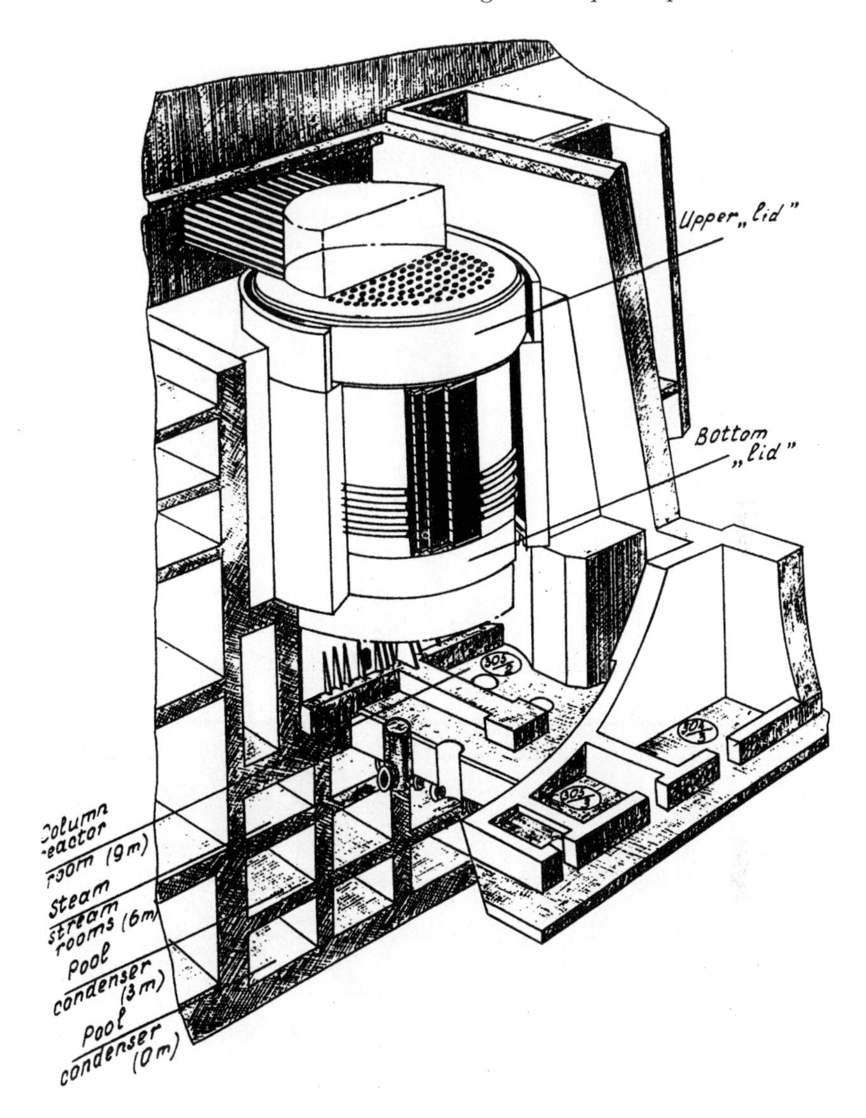

Figure 3.5. Cross-section through the central reactor hall before the accident[17]. (Courtesy: Chernobylinterinform.)

helicopters during the fire extinguishing phase. In some parts the debris is 15 m high.

Investigation of the south pool for spent fuel storage showed that fuel assemblies did not have any noticeable damage within the visible part of the pool. The north storage pool, which was empty, contains some elements

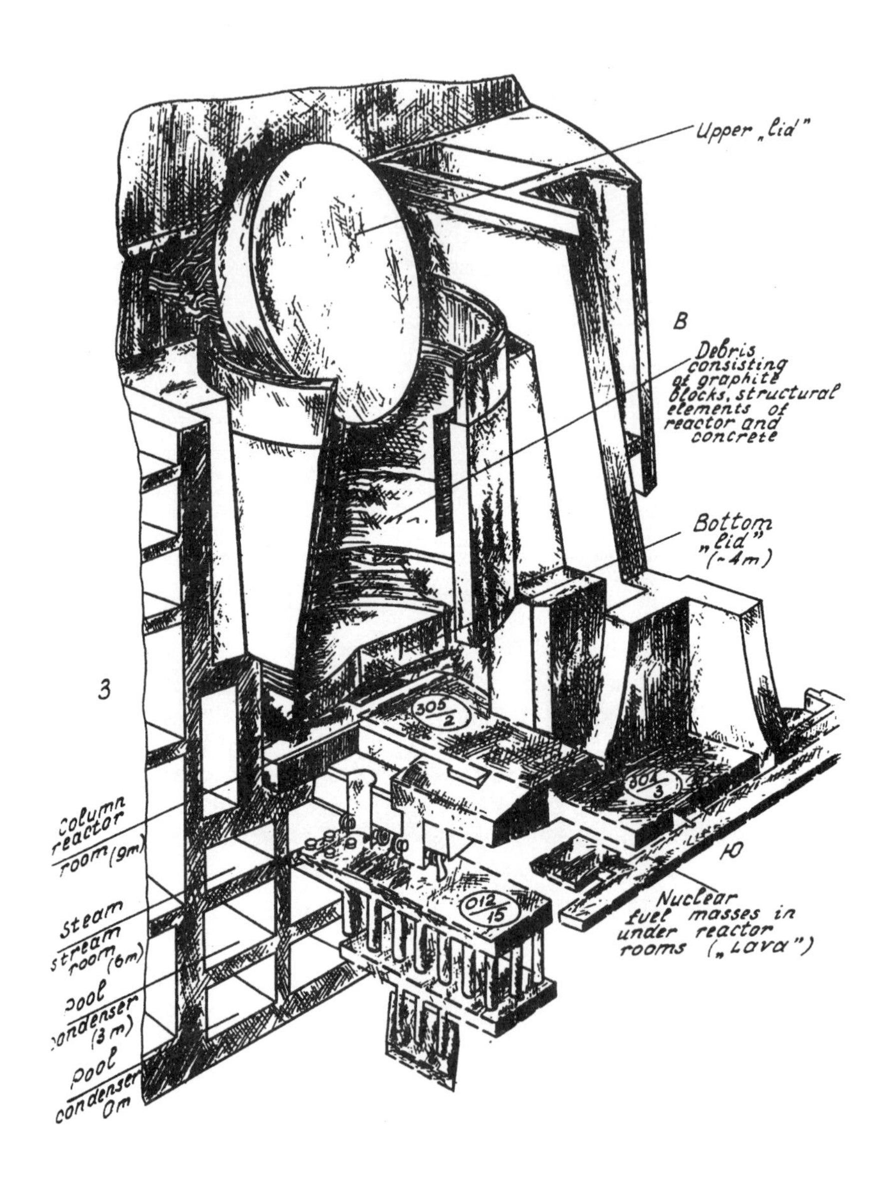

Figure 3.6. Cross-section through the central reactor hall after the accident. The upper lid is seen at an angle of 15° to the vertical[17]. (Courtesy: Chernobylinterinform.)

from the core and materials which were thrown from helicopters. No water was discovered in these pools. Such details of the accident[16] were not available for several years after 1986. Earlier, only exterior photographs of the damaged Unit No. 4 were published.

3.5 Extinguishing the fire

Extinguishing the fire was the first priority and this was achieved not only by the firemen, who worked mainly on the roof of the turbine hall, where the damage is clearly seen in Plate I, but also by helicopter pilots whose task was to put out the fire in what remained of the reactor central hall and to ensure that it did not break out again. This was attempted by dumping 5000 tonnes of boron compounds, sand, clay, dolomite and lead during the period 27 April to 10 May. On 27 April the helicopters flew 93 missions and on 28 April a total of 186 missions. The overflying speed was 140 km/hr.

Their missions continued throughout 1986 and by the end of June they had dumped 14 000 tonnes of solid materials, 140 tonnes of polymerizable liquids and 2500 tonnes of trisodium phosphate[18].

It was extremely hazardous for the helicopter pilots when flying near the electricity pylons, figure 3.7, and there was a fatal accident on 28 October 1986. This was captured on video[19], figure 3.8, and there is a memorial to those who died which incorporates one of the rotor blades. This is situated by the side of the road to Chernobyl and also includes a helicopter of the type which was used. The pilots were well aware of the dangers and an Afghan war veteran is on record[1] as saying 'When we heard that soldiers were being sent to Chernobyl as liquidators, we all felt we were better off fighting the war'. By 1991 it was reported that an unspecified number, *some*, of the helicopter pilots had died[18] and that in spite of their efforts no neutron absorbers reached the reactor core.

3.6 Initial reports of the accident

3.6.1 In the USSR

The Soviet authorities through TASS and the Novosti Press Agency informed the rest of the world about the accident before their own population. This was on 28 April 1986, two days after the accident had occurred. The first communication to reach the United Kingdom from Moscow TASS was terse:

> An accident has occurred at the Chernobyl Atomic Power Plant as one of the atomic reactors was damaged. Measures are being undertaken to eliminate the consequences of the accident. Aid

Figure 3.7. Helicopter flying near power lines and the ventilation stack[18]. (Courtesy: Chernobylinterinform.)

Figure 3.8. The fatal accident when the rotor blades crashed through a power line[19]. (Courtesy: Chernobylinterinform.)

> is being given to those affected. A Government Commission has been set up.

A second communication was later issued on 28 April which attempted, with anti-American rhetoric, to play down the effects of the Chernobyl accident.

> The accident at the Chernobyl Atomic Power Station is the first one in the Soviet Union. Similar accidents happened on several occasions in other countries. In the United States 2300 accidents, breakdowns and other faults were registered in 1979 alone. The atomic power station North Anna-1, Virginia, near Washington DC is topping the list of accident prone stations. A major accident occurred in 1979 at the atomic power station in Harrisburg, Pennsylvania, where radioactive substances leaked due to a reactor breakdown ... etc.

3.6.2 In the West

The USSR also, unsurprisingly, did not admit to any previous accidents, such as that at Kyshtym[20]. Forsmark nuclear power station in Sweden, 130 km north of Stockholm was the site at which the radioactive cloud was first detected outside the borders of the Soviet Union and the events have been related by a Swedish physicist[21].

> Radioactivity was measured on workers passing through the entrance gate to the power station at 7am on 28 April. High levels were measured outside the station and the Swedish authorities were informed at 9.30am. Evacuation of the station began at 11am. About 1pm the indications were that the activity did not come from the Forsmark station and that it was coming from the east, as that was the direction of the prevailing wind. Confirmation came from the Soviet authorities in the late evening of 28 April that an accident had taken place early in the morning of 26 April.

Several satellite photographs were published at this time and, for many years, it was generally assumed that the first of these was taken only after 28 April after the accident was noted at Forsmark. However, this was not correct. An American satellite had passed over the Chernobyl area only 28 s after the accident on Saturday 26 April 1986. This was pure chance. The reason for such a monitoring orbit was to take in a nuclear missile site. An early warning radar screen 132 m high by 96 m wide can still be seen on the road to Chernobyl NPP.

America's initial assessment was that a nuclear missile had been fired, then when the image remained stationary, opinion changed to a missile had blown up in its silo. It was only when a map of the area was consulted that it was realized that it was the Chernobyl NPP.

There are various confirmations of this story, one of the most interesting being that of an IAEA official in Vienna who was attending a British Embassy reception on the Sunday evening being asked about the nuclear accident which had just occurred. 'What nuclear accident?' 'You don't know, well go and check at the Agency'. This he did early on the Monday morning of 28 April to find that there was no knowledge of the accident. It was only later that day that the Forsmark radiation measurements were reported to the IAEA[22].

Once the accident had been confirmed in the West, the press ran riot with various exaggerated claims such as the following: 2000 dead in atom horror: reports in Russia danger zone tell of hospitals packed with radiation accident victims[23]; Please get me out Mummy: terror of trapped Britons as 2000 are feared dead in nuclear horror[24]; 15 000 dead in mass grave[25].

Many cartoons were also published and there were also some spurious photographs. One such series on American and Italian TV networks showed on their screens what purported to be the Chernobyl NPP burning. The truth was that these were pictures of a burning cement factory in Trieste[26,27]. The instigator, a Frenchman Thomas Garino collected a fee of US$20 000, and an ABC TV newscaster later told his audience that 'This is one mistake we will try not to make again'.

Another fraudulent photograph was published by *The Sunday Times*[28] in the United Kingdom on 11 May and in *Time* magazine[29] on 12 May, the former in black and white and the latter in colour. This photograph when in *The Sunday Times* was beneath the headline *Cloud over Kiev* with the quote: 'It was 3 pm on a sunny day when a tourist took this picture of the nuclear cloud, a cloud whose effects fill the residents of Kiev with fear'. The skyline is that of Kiev but it defies all credibility to believe that a black cloud of soot and smoke could travel the 146 km from the NPP and remain intact about a week after the explosion. The photograph eventually located[30] in the John Hillelson photographic agency was found to be only black and white. Journalistic license had added the orange tint to represent the sunset over Kiev.

Chapter 4

Radionuclide Releases

Introduction

The radionuclide releases into the atmosphere reached a total of 1–2 EBq, extended over ten days from the date of the explosion[1,2], and were the cause of the widespread contamination of the environment which depended on wind and rainfall patterns. The contamination remains a problem, especially from ^{137}Caesium with its 30.1 year half-life, although the risks associated with the short-lived 8.0 day half-life ^{131}Iodine have now passed. In some areas relatively near to the NPP the contamination from isotopes of plutonium will continue to exist into the indefinite future. The half-lives of ^{239}Pu, ^{240}Pu and ^{241}Pu are respectively 24 100 years, 6560 years and 14.4 years.

The heat from the Chernobyl reactor core made the radioactive plume rise, and the dry weather at the time over Pripyat was such that it had risen to a height of 1200 m on 27 April. The plume travelled towards Finland and Sweden, becoming stagnant for some time over the Ukraine, Belarus and north-eastern Europe, with some rainfall over Scandinavia. It also rained over the small town of Gomel in Belarus, and the incidence of thyroid cancer amongst Gomel's children is correlated with this event. The plume then spread widely over neighbouring countries.

The Chernobyl releases can be compared with those from Three Mile Island and also from atmospheric nuclear weapons testing. However, Chernobyl cannot be directly equated to the effects of the atomic bombs of Hiroshima and Nagasaki: these were very different types of event. Nevertheless, because the media often erroneously link these together, this chapter includes an explanation of why this should not be done. In addition, comparisons are also made between the radionuclide releases from Chernobyl and those from nuclear weapons testing and from the Mayak plutonium production facility in the southern Urals which has so badly contaminated the Techa river area. Chernobyl is not alone in contaminating to a high level the environment of the former USSR, nor is it alone in causing the evacuation of populations.

4.1 Release sequence and composition

The entire radionuclide release did not occur all on the first day of the accident, but was over a period of ten days with only 25% released on the first day. Temperatures[3] rose well above 2000 °C to a maximum of 2760 °C and the releases consisted of two different components.

- Radionuclides included in the matrix of the dispersed fuel and released as radioactive dust.
- Volatile radioactive substances in the form of gases or aerosols which evaporated from the hot fuel.

Figure 4.1 shows the daily pattern of the releases which have been categorized into four phases, table 4.1. The core inventory and total releases of the various radionuclides are given in table 4.2. At the time of the accident the total core inventory was about 33 EBq. It must be realized though that the range of uncertainty for all estimates of releases is ±50% except for the noble gases xenon and krypton where a 100% release occurred. About 10–20% of the volatile radionuclidens iodine, tellurium and caesium were released, and about 3–6% of other more stable radionuclides such as barium, strontium, plutonium and cerium.

Table 4.1. Release phases with time[2].

- The initial intense release on the first day of the accident.
- A period of five days over which the release rate declined to a minimum value six times lower than the initial release rate.
- A period of four days over which the release rate increased to a value which is about 70% of the initial release rate.
- A sudden drop in the release rate nine days after the accident to less than 1% of the initial rate and a continuing decline in the release rate thereafter.

The initial release was due to the explosion and the second phase decrease was due to the measures taken to extinguish the fire by dropping 5000 tonnes of materials onto the core. The third category, termed a heat-up period, was attributed to heating of the fuel in the core to above 2000 °C owing to residual heat release. The final category was characterized by a rapid decrease in the escape of fission products.

The total release was 3–4% of the core material of which 0.3–0.5% fell on site, 1.5–2% fell within 20 km of the NPP and 1–1.5% fell outside 20 km.

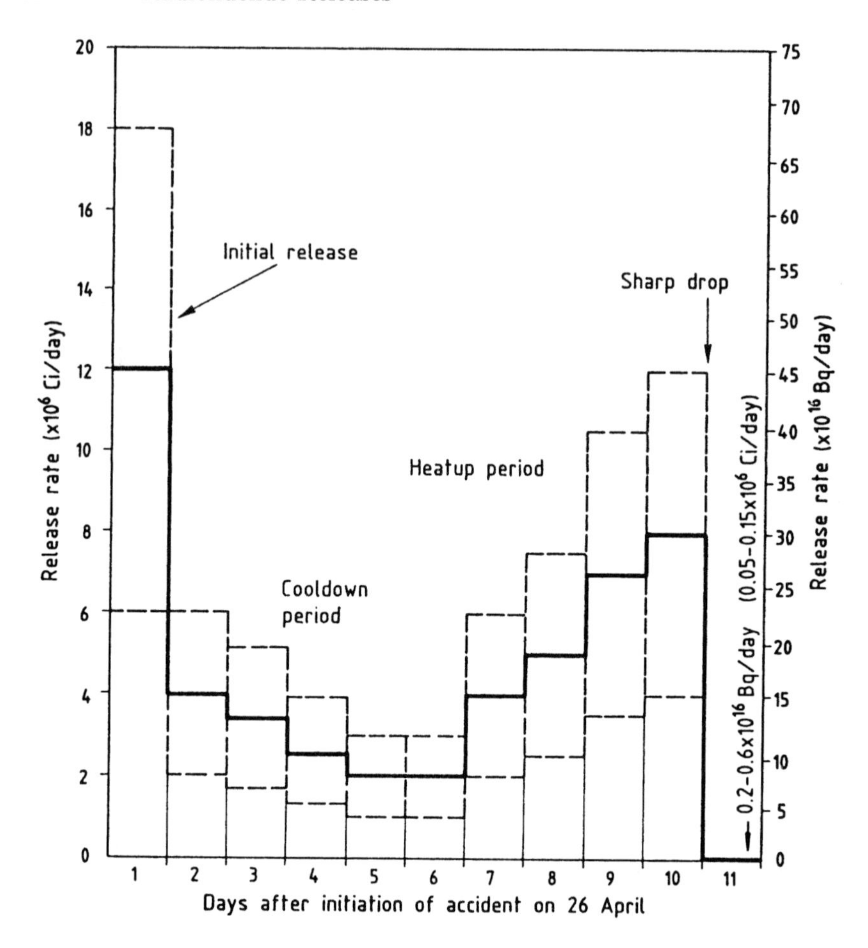

Figure 4.1. Daily releases of radionuclides into the atmosphere, excluding noble gases[2]. The values are calculated for 6 May 1986 taking into account radioactive decay up until that date. (Courtesy: IAEA.)

The proportional amounts dispersed beyond the USSR[1] were 34% for ^{131}I and 56% for ^{137}Cs.

A revision in the releases of ^{131}I, ^{137}Cs and ^{134}Cs from data published by UNSCEAR[4] in 1993 which was based on updated USSR 1990 estimates[5] were, respectively, 0.63 EBq, 0.07 EBq and 0.035 EBq. The earlier data[1,2] were based on estimates made in 1986.

A decade after the accident the release figures have been further revised upwards[6,7] to ~1.8 EBq for the ^{131}I release, and the release of ^{132}Te to ~1.2 EBq. This tellurium radionuclide decays in a few days to the very short-lived radionuclide ^{132}I. The last column in table 4.2 gives the latest estimates[8,9] where the total release excluding noble gases is estimated as 8 EBq, backdated to 26 April 1986.

This is considerably higher than the earlier estimate of 1–2 EBq because it includes more short-lived radionuclides. However, it should be considered as a probable over-estimate[8] for the total release since many of these radionuclides would have decayed inside the damaged core before any release to the atmosphere could occur. Another set of recent estimates[6,8] is of the same orders of magnitude as those[8,9] in table 4.2 but as with the earlier estimates[2] they do not quote the release estimates for some of the short-lived radionuclides.

4.2 Atmospheric transport

At the time of the accident the surface winds at the Chernobyl site were very weak and variable in direction. However at 1500 m altitude the winds were 8–10 m per second from the south-east[1]. This caused the radioactive plume to flow along the western parts of the USSR toward Finland and Sweden where it was detected at Forsmark: see section 3.3.2.

Figure 4.2 shows the behaviour of the radioactive plume and reported initial arrival times of detectable activity in air[1]. Plumes A, B and C correspond to air mass movements originating from Chernobyl on 26 April, 27–28 April and 29–30 April respectively. The numbers **1**–**8** in figure 4.2 indicate initial arrival times: **1**, 26 April; **2**, 27 April; **3**, 28 April; **4**, 29 April; **5**, 30 April; **6**, 1 May; **7**, 2 May; and **8**, 3 May.

Long-range atmospheric transport spread activity throughout the northern hemisphere and reported initial arrival times were 2 May in Japan, 4 May in China, 5 May in India and 5–6 May in Canada and the USA. The simultaneous arrival at both western and eastern sites in Canada and the USA suggest a large-scale vertical and horizontal mixing over a wide area. No airborne activity from Chernobyl has been reported in the southern hemisphere[1].

4.3 Releases and exposed population groups

Figure 4.3 summarizes over time[10] the radiation exposure of defined population groups. The sub-groupings are also given in table 4.3. ^{241}Am is a daughter product of ^{241}Pu and therefore its activity will increase from its 1986 level.

Table 4.2. Core inventory and total release of four groups[8] of radionuclides as estimated from various sources. Figures in columns 3–5 contain decay corrected to 6 May 1986; the last column is the decay corrected to 26 April 1986.

Radio-nuclide	Half-life[1] (y = year, d = day)	Inventory[2] (EBq)	Release[2] (% of inventory)	Release [2] (EBq)	Revised release[8,9] (EBq)
Noble gases					
^{85}Kr	10.72 y	0.033	~100	0.033	0.033
^{133}Xe	5.25 d	1.7	~100	1.7	6.5
Volatile elements					
^{131}I	8.04 d	1.3	20	0.26	1.2–1.7
^{132}Te	3.26 d	0.32	15	0.048	1.0
^{137}Cs	30.0 y	0.29	13	0.0377	0.074–0.085
^{134}Cs	2.06 y	0.19	10	0.019	0.044–0.048
^{129m}Te					0.24
^{133}I					2.5
^{136}Cs					0.036
Intermediate					
^{89}Sr	50.5 d	2.0	4.0	0.08	0.081
^{90}Sr	29.12 y	0.2	4.0	0.008	0.008
^{103}Ru	39.3 d	4.1	2.9	0.119	0.17
^{106}Ru	368 d	2.1	2.9	0.061	0.03
^{140}Ba	12.7 d	2.9	5.6	0.162	0.17
Refractory: including fuel particles					
^{95}Zr	64.0 d	4.4	3.2	0.141	0.17
^{99}Mo	2.75 d	4.8	2.3	0.110	0.21
^{141}Ce	32.5 d	4.4	2.3	0.101	0.2
^{144}Ce	284.9 d	3.2	2.8	0.090	0.14
^{239}Np	2.36 d	0.14	3	0.0042	1.7
^{238}Pu	87.74 y	0.001	3	0.000 03	0.000 03
^{239}Pu	24 065 y	0.0008	3	0.000 024	0.000 03
^{240}Pu	6537 y	0.001	3	0.000 03	0.000 044
^{241}Pu	14.4 y	0.17	3	0.0051	0.0059
^{242}Pu					0.000 000 09
^{242}Cm	163 d	0.026	3	0.000 78	0.000 93
Totals		30.537 EBq¶		1.235 EBq¶	8.0 EBq¶

¶ Excludes the noble gases ^{85}Kr and ^{133}Xe.

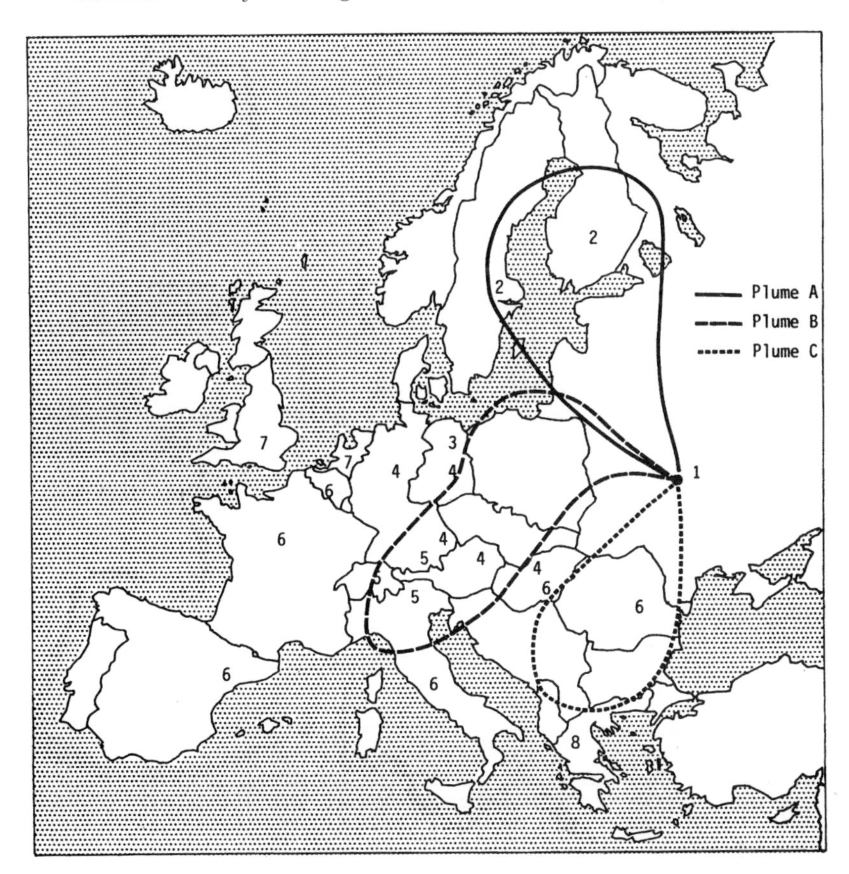

Figure 4.2. Radioactive plume behaviour and reported initial arrival times of detectable activity in air[1]. (Courtesy: UNSCEAR.)

4.4 Residual activity in the global environment after 70 years

An estimate of the residual activity after a lifespan of 70 years from 1986 is given in table 4.4.

4.5 Chernobyl deposition compared to background deposition

Figure 4.4 gives a comparison of the average effective doses in European countries, divided into the contribution from Chernobyl and the contribution from natural background exposures. The Chernobyl contribution

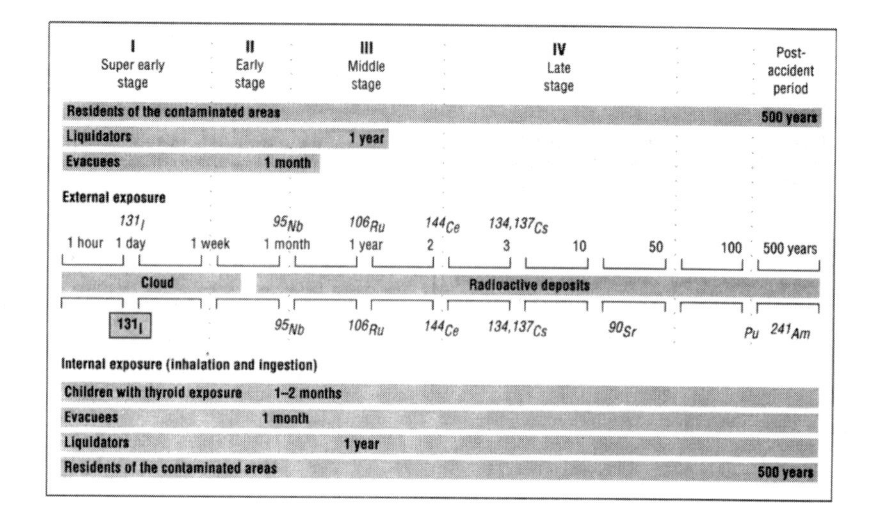

Figure 4.3. Time evolution of important exposure pathways for the sub-populations under exposure[10]. (Courtesy: WHO.)

Table 4.3. Evolution over time for Chernobyl releases and exposed population groups[10].

Irradiation stage	
First. Super-early	0–1 week
Second. Early	1 week–1 month
Third. Middle	1 month–2 y
Fourth. Late	2–50 y
Post-accident	>100 y
Exposed populations	
Residents of the contaminated areas	
Liquidators	
Evacuees	
External exposure	
Cloud	
Radioactive deposits	
Internal exposure: inhalation and ingestion	
Children with thyroid exposure	
Evacuees	
Liquidators	
Residents of contaminated areas	

Table 4.4. Estimate of residual radioactive material in the global environment due to the Chernobyl accident[8]. 1 EBq = 1000 PBq.

Radionuclide	Release in 1986 corrected to 26 April (PBq)	Release remaining in 1996 (PBq)	Release remaining in 2056 (PBq)
^{131}I	1200–1700	0	0
^{90}Sr	8	6	1.5
^{134}Cs	44–48	1.6	0
^{137}Cs	74–85	68	17
^{238}Pu	0.03	0.03	0.02
^{239}Pu	0.03	0.03	0.03
^{240}Pu	0.044	0.044	0.03
^{241}Pu	5.9	3.6	0.2
^{241}Am	0.005	0.08	0.2

ranges from very low to 0.76 mSv with the highest estimated values for countries in south eastern, central and northern Europe. In figure 4.4 the countries are ranked from the highest to the lowest Chernobyl effective doses[11]. In Belarus the first year contribution from Chernobyl was 2 mSv which was comparable to the natural background in Belarus.

4.6 Comparison with Hiroshima and Nagasaki

It is said in the media and in some popular scientific magazines[11,12] that 'the Chernobyl accident was a disaster 200 times greater than the combined releases of the atomic bombs dropped on Hiroshima and Nagasaki'. Actually these were entirely different events.

Chernobyl was a thermal and steam explosion that released fission products accumulated in the reactor core. The Hiroshima and Nagasaki bombs were nuclear explosions that caused a large amount of direct radiation together with thermal and blast effects to devastate large urban areas. There was little production of fallout radionuclides at Hiroshima and Nagasaki. A comparison on this basis is therefore entirely misleading[11].

The figure of '200' does not mean that the explosion was 200 times more powerful, but that it spread radioactive caesium over a large area and that the total amount of caesium was equivalent to some 200 times the combined caesium contamination at both Hiroshima and Nagasaki[12].

The Hiroshima atomic bomb employed ^{235}U and released energy equivalent to 15 ± 3 kilotonnes of TNT and the dissipation of energy is believed to have been in the following ratio: bomb blast (50%), thermal rays (35%)

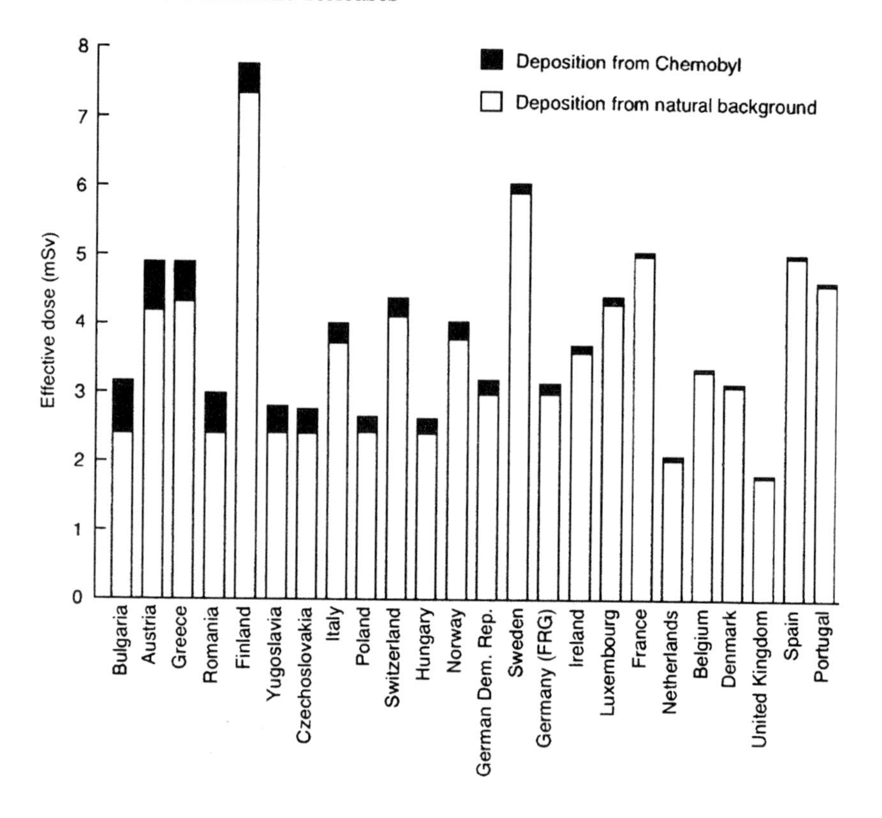

Figure 4.4. First year post-accident effective dose from Chernobyl and from natural background[11]. (Courtesy: IAEA.)

and radiation (15%). The Nagasaki bomb contained ^{239}Pu and released energy equivalent to 21 ± 2 kilotonnes of TNT.

Atomic bomb radiation can be classified into initial radiation, which was released within one minute after detonation, and residual radiation which was released subsequently. The initial radiation accounted for one-third of the total radiation energy released and was composed of gamma rays and neutrons. The residual radiation accounted for two-thirds of the total and can be classified into radioactive fallout, which included gamma, beta and alpha rays, and induced radiation, which was released due to the activation of objects on the ground by neutrons from the initial radiation[13,14].

4.7 Comparison with Three Mile Island

The accident at Three Mile Island severely damaged the core of TMI-2, one of a pair of reactors. The upper portions of the uranium fuel assemblies became molten and flowed down through the core of the reactor vessel before re-solidifying. Despite the destruction, little radiation escaped and the 15.2 cm thick steel reactor vessel did not fail. Its ability to withstand the intense heat of the molten fuel surprised scientists and engineers[15].

The accident, which took place on 28 March 1979 in Pennsylvania, near Harrisburg, in the USA, was initiated at 4am by a failure which had occurred in part of the plant's pumping apparatus. Water in the secondary cooling system (which removes heat from the primary cooling system) stopped flowing. Because of the design of the TMI reactors, if the secondary cooling system fails the entire core can become dangerously hot. The failure was compounded by the backup system not coming into operation because some valves which should have been left open had been closed. By the time this had been discovered the water in the steam generator had boiled away and so heat was not being removed from the core. This rise in the core temperature triggered the opening of a pressure-relief valve at the top of the reactor and this also triggered the mechanism that should have automatically shut down the reactor. By this stage, major damage may have been suffered by the reactor. The pressure-relief valve did not close as it should have done when the pressure dropped and water was being pumped out of the primary cooling system through the faulty valve. Some 250 000 gallons of water poured out onto the floor of the containment building and this caused considerable damage to the core. It was later estimated by the NRC that 60% of the fuel rods had been damaged[16].

The cause of the partial meltdown was thus a combination of operator error, poor design and flawed procedures. TMI-2 is permanently shut down but TMI-1 began generating power again on 9 October 1985.

During the accident the 250 000 gallons of water were automatically pumped into a tank in an auxiliary building but this tank overflowed and the water gave off ^{133}Xe and ^{95}Kr gases into the building and the ventilation system pumped these gases into the atmosphere[16]. The environmental release of these noble gases into the atmosphere was[4] about 0.37 EBq compared to 6.53 EBq at Chernobyl, see table 4.2. For ^{131}I the environmental release was[4] 550 GBq at TMI whereas at Chernobyl it was in the range 1.2×10^9 to 1.7×10^9 GBq, see table 4.2. Because of the containment vessel at TMI ^{137}Cs was not released into the environment.

The most recent news concerning Three Mile Island was in November 1999 when a three judge panel of the United States Third Circuit Court of Appeals allowed 1990 people who claimed they were harmed by radiation from the accident, to proceed with lawsuits against eight defendents, including the plant operator: the Morristown, New Jersey based holding company General Public Utilities (GPU) Inc. and its utility unit, Metropolitan Edison[17]. This ruling came more than two years after attorneys had argued the case before the Court of Appeals. The November 1999 ruling stated that the plaintiffs' cases need not be bound to the fate of ten others that were chosen to represent all cases and finally rejected by a United States District Court judge, Sylvia Rambo, in 1996. Judge Rambo had ruled that there was insufficient evidence to link the plaintiffs' various claims of cancer and birth defects to exposure to the radiation leak at the plant. The 1990 plaintiffs are now being given a chance to object to Judge Rambo's decision, although the original ten *typical* plaintiffs selected for the *mini-trial* cannot revive their cases. The Appeal Judges said that the dangers from lower doses should not be ruled out in the 1990 cases and the response by a GPU spokesman said that the most important aspect of this long-awaited option was its affirmation of 'a lower Court ruling that basically determined that the TMI accident did not cause the illnesses claimed by the plaintiffs' and that 'this is going to reflect on any further lawsuits'[17].

Reports earlier[18] in 1999 also stated that insurers for the parent company GPU Inc. had paid at least $3.9 million to settle lawsuits by residents, although the terms of these settlements are mostly secret, and that the TMI-1 functioning plant is to be sold for $100 million, which is one-seventh of its book value. TMI-2 has been dismantled just enough to reveal a steel skeleton under its cooling towers, and it will never reopen: it cost $700 million to build and to date $973 million to mothball.

4.8 Comparison with nuclear weapons testing in the atmosphere

In a nuclear test, a nuclear device is exploded with large releases of energy and the explosion is caused by nuclear fission, nuclear fusion or a combination of both. In a *fission device*, two sub-critical masses of fissile material, such as ^{235}U and ^{239}Pu, are brought together to produce a supercritical mass. The heavy nucleus is spliced into two parts (the *fission products*) which subsequently emit neutrons, releasing energy equivalent to the difference between the rest mass of the fission products and the neutrons. In a *fusion device*, atomic nucleii of low atomic number fuse to form a heavy nucleus with the release of large amounts of energy. The reaction becomes self-sustaining at very high temperatures with the help of an inner fission device surrounded by light hydrogenous material such as deuterium and lithium deuteride[19].

A third type of nuclear weapons experiment was also undertaken; this was termed a *safety trial*. In such an atmospheric experiment a more or less fully developed nuclear device is subjected to simulated accident conditions. The weapon is destroyed by conventional explosives with no or, in some instances, very small releases of fission energy. While the radioactive residues of a nuclear test are the fission and fusion products, the radioactive residue of a safety trial is the fissionable material itself[19].

Most nuclear explosions in the atmosphere occurred before 1963 and their total yields for the period 1945–80 in equivalent amounts of megatonnes of TNT were estimated by UNSCEAR and are given in table 4.5. Large yield explosions carry radioactive debris into the stratosphere from where it is dispersed and deposited around the world. The radiation doses are mostly due to the ingestion of radionuclides that have become incorporated into foodstuffs, and to external irradiation from ground deposition.

The most significant radionuclides contributing to dose commitments are, in decreasing order of importance, ^{14}C, ^{137}Cs, ^{95}Zr, ^{90}Sr, ^{106}Ru, ^{144}Ce and tritium. Many other different radionuclides were produced in weapons testing, but because of the delayed deposition from high altitudes, some of the prominence of short-lived radionuclides was reduced by decay.

Table 4.5. Numbers and yields of atmospheric nuclear explosions during the period 1945–80[1]. The yields are given in megatonnes.

Years	No. of tests	Fission yield	Total yield
1945–51	26	0.8	0.8
1952–54	31	37	60
1955–56	44	14	31
1957–58	128	40	81
1959–60	3	0.1	0.1
1961–62	128	102	340
1963	0	0	0
1964–69	22	10.6	15.5
1970–74	34	10	12.2
1975	0	0	0
1976–80	7	2.9	4.8
1981–87	No tests		

The greatest *global* release of radionuclides from a man-made source was atmospheric testing of nuclear weapons, but the greatest *local impact* of man-made releases was the Chernobyl accident[11]. Figure 4.5 shows the annual deposition of ^{137}Cs in the mid-latitude band of the northern hemisphere from atmospheric nuclear weapons testing for the period 1953–82: fallout deposition continued for several years after a test. Also shown in figure 4.5 is the average regional ^{137}Cs deposition in east-central Europe[11] in 1986. However, there were local areas in Europe where the deposition was ten times that in figure 4.5 and in areas surrounding the Chernobyl NPP the deposition was 100 times greater or more[11].

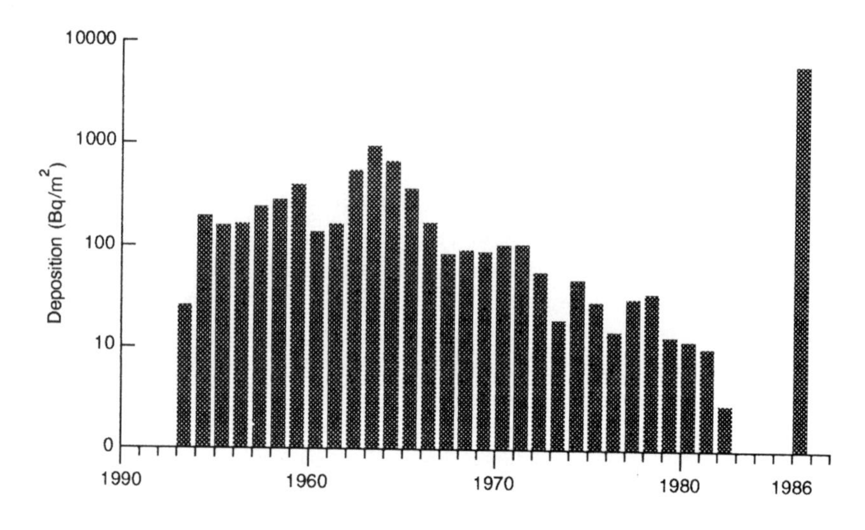

Figure 4.5. ^{137}Cs deposition in east-central Europe from atmospheric nuclear testing (1953–82) and from the Chernobyl accident (1986)[11]. (Courtesy: IAEA.)

Table 4.6 lists[19] the total worldwide activity for 19 radionuclides from atmospheric nuclear tests that have been conducted since 1960 and compares these data with those for the Chernobyl releases taken from the last column of table 4.2.

The main USSR nuclear weapons test site is about 100 km from the city of Semipalatinsk in Kazakhstan, in a very approximately rectangular area of some 19 000 km^2 known as Polygon and including the city of Kurchatov on the river Irtysh and three nuclear reactor sites. Within Polygon there were 26 above ground and 87 in-atmosphere tests in 1949–62 and 346 underground tests in 1961–89. The cumulative external dose at the boundary of this 19 000 km^2 area of Polygon was 2 Gy. The cumulative external dose patterns for the area around Semipalatinsk, which is near the border with Russia and China, were published[20a] in 1995.

Some of the most recent work on radiological conditions at the Semipalatinsk test site have been undertaken by the IAEA[20b] and currently an IAEA videofilm[20c] with the provisional title of *The Polygon* is in production for distribution in the year 2000. One major conclusion of this IAEA preliminary assessment was that 'there is sufficient evidence to indicate that most of the area has little or no residual radioactivity from the nuclear tests. The Ground Zero and Lake Balapan areas, both of which are heavily contaminated, are clear exceptions'.

The centre of the first surface nuclear weapons explosion in the former USSR is historically referred to as *Ground Zero* and figure 4.6 shows the

Table 4.6. Worldwide activity releases from atmospheric testing of nuclear weapons[19] from 1960 and from the Chernobyl accident[8,9].

Radionuclide	Half-life[19] (y = year, d = day)	Nuclear weapons testing (EBq)	Chernobyl accident (EBq)
Tritium	12.32 y	240	—
^{14}C	5730 y	0.22	—
^{54}Mn	312.5 d	5.2	—
^{55}Fe	2.74 y	2	—
^{89}Sr	50.55 d	91.4	0.081
^{90}Sr	28.6 y	0.604	0.008
^{91}Y	58.51 d	116	—
^{95}Zr	64.03 d	143	0.17
^{103}Ru	39.25 d	238	0.17
^{106}Ru	371.6 d	11.8	0.03
^{125}An	2.73 y	0.524	—
^{131}I	8.02 d	651	1.2–1.7
^{137}Cs	30.14 y	0.912	0.074–0.085
^{140}Ba	12.75 d	732	0.17
^{141}Ce	32.50 d	254	0.2
^{144}Ce	284.9 d	29.6	0.14
^{239}Pu	24 100 y	0.006 52	0.000 03
^{240}Pu	6560 y	0.004 35	0.000 044
^{241}Pu	14.40 y	0.142	0.0059

building which was used to monitor the radiation effects of the explosions carried out on this site: some 116 between 19 August 1949 and 25 December 1962, including 30 that were carried out on the surface.

Figure 4.7 shows a field within Ground Zero with the debris of a specially designed *drum bomb* that was tested for its potential in warfare. This was not a nuclear device, but was a drum containing high level liquid radioactive waste which was dropped from a plane to burst open on the ground and spray enemy troops with the radioactive liquid contents.

Lake Balapan or the *Atomic Lake*, figure 4.8, was formed by a cratering nuclear explosion where the explosive charge was placed at a shallow depth below ground. This lake was formed by the first and largest of four such tests, and is 0.5 km in diameter and about 100 m deep with cliffs up to 100 m high. This is a contrast to the normal terrain of the Kazakh steppes which is flat, as in figure 4.7, and where sometimes camels and herds of cows can be seen, figure 4.9.

Figure 4.6. Building at Ground Zero which was used to monitor radiation effects: colloquially known as the *Goose Neck*, May 1999. (Courtesy: V Mouchkin.)

4.9 Comparison with the Techa river area

4.9.1 Mayak

The Mayak Production Association which is located on the eastern shore of lake Irtyash about 70 km north of Chelyabinsk, covers an area of some 200 km^2 and is the site of the first plutonium production uranium–graphite reactor in the USSR, 1948. The complex includes what is termed Facility A, the reactor, and Facility B, a radiochemical plant. For over 40 years the site and surrounding area have been significantly contaminated by the direct discharge of radioactive waste into the environment[19,21].

The principal source of exposure at Facility A was gamma radiation and the main contribution occurred in the central hall of the Mayak reactor, the storage basins and during transportation of radioactive materials. The average annual external gamma-ray doses were 936 mSv in 1949 and were only reduced to 50 mSv by 1957. The average total doses for workers in the period 1949–54 was 1220 mSv[22].

The major factors determining radiation exposure at Facility B were external gamma irradiation and ^{239}Pu aerosols, and practically every room was a radiation hazard. The average annual doses in 1950 and 1951 were, respectively, 940 mSv and 1130 mSv and in the years to 1963 some 30% of the workers were irradiated with doses exceeding 1000 mSv. The average total doses for workers in the period 1949–53 was 2450 mSv[22].

There were three circumstances that led to major exposures[23].

- Large releases into the Techa river system, which was used as a source of drinking water in downstream communities.
- Explosion of a high level nuclear waste tank in 1957, the Kyshtym accident.
- Resuspension from an open intermediate level nuclear waste repository, lake Karachai, in 1967.

These major contaminations in the Urals are summarized in figure 4.10.

During the period 1949–56 liquid radioactive wastes, some 76×10^6 m^3, were dumped into the open river system Techa–Iset–Tobol, figure 4.11, with a total activity of 0.1 EBq (2.75 MCi). The average contents were ^{90}Sr and ^{89}Sr (20.4%), ^{137}Cs (12.2%), rare earth elements (26.8%), ^{103}Ru and ^{106}Ru (25.9%) and ^{95}Zr and ^{95}Nb (13.6%). The major sources of gamma radiation were the contaminated river bed deposits and the soils of the Techa floodlands and water bodies. The highest gamma-ray dose rates were observed during the period of maximum releases, 1950–51, and ranged from 180 R/hr at the release site to 5400 mR/hr on the shores of the Metlinsky village pond to 360 mR/hr on the river bank near the village of Techa-Brod[22].

Figure 4.7. Remains of a *drum bomb* in Ground Zero, May 1999. (Courtesy: V Mouchkin.)

Figure 4.8. Lake Balapan, May 1999. (Courtesy: V Mouchkin.)

Figure 4.9. The Kazakh steppes, May 1999. (Courtesy: V Mouchkin.)

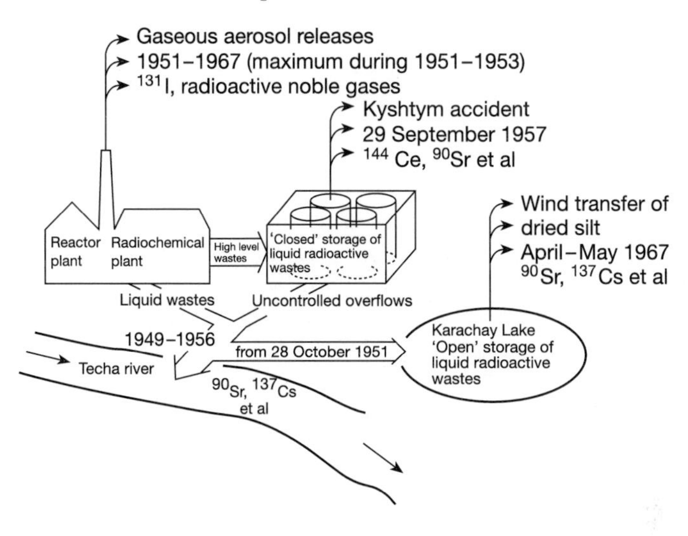

Figure 4.10. Schematic diagram of the contamination in the eastern Urals due to Mayak and Kyshtym facilities. (Courtesy: A Romanyukha.)

The residents of the riverside villages suffered both internal and external exposure, due in part to their irrigated kitchen gardens, and in all 124 000 villagers were exposed. The highest radiation doses were received by 28 000 villagers for whom the river was their main, and sometimes only, source of water supply for drinking and for other domestic purposes. The river water was given to cattle to drink, was used for watering vegetation, breeding waterfowl, fishing and bathing[22,24].

About 7500 people evacuated from 20 settlements received average effective doses in the range 35–1700 mSv. Especially high doses were received by those in the village of Metlino[22], figure 4.11, where certain individuals received red bone marrow doses of 3000–4000 mSv.

4.9.2 Lake Karachai

Lake Karachai is a small natural lake near the Mayak area and in 1951, when large discharges into the Techa river ended, this lake was used for radioactive waste disposal and is currently believed to contain 120 MCi (4.44 EBq) which is primarily ^{90}Sr and ^{137}Cs. In 1967, winds carried about 600 Ci of radioactive particles, mostly associated with dust from the dried exposed shoreline of the lake, up to a distance of 75 km from the site[21].

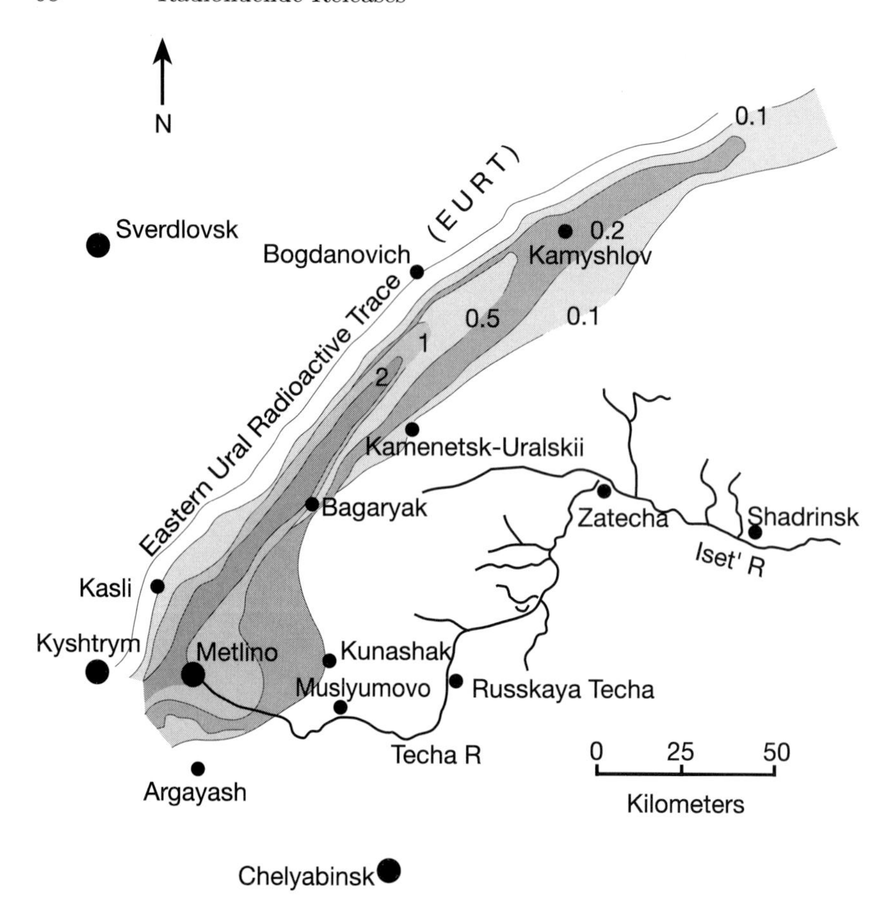

Figure 4.11. The Techa river area and the EURT[2]2. Areas of contamination are shown for 0.1, 0.2, 0.5, 1.0 and 2.0 Ci/km^2. (Courtesy: A Romanyukha.)

4.9.3 Kyshtym

The Kyshtym explosion occurred on 29 September 1957 in a liquid high-level radioactive waste storage tank that contained 20 MCi (0.74 EBq): 90% of the release settled near the site but 2 MCi formed a radioactive cloud that drifted towards the provinces of Chelyabinsk, Sverdlovsk and Tyumen. This is the so-called East Urals Radioactive Trace (EURT). The radionuclides released during this accident consisted mainly of ^{144}Ce and ^{144}Pr (66.0%), ^{90}Sr (5.5%), ^{95}Zr and ^{95}Nb (24.9%) and ^{106}Ru and ^{106}Rh (3.7%).

The maximum length of the EURT, figure 4.11, was 300 km and in terms of ^{90}Sr contamination the deposition was greater than 0.1 Ci/km^2

(3700 Bq/m^2) was 23 000 km^2, greater than 2 Ci/km^2 was 1000 km^2 and greater than 100 Ci/km^2 was 117 km^2. Approximately 10 200 people were evacuated from the contaminated areas at different times after the accident.

4.10 Comparison with Tokaimura

At 01:35 GMT (which was 10:35 local time) on 30 September 1999 a criticality accident occurred at the Tokaimura nuclear fuel fabrication plant owned by JCO Ltd, and which is 160 km north of Tokyo. Three workers were significantly overexposed, one very seriously; and in addition it was reported that a further 66 workers were also exposed, including three emergency service workers and seven local workers assembling scaffolding on a construction site just beyond the JCO site. The cohort of 27 workers who were engaged in the operation to drain water from the cooling jacket received doses which are preliminarily estimated[30b] to be in the range 0.04–119 mGy

A radioactive release occurred and for the local population up to 350 m from the plant evacuation was mandatory, and within 10 km distance the population was advised to remain indoors. This advice remained in force until 1 October[25,26]. The severity of the accident was classified as Level 4, table 4.7.

Table 4.7. Major nuclear accidents in terms of the IAEA International Nuclear Event Scale[30a]. Level 3—Serious incident, Level 2—Incident, Level 1—Anomaly.

Accident	Year	Severity classification on scale 0–7	
Chernobyl	1986	Level 7	Major accident
Kyshtym	1957	Level 6	Serious accident
Windscale pile	1957	Level 5	Accident with off-site risk
Three Mile Island	1979	Level 5	
Tokaimura	1999	Level 4	Accident without significant off-site risk
Windscale reprocessing plant	1973	Level 4	

For some 20 hr after the onset of criticality, radiation could be measured at some distance. However, only trace amounts of noble gases and gaseous iodine escaped from the building itself. After criticality was terminated and shielding was in place, radiation levels beyond the JCO site returned to normal. As the half-lives of the radionuclides detected are relatively short, there is no residual contamination[30b].

There were similarities with Chernobyl in that workers were overexposed, radioactivity was released into the atmosphere, some of the population was evacuated and others advised to remain indoors, but there the similarities end. The differences are that: the scale of the accident was significantly less than Chernobyl, level 4 *versus* level 7, table 4.7; the fact that the accident was brought under control within 20 hr; and that it was a criticality accident and not a nuclear reactor accident.

A *criticality accident* is one in which there is an unscheduled fissioning, a chain reaction, of ^{235}U. This generates a strong external radiation field comprising neutrons and gamma rays and the release of radioactive materials in the form of fission products.

Accidents involving ionizing radiation are divided into nuclear accidents and radiological accidents. The former involves a nuclear facility, especially a reactor, but also a critical assembly, fuel element manufacture, radiopharmaceutical manufacture or a fuel reprocessing plant[27]. The latter involves sealed or unsealed radiation sources and leads to an uncontrolled release of ionizing radiation or radioactive materials into the environment. Such radiation sources include x-ray equipment, sealed sources of ^{60}Co, ^{137}Cs or ^{192}Ir which are mostly used in medicine for the treatment of cancer or used in industry; and unsealed sources used in nuclear medicine and scientific research. Other radiation sources are radioactive wastes.

The accident at Tokaimura occurred because of human error in a process involving adding an oxide of uranium, U_3O_8, to nitric acid. This was one step in the process of making nuclear fuel rods. The solution would then be mixed in a sedimentation tank with a solution of ammonium salt to form a precipitate of ammonium diuranate which is later processed to form uranium dioxide fuel.

The U_3O_8 and nitric acid were supposed to be mixed in an elongated container (a shape designed to prevent a critical mass of uranium coming together) before being fed into the sedimentation tank. The workers had actually mixed the U_3O_8 and then, incredibly, poured the mixture into the tank using buckets.

After the seventh bucket, the sedimentation tank contained 16 kg of uranium, enriched so that 18.8% was the fissile radionuclide ^{235}U, this formed a critical mass and started a chain reaction. It was reported that the most seriously irradiated worker, who received an estimated dose in the range 10–20 GyEq, also quoted in some sources as 17 Sv, saw a flash of blue light as the air was ionized by the radiation release. This 35-year-old male, Hisashi Ouchi, who was unconscious when he was evacuated, was treated on 5 October not by bone marrow transplant, but by peripheral blood stem cell transplantation (PBSCT) with his sister as the donor who was able to provide a perfect match[28,30]. At the time of the IAEA preliminary fact finding mission[30] he had survived 16 days which was unexpected because previous experience of irradiation at this dose level led to a fatal outcome within two weeks. He eventually died on 21 December of multiple organ failure, a total of 83 days post-accident. His doctors at the University of Tokyo Hospital reported that they had raised the oxygen concentration in his respirator to 100% after his breathing had worsened a few days before he died. For two months he had been receiving 15 litres of blood daily.

The second most highly irradiated worker, Masato Shinohara, aged 39 years, who received 6–10 GyEq, had severely damaged bone marrow and it was planned for him to receive a PBSCT but no appropriate related donors could be found. The decision on an alternative treatment was to perform a transplantation of umbilical cord blood cells. On 8 October foetal stem cells were taken from his umbilical cord and were infused. By the end of 1999 he remained hospitalized after undergoing skin grafts on his hands and forearms on 20 December. His prognosis remains uncertain.

The third worker, Yutuka Yokokawa, aged 54 years, who was at a distance of 5 m from the tank, received 1.2–5.5 GyEq, was almost asymptomatic after exposure. He has received cytokines for bone marrow stimulation and supportive care. His prognosis for full recovery is good and on 20 December he left the Chiba Radiation Research Centre where he was taken after the accident. However, he will be expected to be subjected to an increasing risk of incurring cancer or leukaemia at a later date[30b].

Table 4.8. Advantages of PBSCT over autologous BMT[29].

- No need for general anaesthesia.
- No need for an invasive operation.
- Applicable for those over 60 years of age.
- Applicable after irradiation of the pelvis.
- Significantly shorter haematological recovery.
- Significantly less aplastic morbidity and mortality.
- Reduced quantity of antibiotics.
- Reduced demand for red blood cell and platelet transfusion.
- Reduced isolation and hospitalization times.
- Dramatically reduced cost of procedure.
- Reduced probability of relapse.
- Reduced long-term cost.

The doses were estimated by four methods: from blood using ^{24}Na activation measurements; analysis of chromosomal aberrations; lymphocyte counting; and, in the case of the patient exposed to 1.2–5.5 GyEq, by ^{24}Na whole body counting. The dose measurements made using ^{24}Na in blood gave 18 GyEq, 10 GyEq and 2.5 GyEq but they must be considered to be preliminary estimates, owing, among other factors, to the inhomogeneous exposures of the workers' bodies[30b].

It is mentioned elsewhere, section 5.6, that BMT was largely unsuccessful in the treatment of the Chernobyl casualties and that in the 1990s it is considered only to have a limited role in the treatment of radiation accident victims; that is, for victims receiving doses in the range 8–12 Gy where the irradiation is uniformly distributed but the patient is without serious skin injuries and there is an absence of severe internal contamination and conventional injuries[29]. As previously mentioned, none of the Tokaimura accident victims received a BMT. The advantages of PBSCT over autologous BMT are given in table 4.8.

Figure 4.12. Cartoon circulating in Vienna in August 1986 at the time of the post-accident review meeting at the IAEA

Chapter 5

Early Medical Response and Follow-up of Patients with Acute Radiation Syndrome

Introduction

The medical response in this chapter title is defined as that given to the power plant workers on duty at the time of the accident, to some of the early liquidators such as the firemen, six of whom died, and to those who were diagnosed with acute radiation syndrome. The initial response also includes the bone marrow transplants which were carried out in Moscow, the measures taken in Pripyat and Kiev to limit ^{131}I absorption in the thyroid gland, and the medical screening of the evacuees, particularly the children. The establishment of the emergency medical teams is also detailed, some case histories are documented and, as in chapter 3, some eyewitness accounts are given. The current medical department in Chernobyl town is also described. Subsequent medical effects such as the incidence of thyroid cancer in children, the incidence of leukaemia and of psychosocial illness are given in later chapters.

5.1 The first physician to arrive at the power plant after the accident

The first doctor at the scene of the accident[1] arrived almost immediately. This was Dr Valentin Petrovich Belokon. He was 28 years old, with two young daughters aged five years and one and a half months. He was also a sportsman who specialized in weight lifting and was employed in Pripyat as an accident and emergency physician. By the autumn of 1986, he was working as a paediatrician in Donetsk and suffered breathing problems as an after-effect of his experiences.

> On 25 April 1986 at 20:00 hours I started my work at Pripyat, where there is an accident and emergency brigade consisting of one physician (myself) and a doctor's assistant Sasha (Alexander) Ckachok, and six ambulances. On 25 April, Sasha and I worked separately and my driver was Anatoloy Gumarov. At 01:35 hours on 26 April on my return to the medical centre I was told there had been a call from the nuclear power plant and that two or three minutes earlier Sasha had left for the power plant. At 01:40 hours he telephoned to say that there was a fire, with several people burnt and that they needed a doctor. I left with my driver and arrived in 7–10 minutes.
>
> When we arrived the guard asked: 'Why don't you wear special clothes?'. I did not know they would be needed and was only wearing my doctor's uniform, and since it was an April evening and the night was warm I did not even wear a doctor's cap. I met Kibenok (a fireman lieutenant who later died) and asked: 'Are there patients with burns?'. Kibenok's reply was that: 'There are no patients with burns but the situation is not clear and my boys feel like vomiting'
>
> My talk with Kibenok was near the energy block (unit no. 4) where the firemen stood. Pravik (also a fireman lieutenant who later died) and Kibenok had arrived in two cars and Pravik quickly jumped out of the car but did not come to see me. Kibenok was excited a little and alarmed.

Dr Belokon then described his first patients.

> Sasha Ckachok had already taken Sashenok (the second power plant worker to die, the first was buried in the rubble of unit no. 4 whereas Sashenok died of extensive burns) from the nuclear power plant from which he had been pulled by workers after being burnt and crushed by falling beams. He died on the morning of 26 April in a medical recovery room.
>
> My second patient was a young boy about 18 years old. He had vomiting and severe headache, and as I did not yet know about the high level of radiation I asked him: 'What have you eaten and how did you spend the previous evening?'. His blood pressure at 150/90 was slightly higher than the normal 120/80 for an 18 year old. However, the boy was very nervous. At this time, workers who came out of the nuclear power plant were very disturbed and only exclaimed: 'It is horrible' and that 'The instruments went off scale'.
>
> Three or four men from the technical staff all had the same symptoms of headache, swollen glands in the neck, dry throat, vomiting

> and nausea. They all received medication and were then put into a car and sent to Pripyat with my driver Gumarov. After that, several firemen were brought to me and they could not stand on their feet. They were sent to hospital.

Dr Belokon now begins to feel unwell at 18:00 hours on 26 April, and records his symptoms and thoughts.

> I felt something wrong in my throat and had a headache. Did I understand that it was dangerous? Was I afraid? Yes, I understood. Yes, I was afraid but when people see a man in a white uniform is near it makes them quieter. I stood as all of us stood, without any breathing apparatus, without any other means of protection. When it became lighter on 27 April, there was no fire to be seen in the block (unit no. 4), but there was black smoke and black soot. The reactor was spitting but not all the time, only as follows: smoke, smoke, then belch.
>
> Gumarov arrived back from Pripyat after taking the injured to hospital and I felt weakness in my feet. I did not notice it when I walked, but now it has happened. Gumarov and I waited another five minutes to see if anyone else asked for assistance, but nobody did. That is why I said to the firemen: 'I am going to hospital, if there is a need, call us again'.
>
> I went home, but before I washed and changed my clothes, I passed iodine to those in the hospital, asked them to close all the windows and to keep the children inside. Then I was taken to the treatment department of a hospital, by our Dr Dyakanov, and given an intravenous infusion., I felt very bad and started to lose my memory, at first partially and then totally. Later, in Moscow, in Clinic No. 6, I was in one ward with a dosimetrist. He told me that just after the explosion all instruments were off-scale, that they called to the safety engineer who then answered: 'What is the panic? Where is the shift chief? When he is available tell him to call me, you yourself don't panic, such a report (about off-scale measurements) is not correct'.

5.2 Medical examinations

The main workload was undertaken by institutions under the aegis of the Ukranian Ministry of Health[2]. The personnel included more than 7000 medical staff, 230 mobile dosimetry laboratories and 400 teams of physicians which included 212 specialist teams.

5.2.1 Experience of a medical student

These 7000 were divided into 450 medical teams and included medical students who were drawn from several cities, including Kiev where the medical course takes six years with the final examinations in May–June each year. The fifth-year medical students were therefore drafted as Chernobyl liquidators* rather than the sixth-year students[3].

They arrived at the power station on 3 May 1986 and worked till 17–18 May, on eight-hour shifts. They were sited near the border of the 30 km exclusion zone and, more often than not, worked in tents. The paediatric medical students from Kiev, of whom there were some 25, saw about 100 children per day and divided them into three groups. Group 1 were termed *Healthy* as they had no signs of symptoms or disease. Group 2 were also termed *Healthy* but they had suspected signs or symptoms of disease. Group 3 were termed *Sick*.

Thyroid ^{131}I uptakes were measured in the medical tents, but this was only possible because the Kiev Urology Institute had appropriate counters which they normally used for kidney function studies. If the measured ^{131}I uptake was greater than the threshold ^{131}I uptake for any Group 1 or 2 child, then they were also sent to Kiev with the Group 3 children who were sick.

5.2.2 Experience of a haematologist

A total of 500 000 people were examined including 100 000 children and more than 200 000 of these examinations included uptake measurements for ^{131}I and ^{137}Cs, but the examinations were complicated by the 'extremely low level of knowledge of radiation medicine' of many of the physicians[2].

This was confirmed by the personal experience of one of the physicians[4] from the Institute of Haematology and Blood Transfusions, Kiev, who was in a field team in early May 1986. Each field team consisted of three qualified physicians, a haematologist, an endocrinologist and an opthalmologist. The visits, each over a period of 7–8 days, were to schools to examine the boys and girls. They worked each day from 7am in the morning to 9pm in the evening and they lived in the so-called *Lenin Room* of each school. This was a room in Soviet times for party meetings and there was one in every school, hospital and government office. The walls contained a portrait of Lenin and his *words of wisdom*.

However, in Narodichi village school, there was no hot water for showers or even to wash hands because it had been Pripyat river water which was fed to Narodichi. Water therefore had to be fed from other streams and was filtered through sand and looked brown and rusty. There was in the school a bathroom for the children with a single bath tub and this was

* They are entitled to a category 1 liquidator certificate, see section 5.9.

used as a water reservoir. Only in the morning could the physicians use the water, to which potassium permanganate crystals had been added as a disinfectant. The physicians closed their eyes before using the water to wash their hands!

The types of examination for each child included a complete blood count except for those considered to be at high risk, when basic biochemistry tests were also carried out. Haemoglobin measurements were made by a technician using a primitive machine. No blood counters were available and therefore the haematologist often worked all night looking visually at the blood smear test results, taken from a finger, and manually counting platelets, neutrophils, monocytes, lymphocytes and atypical cells.

The physical examinations, because of lack of equipment, were basic in the extreme. The haematologist palpated lymph nodes, the endocrinologist examined the testicles and (without an opthalmoscope available) the opthalmologist could only look at the eyes visually[4].

5.3 Iodine prophylaxis against thyroid cancer

5.3.1 Radiation-induced thyroid cancer

External radiation is known to cause thyroid cancer such as the case in a series described by Sir Stanford Cade[5,6] of a man treated in 1912 for enlarged cervical lymph nodes who, 42 years later, developed carcinoma of the thyroid, although the latent period for radiation-induced cancers is not usually more than 40 years. Indeed, it is more likely to be some ten years post-irradiation[7], for some non-malignant condition such as goitre, but this was in the early years of the 20th century and also for a small proportion of those treated. It should also be noted that most of the documented cases are of skin cancers caused by the use of x-rays in the early treatment of ringworm or for removal of facial hair as a beauty treatment[6]. Radioactive ^{131}I which has been used for many years in the treatment of thyrotoxicosis has never been proven to cause thyroid cancer[7,8].

5.3.2 Iodine metabolism

The relatively high thyroid doses due to the Chernobyl accident in areas of high deposition of ^{131}I such as Gomel in Belarus are due to the thyroid's ability to concentrate iodine, which enters the blood stream through inhalation or ingestion, of which the former is likely to have been particularly important near the Chernobyl NPP.

Iodine's metabolism is such that it is rapidly absorbed from the gastrointestinal tract: some absorption takes place in the stomach but the greater part occurs in the small intestine. After absorption, iodine, whether radioactive or not, will become distributed throughout the extra-cellular fluid. It also diffuses into red cells where it is present at a concentration of about 65% of the plasma concentration. Iodine is removed from the plasma almost entirely by the thyroid and kidneys and the normal value for the thyroid clearance rate[9] is 20 ml/min.

The high propensity of the thyroid for iodine is increased in people with iodine deficiency, the latter causing endemic goitre. This is important because it has been recognized that some of the populations in Belarus and Ukraine which were affected by the Chernobyl accident's ^{131}I deposition, included a high proportion of persons with iodine deficiency. This, therefore, caused a larger than normal uptake of ^{131}I.

If, on the other hand, there had been little or no iodine deficiency, and prophylactic iodine had been administered to block the thyroid from absorbing more iodine, then the radioactive ^{131}I would not have been absorbed. The underlying concept for iodine prophylaxis was therefore to limit the absorption of ^{131}I and hence the incidence of thyroid diseases, possibly including cancer. This concept was well recognized throughout the world, but the problem in late April and early May 1986 was to have enough stable iodine available for the populations, especially the children and adolescents whose thyroids were more susceptible to irradiation than those of adults.

5.3.3 Distribution of stable iodine in the USSR immediately after the accident

Potassium iodide or iodate tablets were distributed, with a considerable delay due to a lack of reserves of stable iodine, (given only from 7 May), to some 5.3 million people which included 1.6 million children: but even so, there were large numbers of children who did not receive any tablets[10]. This was due in part to the major fault in the immediate phase following the accident of a failure of communication, not only with the local population but also to the governmental bodies of Ukraine, Belarus and Russia[2].

A second major fault was the absence of data on radiation doses received by both the local population and the liquidators. This led not only to an inadequate distribution of stable iodine, but also to delays in the evacuation of populations[2].

The inadequacy of the distribution is well encapsulated in the experience of a haematologist who was also a liquidator in early May[4], see also section 5.2.2.

> When parents realized that no tablets were available early on, they bought 50 ml bottles of 1% or 5% potassium iodide liquid *over the counter* from a pharmacy and then purchased separately an eye dropper. Using the dropper they made their young children take the iodine liquid and the result was seen in cases of gastric burning and gastric irritation and of vomiting.

Another course of action taken by parents was to use a medicine that had been taken widely before Chernobyl, a *Russian sedative mixture* which contained five ingredients, one of which was potassium iodide. The children who had been given this sedative in relatively large doses were found later to have ^{131}I uptakes much lower than in other children.

The distribution of stable iodine in Kiev was late, and only 10–20% of children received this prophylaxis[11]. However, it was reported by the Soviet Delegation[12] that potassium iodate tablets were distributed to the NPP workers at 03:00 hours on 26 April and in the town of Pripyat later that day at 20:00 hours. This was undertaken by medical staff and local volunteers, who organized a door-to-door visiting schedule.

5.4 Acute radiation syndrome initial diagnoses

The problem immediately following the accident was to diagnose at the NPP the most severely injured cases. Essentially this meant the diagnosis by degree of acute radiation syndrome (ARS), see section 1.2.1, but not surprisingly some cases had their degree of ARS re-graded on arrival at hospital when there was more time for assessment. The August 1986 report of the USSR authorities[12] is summarized in table 5.1.

No clinical symptoms of ARS were seen in any of those evacuated from the 30 km zone and therefore the cohort of ARS cases only contained NPP workers and some of the firemen who fought the original blaze.

The definitions of the first and second degrees were not always consistent and the first degree was also defined as less than 1 Gy and the second degree as 1–4 Gy. The upper and lower limits of ARS were also questioned following a re-analysis of the data from Hiroshima and Nagasaki and table 5.2 gives the five grades of dose–effect relationship for clinical illness which were published by Geiger in JAMA[13].

5.5 Follow-up of patients with acute radiation syndrome

There were 444 people working at the NPP and some 300 were admitted to hospitals. The figure of 203 in table 5.1 was later revised downwards to 134 with a diagnosis of ARS: 108 in Moscow hospitals and 26 in Kiev hospitals[10]. Of these, by September 1986, 14 remained in hospital, three in Moscow and 11 in Kiev, and by January 1987 there were only five remaining in hospital[14].

Table 5.1. The 203 cases originally diagnosed with acute radiation syndrome[12].

Degree of ARS and dose range (Gy)		No. of patients hospitalized		No. of Deaths
		Kiev	Moscow	
Fourth¶	6–16 Gy	2	20	21
Third	4–6 Gy	2	21	7
Second	2–4 Gy	10	43	1
First§	1–2 Gy	74	31	0

¶ The characteristics of the fourth degree are that the period is short, 6–7 days. Primary reactions are early, in the first 15–30 minutes. The number of lymphocytes is less than 100 per microlitre. On the seventh to ninth days there is vomiting, damage to the digestive tract and the granulocytes are less than 500 per microlitre. Thrombocytes from the eighth to ninth days are less than 40 000. General intoxication is clearly shown and there is fever. In 18 cases there were great beta-ray burns on large areas of skin and in two patients there were also heavy thermal burns. The lethal outcome commenced from the ninth day and 21 patients, all with fourth degree ARS, were dead by the 28th day[12].
§ Confirmation of a diagnosis of first-degree ARS requires a much longer observation period for the patient[12].

5.5.1 Deaths

Out of the 134, 28 died within the first three months and the survival times as reported by Guscova[15] are given in table 5.3. Of the 28, a total of six were firemen and the remainder included engineers, technicians, power plant operators and other NPP workers.

Photographs of four of the victims who did not survive are given in Plate III. These were shown publicly by the USSR only once[12] and are reproduced here from photographs taken directly from the projection screen. The small red artefact on three of the photographs is the image of a laser pointer. These figures represent reality and are very different from the *sanitized* photographs of smiling patients which were distributed by TASS to the media.

All firemen were in their twenties although Plate III (*top left*) could be thought to be of a 70 year old. This victim and his colleagues did not have adequate protective clothing; for example his fire helmet had no backpiece

Table 5.2. Dose–effect relationships for ARS[13] (1 Sv = 100 rem). The ICRP recommended dose limits for those occupationally exposed are an annual effective dose of 20 mSv (2 rem) and an annual equivalent dose to the skin of 500 mSv (50 rem), see table 1.1 in section 1.1.2.

Dose range (Sv)	Clinical Illness	Percentage surviving if treatment given
10–50	Acute encepalopathy and cardiovascular collapse	0
6–10	Gastrointestinal syndrome	0–10
3–4	Severe leukpenia, thrombocytopenia and epilation	50
1–2	Nausea and vomiting, bone marrow suppression	100
0.15–0.5	Asymptomatic, maybe chromosome aberrations	100

Table 5.3. Survival times of the 28 NPP workers and firemen who died within three months.

Degree of ARS	No. of cases	Survival times (days)
Fourth	20	10, 14, 14, 14, 15, 17, 17, 18, 18, 18, 20, 21, 23, 24, 24, 25, 30, 48, 86, 91
Third	7	16, 18, 21, 23, 32, 34, 48
Second	1	96

to prevent radioactive particles from dropping down between his shirt collar and his skin. When attempts were made to decontaminate the skin, some of the beta-ray emitting particles were inevitably pushed deeper into the tissue, resulting in a second wave of beta-ray burns. The red colour burn on his neck is clearly seen.

Using Guscova's commentary, Plate III (*bottom right*) shows 'epilation of the scalp, and the blue skin is where there was total ulceration', Plate III (*bottom left*) shows 'a heavy burn to the thigh, which has deep damage with very painful sections: scabs are also forming'. Plate III (*top right*) was described as follows: 'patients have very different kinds of burns. This one is another characteristic dark pigmentation. You have these waves which are moving into the skull of this patient, into the ear and in the edges of the eye.' This patient was further described as having 'a viral infection of the herpes type: herpes simplex of the facial skin, lips and oral mucosa'.

After the acute phase of deaths within three months post-accident, there had, by the end of 1995, been an additional 14 deaths among those diagnosed in 1986 with ARS. However, these deaths are not correlated with the original severity of ARS and, in some cases, are certainly not directly attributable to radiation exposure. The 14 causes of death were[30]: four due to coronary heart disease, two due to myelodysplastic syndrome, and one death due to each of the following causes: car accident, hypoplasia of haematopoiesis, lung gangrene, encephalitis/encephalomyelitis, sarcoma of the thigh, lung tuberculosis, liver cirrhosis and fat embolism.

Finally, the individual doses to thyroid and lungs accumulated by the time of death among 21 of those who died are given in table 5.4. These doses were reconstructed using *in vivo* monitoring of gamma-emitting radionuclide clearance in the urine, as well as from post mortem studies of the radionuclide distribution in thyroid and lungs[16].

Table 5.4. Equivalent doses (in Sv) to thyroid and lungs of 21 liquidators who died from acute radiation sickness[16].

Thyroid doses	Lungs doses
21	0.26
24	2.8
54	0.47
62	0.57
77	0.68
130	1.5
130	2.2
210	3.5
310	2.3
340	8.7
320	27
470	4.1
540	6.8
600	120
640	34
890	9.4
740	29
950	20
1900	19
2200	21
4100	40

5.5.2 Skin burns

Obvious signs of radiation-induced skin injuries did not develop until three days after the accident, when a transient skin erythema developed which lasted no longer than one day. One group of patients then developed widespread erythema 5–10 days later with areas of skin breaking down, some of which required surgery. It is thought that this might have been caused by non-uniform exposure greater than 80 Gy to these areas: some of which were exposed and some of which were covered by clothing.

Another group developed moderate to severe reactions 21–24 days after the accident and these particular skin reactions were similar to those which can develop after radiotherapy and therefore suggest surface skin doses in the range 20–80 Gy.

The most severely affected patients were those who had remained in the area of the accident for up to five hours and, in these cases, even their clothing had become sufficiently contaminated to expose covered skin areas to a radiation reaction. In addition some 28–30 patients also experienced a late wave of radiation erythema two to four months after the accident. This late stage occurred after the acute skin reaction had healed and recovery from bone marrow damage[12] had occurred .

The grouping by Guscova[12] of 56 patients with a history of radiation burns is given in table 5.5. If the burns covered more than 40% of the body surface then the patient would inevitably die. One of the less severe cases has already been described, see section 3.1.5, and Plate II shows the late effect of this NPP control room operator's burns.

Table 5.5. Grouping of 56 patients by medical history of radiation burns[12].

Group	No. of patients
Burns three weeks after the accident	48
Non-compatible with life (40–100% body surface burns)	20
Life threatening (1–40% body surface burns)	9
Non-life threatening (1–40% body surface burns)	19
Burns later than three weeks after the accident	9
Non-life threatening (1–40% body surface burns)	9

A special meeting[17] was organized by the IAEA in 1987 to discuss the problems of the skin burns and the aim of this meeting was described in the following terms.

> The accident at Chernobyl caused an unexpected high frequency of skin lesions in various combinations, that is, thermal and chemical burns with contamination, or beta-ray burns with thermal

and chemical lesions, skin burns with various degrees of external irradiation. New methods have recently been applied in the treatment of skin lesions, for example the use of artificially produced human skin which has enabled brilliant successes even in cases in which the area involved in the burns is 90% of the body surface.

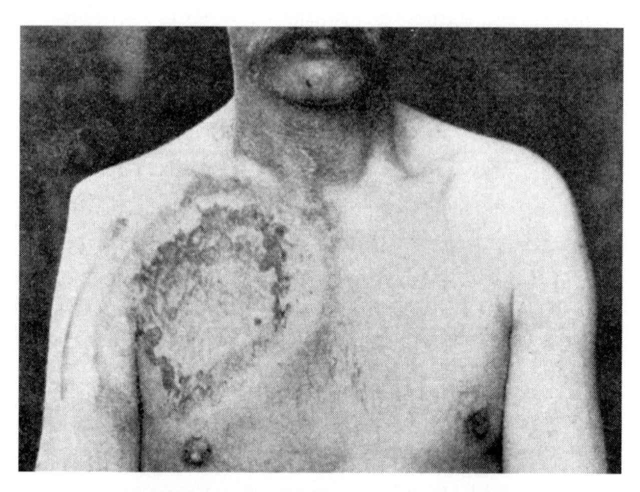

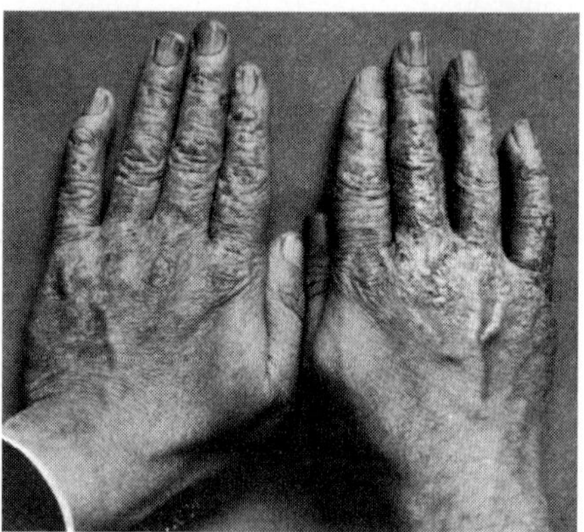

Figure 5.1. X-ray burn to an American soldier, 1898 (Courtesy: Otis Archives of the Armed Forces Institute of Pathology, Walter Reed Medical Center, Washington) and the hands of a pioneer radiologist in Philadelphia, Mihran Kassabian, showing the result of chronic x-ray dermatitis[18a]. He began x-ray work in 1899, the photograph was taken in 1903, and he died in 1910, due to his x-ray injuries, at the early age of 40. Such wrinkled and shrivelled hands were often termed *Röntgen hands* with the later injuries becoming progressively worse and turning into radiation induced ulcers which eventually could develop into skin cancers[18b].

On a historical note[18b], radiation-induced skin burns were first documented soon after the discovery of x-rays by Wilhelm Röntgen in November 1895 and for comparison with the Chernobyl burns figure 5.1 shows one of the earliest photographs of an x-ray burn. This dates from 1898 in the Spanish–American war. A private soldier in the sixth United States Infantry had received a gunshot fracture of the upper third of the humerus for which an excision of part of the humerus was made. In December 1898 using a 20 minute exposure at 10 inches from the shoulder an attempt was made at radiography, but this was unsuccessful. The second and third attempts also did not work. Six days after the last exposure a slight erythema appeared, later forming small coalescing ulcers and tissue necrosis. The burn showed no sign of healing for four months and was not entirely healed until 11 months after the x-ray exposure[19]. Although this was not a beta-ray burn, it would have been from relatively low energy x-rays as the year was 1898. It is also of interest to record that as early as 1904 a scale of x-ray burn severity had been defined[20], table 5.6, which can be compared with those of tables 5.1 and 5.2.

Table 5.6. Degrees of x-ray burns defined in 1904 by a New York physician[20].

Degree	Characteristics
Third	This is the gravest degree and is characterized by the escharotic destruction of the irradiated tissues. They show signs of dry gangrene and their appearance is brownish-black. The necrotic area should be surgically removed. An ulcer remains which may take months to heal.
Second	Main feature is the formation of blisters containing clear or yellowish contents. Inflammatory signs are well pronounced. Pain is intense.
First	Characterized by the symptoms of hyperaemia. The most pronounced subjective symptom is a tormenting itching of the skin.

It has already been noted in section 4.6 that the radionuclide releases were very different for the atomic bomb explosions and Chernobyl, and so were the injuries received. Table 5.7 summarizes the acute (early) and late effects[21,22] and figure 5.2 includes an example[22] of late effects following burns which, healing once, formed keloids with the typical raised scars. For further reading of the medical effects of the Hiroshima atomic bomb the books by John Hersey[23], a *New York Times* journalist who visited in May 1946, and by Anne Chisholm[24] who followed up young girls who survived the explosion, are recommended.

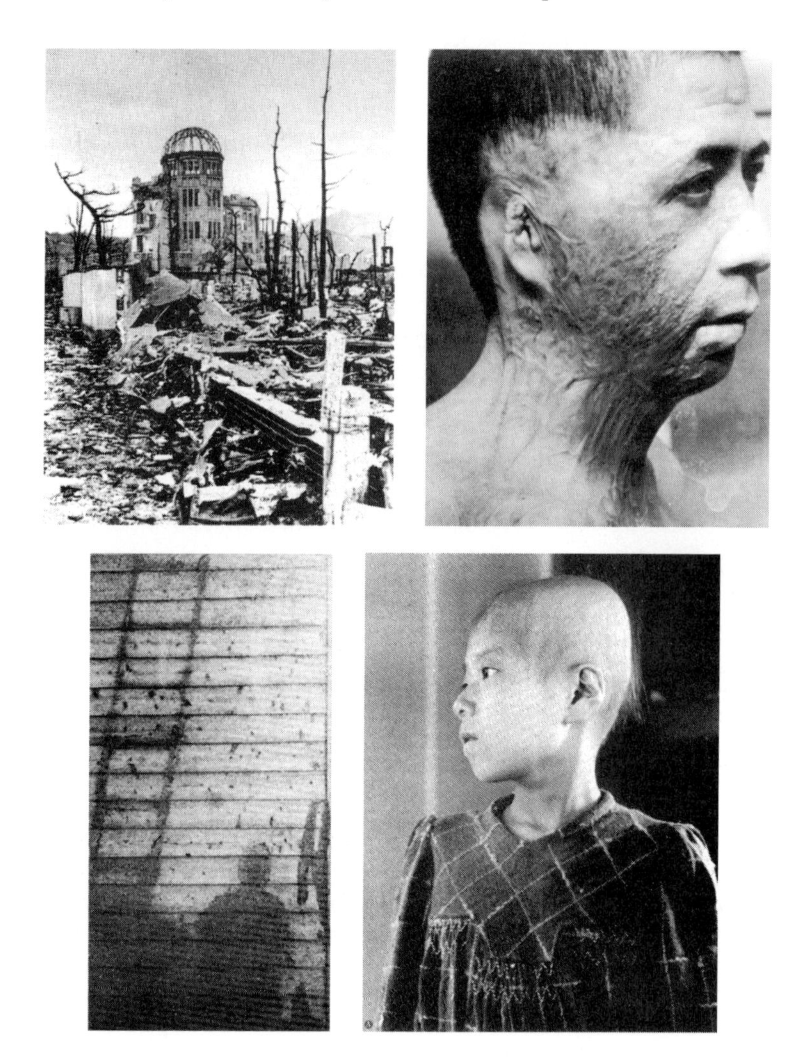

Figure 5.2. Images of Hiroshima and Nagasaki[21,22]. (Top left) The Hiroshima A-bomb dome, October 1945. The bomb exploded at about 160 m south-east of, and in the air at some 580 m above this dome. It now stands as it did at the time of the explosion, and acts as a permanent memorial. (Bottom left) Shadow of a man and ladder imprinted upon a wooden wall, 3.5 km from the hypocentre, Nagasaki. (Top right) Burns due to the atomic bomb, healing once, formed keloids with the protuberance of these scars, Nagasaki. (Bottom right) Epilation due to acute radiation effects. Epilation markedly appeared up to the 8th week after the bomb, and to the 10th week at the latest. Hiroshima, about 1 km from the hypocentre. (Courtesy: The cities of Hiroshima and Nagasaki.)

Table 5.7. Some examples of the acute and late effects of the atomic bomb[22].

Acute effects
• Effects were caused by the complicated involvement of the following factors: burns due to thermal radiation and fire, injuries due to the blast, and radiation effects.
• At distances within approximately 1 km from the hypocentre, the effect of any one factor, be it thermal radiation, blast or initial radiation, alone was fatal in many cases.
• Epilation markedly appeared up to the eigth week after the bomb, and to the tenth week at the latest, see figure 5.2.
Late effects
• Burns, healing once, formed keloid scars, see figure 5.2.
• Besides keloids, the following late effects also occurred: opthalmological disorders (e.g. cataract), blood dycrasias, malignant tumours, psycho-neurological disturbances.
• Of the malignant tumours the following are considered to have been radiation exposure related in some of the survivors: leukaemia and cancers of the thyroid, breast, lung and salivary glands.
• Due to *in utero* exposure (e.g. seven weeks gestation) delayed growth and microcephaly.
Number of deaths
The damage being enormous and caused instantaneously, an accurate number of casualties is not available. However, it is estimated that deaths by the end of December 1945 were approximately (±10 000) 140 000 in Hiroshima and 70 000 in Nagasaki. The sociological destruction was also enormous for the survivors and as regards buildings, about 76 000 in Hiroshima and about 51 000 in Nagasaki were either burnt down or destroyed. It required almost a decade for Hiroshima and Nagasaki to be restored from the massive destruction.

5.5.3 Survivors

Among the remaining 106 who recovered from ARS (28 died, table 5.3), most had after-effects of emotional and sleep disturbances and some 30% suffered various disorders, including gastrointestinal, cardiovascular and immune function illnesses, that reduced their ability to work.

Most of the first wave of liquidators have developed symptoms of premature ageing such as lung disease, heart problems and arthritis. In addition, some have committed suicide, others have turned to drink, their wives are afraid to have more children and depression is common[25]. In the Ukraine about 4000 of the 173 416 registered liquidators have died and the

Ukranian Ministry of Health claims that 77% of the deaths which occurred in 1994 were Chernobyl related. It is also estimated by the organization called the *Chernobyl Union* that about 10% of the 350 000 Russian liquidators are now disabled and that 2% have died from various causes including suicide. Fuller details of the illnesses suffered by these liquidators are given in subsequent chapters.

Liquidators working in later years, after the first wave, and taking part in clean-up operations, are healthier and usually still able to work after completing their Chernobyl duties.

A special group of liquidators were the firemen, and in March 1987 the chief of the Chernobyl NPP fire service at the time of the accident, Major Leonid Telyatnikov, visited the United Kingdom and recorded the treatment and rehabilitation schedule for the firemen who survived[26], table 5.8. However, 'resumption of duties' did not mean full duties but only desk jobs, because those who survived usually suffered permanent lung damage.

The bravery of the firemen is remembered by a memorial at the side of the road to the NPP past the control and passport point at Ditjaki village on the border of the 30 km zone and between Chernobyl town and the NPP, figure 5.3. The tall column of the sculpture is topped by a cross and the hollow spherical form symbolizes an atom's orbiting electrons. Further down is the ventilation stack of unit no. 4. and a series of firemen.

They are also remembered in a museum in the Kiev City Fire Department building near Contractoviya Square, formerly named Red Square, which houses many items of interest, including fire suits donated by Germany, the United Kingdom and France, and many photographs. It is the only museum devoted to the Chernobyl accident in the former USSR.

Table 5.8. Treatment and rehabilitation of the Chernobyl firemen.

- 2 months of treatment.
- 1 month of rehabilitation.
- 1–2 months health resort treatment at a sanatorium.
- 2–3 months of general medical checks.
- Resumption of duties.

The firemen, together with other liquidators, see for example table 5.10, after completing their work were issued with a special passport entitling them to a range of benefits, see section 5.9. In addition there are benefits in terms of wages for the current workers at the Chernobyl NPP and these include special bonuses, which can double their pay packet when they work in a contaminated area: they term these *coffin bonuses*[25].

Figure 5.3. Memorial to the firemen of Chernobyl. (Photograph: R F Mould.)

Table 5.9. Classification of liquidators and their type of work[27].

Class of liquidator
• Gas protection regiment.
• Motor battalion.
• Composite regiment on guard in the 30 km zone.
• Fire brigade.
• Medical battalion.
• Other detachment.
Class of work¶
• Collection of fragments on the NPP roof.
• Decontamination of buildings.
• Decontamination (i.e. removal of soil).
• Evacuation of people.
• Services (i.e. guard, communications, etc.).
• Other job.

¶ Subsidiary data for class of work was 'Were dosimetrists present at place of work?'

5.6 Transplantation of bone marrow and embryonic liver cells

5.6.1 Radiation damage to bone marrow

Blood cell production is maintained from a number of so-called stem cells which reside in the bone marrow and generate some 1000 or about 100 g of new blood cells daily. As bone marrow stem cells are readily transplantable if injected into the blood stream, donor bone marrow transplants had by 1986 long been considered the treatment of choice for the severely injured radiation accident victim who had undergone whole body irradiation[30].

This was even though donor bone marrow transplantation has several risks. For acceptance of all types of transplants, the immune functions of the recipient need to be suppressed to prevent the immune cells, the lymphocytes, of the recipient from recognizing the graft as *foreign* and launching an immune attack. In the case of bone marrow transplantation, the immune suppression required is extreme compared with that in the case of kidney and heart transplants. In clinical medicine, such an immunosuppression can only be achieved by relatively high doses of homogeneous whole body irradiation.

However, in radiation accidents exposures are always heterogeneous, which means that some of the circulating lymphocytes can escape death and may reject the graft. It has been calculated that even a one-thousand-fold reduced immune system still has the capacity to reject a donor bone marrow graft[30–32].

Bone marrow grafts are singular in that a significant proportion of the cells are similar lymphocytes, but are those of the donor. These cells may, in turn, recognize the tissues of the recipient as *foreign* and try to reject them. This reverse reaction is one of the characteristic and either lethal or disabling risk for recipients of bone marrow transplants.

In the event of insufficient immune suppression of the recipient or insufficient removal of the immune cells from the graft, these cells can also recognize each other as foreign and start a deadly mutual attack. This will result in both rejection of the bone marrow graft and elimination of the residual stem cells of the accident victim. These mechanisms played a significant part in the failure of bone marrow transplants in the Chernobyl recipients[30].

This failure is described in section 5.6.2 and Guscova has summarized[33] the two main lessons which were learned from these failures.

- Highly exposed patients need to be treated in one or two specialist centres with a uniform approach to the diagnosis and treatment of the patients.
- The greatest difficulties occur when the total body doses exceed 8–13 Gy, when localized injuries occur, and when radionuclides are ingested in large concentrations.

The European Commission has also published an up-to-date report on the diagnosis and treatment of patients with ARS[34].

There is little doubt that the patients in 1986 received the best possible treatment in line with the state of knowledge at the time in the most experienced hospitals. However, it has now been recognized that immature haemopoietic stem cells are heterogeneous with respect to radiation sensitivity, the most immature stem cells, which are responsible for long-term haemopoietic and immune reconstitution, being less sensitive to radiation than was previously thought on the basis of lethality data[30,35–37]. The recent reappraisal of the radiosensitivity of immature haemopoietic stem cells[38–43], which was not recognized at the time of the Chernobyl accident, simply tells us that the bone marrow stem cells will ultimately regenerate the blood cell production tissues following radiation doses that are not otherwise incompatible with the survival of the human organism[30].

5.6.2 Review of results of transplants for the Chernobyl liquidators

The physician in charge of the patients in Moscow was one of the most experienced in the world with regard to these types of victim: Dr Angelina Guscova had treated over 1000 victims from the radiation accidents and emergency situations that had occurred in the USSR over a period of 45 years[28]. Her experience has been reported at several conferences in the last decade[12,15,29,30] and is summarized here with regard to transplantation of allogenic bone marrow (TABM) and transplantation of human embryonic liver cells (THELC).

> The indication used for TABM and THELC was a whole-body gamma radiation dose estimated to be of the order of 6 Gy or above. TABM was undertaken 30 times and THELC was carried out six times.
>
> After THELC, all patients died in the early stages (14–18 days post-irradiation) from lesions of the skin and intestines. The only exception was a woman aged 63 years who survived 30 days post-irradiation after receiving a dose estimated as 7–10 Gy. Her death occurred 17 days post-THELC.
>
> Seven patients died at 2–19 days post-TABM (15–25 days post-irradiation) from acute radiation lesions of the skin, intestine and lungs. Of six patients who had no lesions of the skin and intestine, which were considered to be incompatible with life, and whose doses were estimated to be 4.3–10.7 Gy, two survived after TABM. Their doses were 5.8 Gy and 10.7 Gy and in both cases their donor was their sister.
>
> Four patients died 27–79 days post-TABM from mixed viral–bacterial infections. Two of these four cases received doses in the ranges 5.0–7.9 Gy and 5.8–6.0 Gy, against a background of a well-functioning graft. The remaining two patients, who had received doses of 4.3 Gy and 10.7 Gy, died after early rejection.

As a footnote, the name of Robert Gale should be mentioned because at the time these TABM treatments were being given, the press in western countries, including TV, were giving the major credit for the care of the patients to Armand Hammer, the entrepreneur and Chairman and CEO of Occidental Petroleum and to Gale. Certainly Hammer, as he states in his autobiography[44], supplied medical aid and facilitated the travel of Gale to the USSR. However, it turns out from a biography[45] of Hammer after his death that he was, to say the least, a rather dubious character whose interests included espionage for the Russians, having first met Lenin in 1921. His protégé Gale, although being of little help to Guscova and her team, did later admit that the transplants were a failure[46], but he also to a certain extent, followed the USSR *party-line* and was intentionally seen to take his family around the city of Kiev as if there was no significant contamination problem, apart from that around the Chernobyl NPP, and to endorse this official Soviet view in the media. He even went to the lengths of saying that the only problem he had with staying in Kiev was that he missed his bagels, whereupon Hammer air freighted some from the United States. This was not appreciated by the local population[4].

5.7 Case histories

Late post-irradiation effects on the skin are seen in Plate II and figure 1.1 and acute effects in liquidators who subsequently died, in Plate III. However, very few reasonably detailed individual case histories have ever been described and for that reason the three which I have been able to locate are included in this section. These were described by Soviet physicians at a conference[47] in 1987 and I am indebted to Dr John Hopewell for providing me with these histories. For comparison with the much more extensive injuries of case histories from Hiroshima and Nagasaki see the books by Hersey[23] and by Chisholm[24].

5.7.1 A male nuclear plant worker

This patient had received an estimated average total body dose of 9 Gy and was treated by TABM from a female donor. The transplant was rejected but his haematological status improved due to recovery of his own bone marrow. This indicated a highly non-uniform distribution of the dose. He developed skin lesions five days post-accident and these eventually involved 40% of the body surface area.

He showed epilation of the scalp and eyelashes, but the eyebrows were not affected. Lesions over both buttocks were severe as a result of his sitting on a contaminated surface. These areas of skin developed blisters and foci of ulceration, which required covering with free skin grafts taken from the patient's flank two months after the accident.

These 0.8 mm thick skin grafts took fairly satisfactorily and the patient was released from hospital five months post-accident. Small areas of necrosis developed in the graft areas at seven months but these had healed by 12 months, when he was otherwise well, although asperimia persisted.

5.7.2 A male turbine operator

This case received an estimated whole-body gamma dose of 2.0–2.5 Gy and thus bone marrow syndromes were only slight to moderate. He developed severe erythema and oedema of the skin in the second week after the accident and by the end of the third week had developed widespread erosion of the skin over the wrists and also of the trunk and thighs.

He experienced a protracted and severe fever that appeared to be associated with the severity of the skin lesions. Topical therapies were tried, but he required surgical intervention on the 50th day after exposure. This involved removal of dead tissue from the ulcerated areas of skin on both wrists and grafting with skin taken from the patient's flank.

The graft on the right wrist proved unsuccessful, either because of the severity of the injury to the deeper tissues, the more likely cause, or because of radiation injury to the grafted skin. A pedicle flap taken from the anterior abdominal wall was successful in covering the ulcerated area, becoming established in three weeks.

Although there was no evidence that the tendons were damaged, the patient now had only limited movement of the wrists and is unable to use his hands. He unexpectedly developed an annular ulcer at the base of the fifth finger on the left hand in April 1987 and he requested amputation of this digit because of the associated severe pain.

5.7.3 A female nuclear plant guard

This 58-year-old patient was on duty in a booth about 300–500 m from the reactor site at the time of the accident. She ran several kilometres from the site and as a result had dry radionuclide contamination, in the form of soil dust, on her legs and shoes.

She received an estimated bone marrow dose of 3 Gy from which she recovered. She experienced three waves of erythema over her thighs and lower legs: the third wave developed almost three months after the accident and was accompanied by oedema and severe pain.

Severe lesions also developed later on the feet. The severity of these late developing lesions was accompanied by a deterioration in her general physical condition. She subsequently suffered a cerebrovascular accident which was probably superimposed on a condition of generalized radiation-induced vascular damage.

5.8 Medical centre in Chernobyl town

Dr Belokon's medical centre no longer exists as Pripyat, the dormitory town for the power plant, formerly with a population of 47 000, is deserted except for scientists undertaking environmental measurements; there is, however, still a great need for the medical centre in Chernobyl town. Now in the late 1990s this is called Medical Department (or Point) no. 5 but in 1986 within the USSR framework of hospital centres was called Medical Sanitary Department no. 125. The title Medical Sanitary Department was used throughout the USSR for all medical institutions which belonged to a factory or power plant or some such organization.

Some 130 000 persons were evacuated and by 1990 there were some 600–800 returnees. Medical Department no. 5 serves them and the scientific and administrative workers still in the 30 km exclusion zone, with the medical personnel working shifts of 14 days on duty in the zone followed by 14 days outside the zone in their homes in such as Kiev and Chernigov.

There is now also a Medical Department of a similar size to no. 5, in the new town of Slavutich which was built outside the zone as a permanent town to re-house some of the evacuees. The temporary town of Zeleny Mys (Green Cape), on the edge of the zone and built in 1986, has by 1998 been abandoned, even though the facilities were still perfectly adequate. This is due to the Ukraine being unable to afford the cost of its upkeep; 20% of the gross national product of the Ukraine in 1998 was spent on the cleanup operations.

The staffing levels of Medical Department no. 5 in Chernobyl town (similar in number to that of Slavutich) consists of a total of 132 persons of whom 42 are physicians with the remainder being nurses and junior medical staff. The medical staff includes surgeons, opthalmologists, otolaryngologists, neurophysiologists, dermatologists, dentists, stomatologists and gynaecologists. Medical centres in outlying villages are visited daily, urgent cases for hospitalization are sent to the town of Ivankov, between Kiev and Chernobyl, and when hospitalization is required but it is non-urgent, the patients are sent to Kiev. Several of the dedicated medical team have worked at Medical Department no. 5 since 1986[48].

5.9 Liquidator certificates for benefit entitlements

Just after the accident many people received so-called *liquidator certificates*, some of whom did so illegally when they were not entitled to the benefits and instances have been found of forged applications stating that the person applying had worked as a liquidator only for a single day but was severely irradiated. There were also many cases of actual liquidators not bothering to apply for their benefits until much later, and then finding difficulty in supplying the necessary documentation and supporting witnesses to prove that they were liquidators. Because of such problems the legal requirements for receiving a liquidator certificate were changed on 28 February 1991 when a law was adopted in the USSR, for the Ukraine (which I am using as an example) with the title *On the Status and Social Protection of Citizens who Suffered as a Result of the Chernobyl Catastrophe.* This legal decree defined four categories of liquidator, table 5.10.

In addition, there are certain other groups, not given in table 5.10, who qualify for one or other of the categories. These include Soviet citizens who implemented various Chernobyl-related governmental tasks during the period 1986–90, including some which were undertaken outside the boundaries of the 30 km and other designated zones. The legal decree also stated that 'Soviet citizens who took part in the elimination of the effects of other nuclear accidents and of nuclear tests, including military exercises

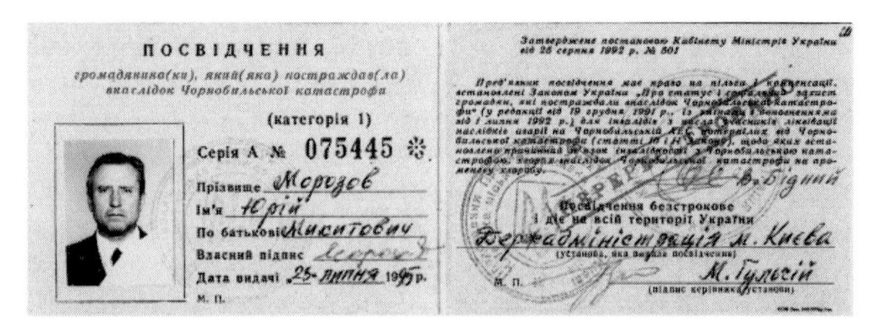

ПОСВІДЧЕННЯ
громадянина(ки), який(яка) постраждав(ла) внаслідок Чорнобильської катастрофи
(категорія 1)
Серія А № 075445
Прізвище Морозов
Ім'я Юрій
По батькові Микитович
Власний підпис
Дата видачі 26 липня 1995 р.
М. П.

Затверджене постановою Кабінету Міністрів України від 25 серпня 1992 р. № 501

Посвідчення безстрокове
і діє на всій території України
Держадміністрація м. Києва
(установа, яка видала посвідчення)
М. П.
М. Гульсій
(підпис керівника установи)

Figure 5.4. Category 1 certificate of a Ukranian liquidator. Iourii Morozov, born November 1936, was a Lieutenant-Colonel of troops within the Ministry of Internal Affairs and took part in construction works for the protective fence encircling the 30 km zone as commander of a military unit. He worked in the 30 km zone twice, 4–16 June 1986 on reconnaissance and the start of construction work, and 23–31 July 1986, also for construction work. He received a dose of 7 rem during his 19 days in the 30km zone. On 31 December 1987 he retired from military service as he had already served the 25 years required of officers in the former USSR, and then worked as a civilian engineer. His illnesses following his work as a liquidator included bronchial asthma and a myocardial infection and in 1992 the Chernobyl Medical Commission recognized these as being Chernobyl related and in 1995 this Commission in the city of Kiev classified him as an invalid. By 1999 he could no longer work because of ill health.

with nuclear weapons' could be classified within categories 1–3. The benefits available to those holding a category 1 liquidator certificate are given in table 5.11, and an example of a certificate in figure 5.4. A translation of the wording in figure 5.4 is also given here.

CERTIFICATE
of the Citizen who suffered as a result of the Chernobyl Catastrophe
(Category 1)
Series A No. 075445
Surname MOROZOV
Name IOURII
Patronymic MIKITOVICH
Personal Signature
Date of Issue 26 July 1995
Adopted by Decree by the Cabinet of Ministers of Ukraine from 25 August 1992 No. 501
The owner of this Certificate has the right of privileges and compensations as established by the Law of the Ukraine 'On Status and Social Protection of Citizens who Suffered as a Result of the Chernobyl Catastrophe'

(edition from 19 December 1991 with changes and additions from 1 July 1992) for invalids from the number of participants of the elimination of the consequences of the Chernobyl NPP accident and who suffered from the Chernobyl Catastrophe (Articles 10 and 11 of the Law) for which it has been established that there is a correlation between their invalidity and the Chernobyl Catastrophe, or an acute radiation syndrome developed as a result of the Chernobyl Catastrophe.
This Certificate is valid for the whole territory of the Ukraine without limitation.
State Administration of the city of Kiev
Signature of the Head of the City Administration
The Circular Seal with the Coat of Arms of the Ukraine
Signature: Representative of the President of the Ukraine
Signature: State Administration of the city of Kiev

Figure 5.5. Dr Angelina Guscova (see Plate III, table 5.5 and the TABM and THELC summary on page 92) and the author at Novosti Press Agency, Moscow, December 1987. (Courtesy: Novosti.)

Table 5.10. Categories of liquidator certificate.

Category	Qualifications for a given category
1	• Invalids among the liquidators and victims of the catastrophe for whom a causal effect for the invalidity can be linked to the catastrophe: for example, cases with acute radiation syndrome.
2	• Those who worked in the 30 km zone until 1 July 1986, irrespective of the number of working days; or for not less than 5 days during the period 1 July 1986 to 31 December 1986; or for not less than 14 days during 1987. • Those evacuated in 1986 from the 30 km zone, including women who were pregnant at the time of the evacuation. • Those who continuously resided from the time of the catastrophe until their relocation, in a region which was not initially, but only later, classified as a zone of compulsory relocation.
3	• Those who worked in the 30 km zone for 1–5 days during the period 1 July 1986 to 31 December 1986; or for 1-14 days in 1987; or for not less than 30 days in 1988–90. • Those working at health centres, or at centres for equipment contamination, or with building construction, for not less than 14 days in 1986. • Those who were guaranteed voluntary relocation and until this occurred, resided on contaminated territories: or who in the period to 1 January 1993 have lived in a zone of guaranteed voluntary relocated for not less than 3 years and were eventually relocated. • Those who constantly worked or studied in the zone of compulsory relocation for not less than 2 years in the period to 1 January 1993, or not less than 3 years in the zone of guaranteed voluntary relocation.
4	• Those who constantly resided, worked or studied on territories of enforced radioecological control for not less than 4 years in the period to 1 January 1993.

Table 5.11. Benefits of persons holding a Category 1 certificate. The numbering 1–32 is within the legal decree. For holders of Category 2–4 certificates the benefits are less and for example where the reimbursement for Category 1 is 100% for Category 2 it may be only 50%. For Category 2 the following clauses are relevant *in full*: 1–3, 5–8, 11–12, 17–18, 20 and 22–31; and for Category 3 the clauses: 1–3, 5–6, 8, 17, 20 and 27; and for Category 4 the clauses: 1, 3, 5–6 and 8.

[1]	Free medicines when prescribed by a physician.
[2]	Free dentistry, excluding the use of precious metals (e.g. gold fillings for teeth).
[3]	Priority service in medical establishments and drug stores.
[4]	Free vouchers, annually, for a holiday in a resort or sanatorium: or the cash equivalent.
[5]	After retirement, or in cases when the place of work changes: use of outpatient medical departments equivalent to permitted use in previous work.
[6]	Annual health check-up and medical treatment in specialist clinics if required.
[7]	Job protection (retention of existing job or right to receive a new job) during any management/worker organization changes which result in job redundancies. The salary level is also guaranteed.
[8]	Cash payment because of any temporary working disability: payment equal to 100% of average salary.
[9]	For working invalids, cash payments if they cannot work the entire year, of 4–5 months salary in a calendar year.
[10]	Priority on the housing list (e.g. for flats) for those who need improved housing (including families of those who have died). Guaranteed housing within 1 year of application. (15% of all new residential premises are allocated for this purpose by Local Councils of People's Deputies). Those requiring improvement in housing include those with living areas smaller than the average provision for Ukranian citizens, of if accommodation is in communal flats¶. A separate room will be provided for all those suffering from acute radiation syndrome. A family which has lost the bread-winner as a result of the Chernobyl catastrophe has the right to additional residential living space. These privileges can be used only once.
[11]	Payment of up to 50% of the cost of a flat for Category 1 liquidators and members of their families and also payment of 50% of communal services (water, gas, electricity and heating) and 50% of telephone charges. For those living in accommodation without central heating, 50% of the cost of fuel is reimbursed.
[12]	Free cost of transfer into private property: flats or houses. This privilege can be used only once.

Table 5.11. (Continued.)

[13] Priority for purchase of a car (type defined by the Cabinet of Ministers of the Ukraine) for those in Category 1 and also, if medical support is given, for those in Category 2. If the car required is of a more expensive type then the liquidator can obtain this car by payment of the difference in cost. This clause is only valid if the invalid has no car or if he did not purchase a car in the previous 7 years.

[14] Provision of food products required medically. 50% of the cost of such food to be reimbursed.

[15] Free use of all urban and suburban transportation (excluding taxis where the number of seats is less than 9) in the territory of the Ukraine[§].

[16] Liquidators to be provided with a hospital certificate[‖] for the entire period of treatment in a health centre or specialist medical establishment: including the time for travel from home to the centre.

[17] Priority for pre-school places for the children of the liquidator.

[18] This clause was subsequently cancelled but originally referred to privileges with regard to customs fees and taxes.

[19] Free travel, once annually, to and from any point in the Ukraine, by automobile, air, railway or navy transportation and including priority for the purchase of tickets.

[20] Guaranteed allocation, within one year of application, of land for building construction for those who need improved living conditions. This to include land for the establishment of a garden or a kitchen garden, construction of garages and of dachas.

[21] Provision of a loan with no interest charges, for individual residential (cooperative) construction, including garden cottages, garden development and garages, calculated as an area of 13.65 metres^2 for each member of a family. Repayment of only 50% of the loan is required, the remainder being paid by the State budget. These privileges can only be used once.

[22] Permission to take an annual vacation for 14 days at a time of their choice with their salary paid for this period.

[23] Priority for joining a residential-construction cooperative if the clause [10] privilege has not been taken. If clause [10] is taken then clause [23] is not relevant.

[24] Priority in establishing enterprises such as communications organizations, technical services, vehicle repair, consumer services, trade, public catering, residential and communal husbandry and intercity transportation.

[25] Priority for the purchase of industrial goods including a car, motorcycle, motor boat, TV set, refrigerator, furniture, washing machine and vacuum cleaner.

Table 5.11. (Continued.)

[26]	Priority for entrance into State higher educational establishments and professional and technical establishments* and into courses for professional training. Guaranteed payment of 100% scholarship fees, and if necessary, of dormitory facilities.
[27]	Priority for entry into social security establishments, and for home help if required, if a sick individual has no relatives living with him.
[28]	Priority for installation of a telephone and payment of 50% of the installation costs.
[29]	Provision of a loan with no interest for business activities or agricultural husbandry. This privilege can be used only once.
[30]	Payment, up to 5000 roubles per family†, of an unpaid part of a loan with no interest which has been granted for husbandry, and which was paid to evacuees from the 30 km zone.
[31]	Payment for those with hospital certificates‖ relating to temporary working disability is calculated on the basis of the average income they received as a liquidator when working on the territory of the Chernobyl NPP.
[32]	Reimbursement of financial losses as a result of disease or mutilation because of work as a liquidator.
	Clauses [1], [2], [17], [26] and [27] are relevant to children of minor age of Category 1 liquidators whose death is connected with the Chernobyl catastrophe. Clauses [5], [7], [8], [11], [12], [20], [23] and [27] are relevant to a wife (or husband) of a deceased liquidator, or to a tutor who is required to teach the children of a deceased liquidator.

¶ During the period when the Soviet Union existed there was a significant lack of flats for the population such that one flat could be provided for each family. Many Soviet citizens lived in large flats with many rooms and each family only had a single room, with the bathroom, toilet and kitchen being shared by all residents.

§ This includes the tram system in the city of Kiev.

‖ This is a special document issued by a hospital which give the patient the right not to work because of his illness, but also the right to receive his salary for the time he is ill, because of his membership of a trade union.

* Technical colleges which provide professional and secondary education.

† These are Russian roubles. The Ukranian currency is now the Gryvna.

Figure 5.6. Major Leonid Telyatnikov and the author at the Soviet Embassy, London, March 1987. (Courtesy: TASS.)

Figure 5.7. Symbol of the Chernobyl Museum in Kiev which is located within the headquarters building of the Fire Department.

Chapter 6

Evacuation and Resettlement

Introduction

The earliest Soviet estimate of the initial evacuation was 135 000 people, but although this has now been revised down to 116 000, it is still an enormous number considering that most of the evacuation occurred within 11 days, from 27 April to 7 May 1986, and that not only had a large city—Pripyat—of almost 50 000 to be evacuated but also the population from a total of 187 outlying settlements. The logistics of such an evacuation, involving transportation and temporary resettlement, followed by permanent resettlement were vast.

6.1 Evacuation zones and populations

Several zones have been defined at various times following the accident: these include the 10 km and 30 km zones and SCZs, the *strict control zones* where populations still reside.

The 30 km zone and the inner 10 km zone centred on the Chernobyl NPP were termed *exclusion* or *alienation zones* and compulsory evacuation of the entire 30 km zone was completed within the first few days following the accident.

Table 6.1 defines the zones classified by the Soviet authorities, based on Government Commission recommendations, after the accident[1] and table 6.2 gives the definitions of the various zones in the Ukraine after the break-up of the USSR[2]. However, there is not always compatibility in the terminology used by the Ukraine and Belarus on whose territories the major portion of the contamination fell. Table 6.3 gives a breakdown in terms of settlements and dates of evacuation as well as number of evacuees[3–6]. For the Ukraine the total number of evacuees was 91 406, for Belarus it was 24 725 and for Russia it was 186. This correlates with the areas contaminated above 1480 kBq/m^2 which were 2100 km^2 in Belarus, 2044 km^2 in

Ukraine and 170 km^2 in Russia[7]. These areas included *hot spots* of ^{137}Cs up to 370 000 kBq/m^2 (10 000 Ci/km^2), of ^{90}Sr up to 185 000 kBq/m^2 (5000 Ci/km^2) and of plutonium up to 555 kBq/m^2 (15 Ci/km^2)[7].

Table 6.1. Zones classified by the Soviet authorities in 1986[1].

Contamination (kBq/m^2)	Description of zone
37–555	Periodic health monitoring, no special measures adopted.
555–1480	Strict control, restrictions imposed on population, restrictions on use of locally produced food, decontamination measures adopted.
>1480	Area classified as unfit for human habitation, population evacuated.

Table 6.2. Evacuation zones in the Ukraine from 1986 onwards[2].

Zone radius (km)	Zone description	Evacuees from zone
30	Exclusion/alienation zone	91 200
60–70	Second zone: compulsory (but planned) resettlement	35 000 (but 50 000 planned)
200	Third zone: voluntary relocation	35 000
>200	Fourth zone: strict supervision by law (Belarus has no such fourth zone)	
Total		161 200

The relocation/resettlement problems were vast and for example in the Ukraine from 1986–96 a total of two million m^2 of housing was completed and schools for 35 000 children and pre-school facilities for 10 000 infants were constructed. In addition, hospitals were built with 2000 beds each and policlinics which enabled a patient workload to be achieved of 7000 cases per shift. This cost some US$2500 million in investment[9].

The evacuation did not cease in 1986 and, for example, for the period 1990–95 the following persons were evacuated: 53 000 from Ukraine; 107 000 from Belarus; and 50 000 from Russia[10]. As with the earlier evacuations and resettlements, these have created a series of serious social problems, linked to the difficulties and hardships of adjusting to the new living conditions. Some of these problems have been due to relocations very far afield. For

Table 6.3. Time frame of the evacuation to September 1986[3–6].

Area	Date of evacuation	No. of evacuees
Ukraine		
Pripyat town	27 April	49 360
Yanov railway station[¶]	27 April	254
Burakovka village	30 April	226
Belarus		
51 villages from the 30 km zone	2–7 May	11 358
Ukraine		
15 villages from the 10 km zone	3 May	9864
Chernobyl town[§]	5 May	13 591
43 villages from the 30 km zone[‖]	3–7 May	14 542
8 villages outside the 30 km zone	14–31 May	2424
Belarus		
28 villages outside the 30 km zone	3–10 June	6017
Ukraine		
4 villages outside the 30 km zone	1 June–16 August	434
Russia		
4 villages in the Bryansk region	August	186
Belarus		
29 villages outside the 30 km zone	August–September	7350
Ukraine		
Bober village	September	711
Total		116 317

¶ The railway station area near to the NPP was so highly contaminated that it was impossible to evacuate any people by rail and therefore all had to be evacuated, first to Kiev, by road.

§ Figures 6.1 and 6.2 are photographs from Chernobyl town, the first showing two elderly ladies and the second illustrating that all that remained were family pets: rabbits in this instance[8].

‖ The cow farms such as in Plate IV and many of the small single-story wooden houses which were to be found in villages, were burnt to the ground and a decade later all that remained were small grassy hillocks covering the charred remains of the homes of the evacuees, figure 14.10.

Figure 6.1. Evacuation of Chernobyl town, May 1986[9]. (Courtesy: V Zufarov.)

Figure 6.2. After the evacuation, only family pets remained in Chernobyl town[9]. (Courtesy: V Zufarov.)

example, some were evacuated relatively close, for example from South Gomel district to North Gomel district as there was a shortage of workers in the north, whereas others were dispersed to Kazakhstan, the Baltic republics and Siberia[11].

Voluntary resettlement is still continuing and as of 1995 the population still living in territories contaminated at the level of 555–1480 kBq m^{-2} totalled 270 000[1].

6.2 Pripyat

Very soon after the explosion, the Pripyat Internal Affairs Department including representatives from the militia and KGB, held their first meeting, at 02:15 hours on 26 April, and their first decision was to ban all unnecessary traffic from the town. An *Izvestia* correspondent later reported that the second priority was to 'maintain order'.

The decision to evacuate Pripyat was agreed by the authorities at 22:00 hours on 26 April and during the night arrangements were made for the provision of 1216 buses from Kiev which were to be required for the evacuation. The announcement to the population was made at 12:00 hours on 27 April, giving the population only two hours before they left their homes for ever. The evacuation started at 14:00 hours and was completed by 17:00 hours: 34 500 of the population left in the buses and the remainder in Pripyat city transport and in their own cars. The column of buses, cars and trucks stretched for 15 km.

The residents were only allowed to take bare essentials with them and had to leave the remainder of their possessions behind in the abandoned town. In December 1987 washing was still to be seen hanging on lines outside the balconies of flats, and in June 1998 the Soviet symbols of red metallic stars, and hammers and sickles still decorated the lamp posts of Pripyat, having been set up for the 1986 May Day celebrations which never took place. The new athletics and football stadium was also scheduled to be opened on 1 May 1986 and figure 6.3 shows the site in the centre of Pripyat. This photograph also includes parts of some of the housing estates for the 49 360 population.

The town was divided into five main sectors for the evacuation, each covering one housing estate. Evacuation workers were distributed according to the number of buildings and the number of doors and were given defined evacuation routes, eventually joining the long column of buses, which included mobile garage facilities and communications, winding its way to Kiev. No assembly points were used at which large number of the population could gather, so as to avoid any possibility of panic. Militia forces organized checkpoints, roadblocks, cordons and control points.

Nowadays, any authorized visitors or environmental scientists enter-

Figure 6.3. View of Pripyat from a helicopter, 1991. (Courtesy: TASS.)

ing the barbed-wire surrounded Pripyat do so through a checkpoint and pass a tall block of apartments with the following slogan from 1986 still visible 'The Party of Lenin The Force of the People is Leading Us to the Triumph of Communism'. All the surfaces of the roads are cracked, the telephone boxes look as if they have been vandalized, buildings such as the apartment blocks, Palace of Culture and hotels are still standing but are of course deserted, and the school looks as if a bomb has fallen with the debris of books, desks and other school equipment lying everywhere. There is one building still occasionally used as a restaurant for workers and visitors, but this only opens for short periods: the monitoring equipment is broken, and heavy duty plastic covers the stairs, dating from 1986 for radiation protection purposes. Views of the abandoned playground are seen in figure 6.4 and Plate VI.

6.3 Evacuation of livestock

The contaminated areas included much prime agricultural land and dairy farms, such as that in Plate V, and were evacuated not only of people but also of livestock of which 86 000 head of cattle were transported in open trucks. Evacuation from the 10 km zone was completed by 2 May and from the 30 km zone by 5 May.

Figure 6.4. The abandoned children's playground in Pripyat, June 1998. (Photograph: R F Mould.)

An eyewitness report[12] from a farmer's wife graphically describes a typical experience of these times.

> We were milking the cows and the authorities came and said 'Don't panic, just carry on milking'. On 2 May people started to run away with their children. We were told to take our cows to the collective farm. At 5am on 4 May a truck came to our house and dropped us in Borodyanka (this is a small industrial town some 45 km from Kiev, with a population of 14 500) at noon the next day: a journey of 31 hours. We had travelled all night. We took some pork fat and eggs for three days. We had to cook in the garden.

Figure 6.5. Painting of how Alena, 11 years old in 1996, visualized her concept of the Chernobyl accident: the evacuation of her family from Pripyat, see also Plate VI.

6.4 Resettlement

The evacuees from Pripyat and the surrounding evacuated areas were taken in by families who lived in settlements in the surrounding districts and most stayed with these families until August when they were resettled in apartments in Kiev.

A complete and detailed set of statistics for re-housing developments for the evacuees is not available but the following Ukranian figures give some idea of the speed with which construction was achieved. By 1 September a total of 8210 houses had been built and 7500 apartments had been allocated to the evacuees. A further 4500 houses were built by 1 January 1987. To accommodate the Chernobyl NPP workers in permanent accommodation, the new town of Slavutich was built, which is outside the 30 km zone, across the river Pripyat and towards the city of Chernigov[13]. By December 1988 a new 685 m road bridge had been constructed across the river to make the journey between Slavutich and the NPP as quick as possible.

Throughout the entire USSR for the period 1986–87 the following accommodation was found or built for the evacuees: about 15 000 apartments, hostel accommodation for over 1000 persons and 23 000 houses. In addition about 800 social and cultural establishments were also built[14].

Slavutich was preceded by the building within two years of the Zeleny Mys (translation: Green Cape) housing settlement for Chernobyl NPP workers. In May 1987 each shift for power plant personnel lasted five days and was followed by six days leave in Kiev. For other workers in the 30 km zone, the schedule was, and still is a decade later, 15 days work followed by 15 days leave. Zeleny Mys with its prefabricated construction, which was nevertheless of good quality, was intended for use for only five years, but in the event was closed in 1998. This was not because the housing had deteriorated but because the Ukranian economy could no longer support the running costs.

As an example of the building effort in 1986: one new village of 150 homes, Ternopolskoye in Makarov district, was constructed in 50 days for evacuated collective farm workers. Exterior decoration of the houses, weather vanes, dovecotes, letter boxes, shelves in the cellars, and food gifts per household of bags of potatoes, cereals, jars of pickled cucumbers and tomatoes were provided for each household. In addition, 10 hens were allotted to each family as well as kitchen utensils and furniture including two beds and a cot. Barns for cattle were also included in the construction.

State compensation cash benefit payments were also made for the evacuees (approximate exchange rate in 1986 was 1 rouble = £1.00 = US$1.5) of 4000 roubles for single persons, 7000 roubles for a family of two and 1500 roubles for each additional family member.

At least 284 000 people were uprooted from their family homes and evacuated to distant communities or new towns, and thousands of farmers have lost their livelihoods[12]. A unique insight into the attitude towards the disaster in the minds of the children, who were either evacuees themselves or who were born to evacuees post-1986, is encapsulated in an exhibition of paintings by these children who were asked to draw a picture of 'How you feel in your life about the accident'. These were drawn for the tenth anniversary and were exhibited in the Chernobyl Museum in the Kiev City Fire Department 1996–98, see figure 6.5 and Plate VI.

The artwork of the younger children was more positive than that of the older children. Brightly coloured clowns and birds, for example, *versus* dark coloured images, many with black trees devoid of leaves. Plate VI shows four examples from the older children.

6.5 Returnees

By December 1987 some of the residents of a very small number of evacuated villages in the least contaminated areas, with typical populations of 50–60 persons, were permitted to return. Following the dissolution of the USSR, the guards, because of their lower number, became less able to prevent elderly ex-residents from returning to their homes. Some of them were formerly partisans in the Second World War and knew the area so well that it was easy for them to evade these guards, as they once evaded the Nazis. The wooded area around Pripyat had been a focus of resistance during the Second World War and an old oak tree known as the *cross-tree* because of its shape, figure 6.6, was used to hang captured partisans.

Figure 6.6. The cross-tree oak World War II memorial with the NPP and Sarcophagus in the background, 1988. (Courtesy: TASS.)

This later became a memorial, after the partisans had hung some of the Nazis from it at the end of the War, and following the Chernobyl accident it was contaminated square cm by square cm and was the only remaining tree in a once densely populated forest. This fact is obvious in figure 6.6 with the NPP and the Sarcophagus appearing in the background. In 1996 this cross-tree fell down because of its great age, but the area is still retained as a memorial, figure 6.7. The trees and bushes have re-seeded themselves and by 1998 the NPP and the Sarcophagus can no longer be seen from the memorial.

By 1998 the number of returnees was between 600 and 700, with many in their eighties and an average age in the range 65–70 years, figure 6.8. From 1992 it was no longer illegal for elderly residents to return to certain areas within the 30 km zone, but the return of children was prohibited. Nobody is allowed to return to the 10 km zone and this requirement is likely to remain for ever. The returnees are, however, not all permanent residents and many come only to plant vegetables in their gardens and to return to Kiev in the winter. The church of St Eliah in Chernobyl town (Eliah is the biblical spelling for the Old Testament Prophet Elijah, figure 6.9, but in ordinary Russian literature it is written Ilya) has been recently redecorated with an attractive exterior of yellow, blue and white and in

Figure 6.7. The memorial in 1998, two years after the oak fell. The view of the NPP is now obscured by re-seeded trees. (Photograph: R F Mould.)

1997 a new younger priest replaced Father Melody, figure 6.10, who was priest at the time of the accident.

Even in the Communist era the Orthodox Church had a significant following, particularly among the older population, and this has continued. In an article entitled *The Time has Stopped: Atlantis of Polesye Leaves for Eternity*[15] where the disappearance of the mythical Greek island of Atlantis is equated to the Ukranian forest area of Polissya within the 30 km zone there is much mention of the abandoned churches. One poignant story was the visit in November 1998 during the feast of St Michael when the temperature was 20 degrees below zero, of a group who entered the Church of St Michael in the village of Krasnoye. The abandoned church had deteriorated so badly that only pieces of canvas with the remains of paintings were still hanging, but there were candles lit and a divine service was apparently being conducted. The congregation were crying as the music and the words were only tape recordings from the Church of St Nicholas in Kiev.

6.6 Radiation phobia

Radiation phobia, as it was termed in the USSR, was one of the problematical factors of the social impact of the accident. The populations

Figure 6.8. Two of the returnees, Olga and Sergei on 7 May 1996. They did not learn any details of the accident until three months after it occurred. This photograph was published on 24 April 1996 in the Örebro *Nerikes Allehanda* under the title *A Laugh from the Zone of Death.* It won the award for the best portrait photograph in all Swedish newspapers in 1996. The village is just within the boundary of the 30 km zone. The photograph was taken by Bernt Larsson who was accompanying a Swedish radiation oncologist on a visit to the Ukranian Research Centre for Radiology and Oncology. (Courtesy: B Larsson and B Johansson.)

most directly affected were first the evacuees and second the liquidators. The general population developed a fear of radiation but were not directly affected by having to be evacuated.

It is sad to relate that the initial compassion felt for the evacuees, and also liquidators, has in the intervening years since 1986 evaporated and been replaced by a certain amount of hostility, so much so, that those resettled from Pripyat and the surrounding contaminated areas try to hide this fact. This was in part due to the evacuees being moved into apartments in Kiev that local families had been waiting to fill for more than ten years. In addition, the local communities thought that the evacuees were radioactive and called them *glow worms* or *fireflies* to signify that they were total outsiders[12].

Figure 6.9. Ukranian icon of Saint Elijah from the Lavra Monastery in Kiev. In the Bible, in chapter 1 of the Second Book of Kings, it is described how the prophets Elisha (seen bottom left in the icon) and Elijah were journeying to Bethel, Jerico and on to the River Jordan, where Elijah smote the waters with his mantle and the waters divided so that the prophets could pass to dry land. (The river is dipicted centrally towards the bottom of the icon). Verse 11 then tells of Elijah's ascent to heaven. 'And it came to pass, as they still went on, and talked, that behold, there appeared a chariot of fire, and horses of fire, and parted them both asunder; and Elijah went up by a whirlwind into heaven.'

To a certain extent this behavioural pattern in the unaffected populations mirrors that noted some 40 years earlier in Hiroshima and Nagasaki. In Japan, the atomic bomb survivors were discriminated against by prospective employers because it was thought that they might suffer from cancer in the future, and the evacuees were also sometimes shunned by the local population because they were provided with new houses and with special pensions[15].

Several anecdotal stories and more scientific studies have addressed the

Figure 6.10. Father Melody the Archpriest of the Autocephalous Orthodox Church in March 1991 with some of the returnees now living in the contaminated territories. Although living in Kiev since the accident he regularly travelled within the 30 km zone. The word *Autocephalous* means *Independent* in that formerly in the Soviet Union the Church was a Russian Orthodox Church but now it is independent. (Courtesy: TASS.)

problem of radiation phobia and in one such study[16] in Kiev two groups of children in the age range 13–15 years, one in this age range in 1987 and the second group in the same age range but in 1996. They were interviewed for their opinions at the time. In 1987 the emphasis from the children was two-fold.

- Stress due to lost homes and the process of evacuation.
- We are victims, the state must help us.

Nine years later in 1996 the attitude of the children had changed dramatically and their overriding opinion was: we want to merge into society and not be different.

In this same study[16] a 14 year old boy's family had to be relocated to Kiev from Pripyat and one of his friend's mother objected to her own son playing with him as the evacuated boy was 'not clean and will have terrible children in the future'.

To end this section on a personal note, when I returned to Moscow from my visit to the Chernobyl NPP and Pripyat in December 1987, my accompanying journalist, Dmitri Chukseyev, walked into the Novosti Press Agency's large cutting room in which there were some 12 journalists. He threw a stone down onto the press cutting table (having picked it up on the Moscow street) and said 'Here is a present from Chernobyl', and the entire room emptied of journalists. Earlier, on the non-stop Moscow–Kiev overnight express train, Chukseyev had become drunk on vodka with a fellow journalist, because of radiation fear, and was so inebriated that the lady conductor in charge of our railway carriage radioed to the militia and stopped the train in the middle of the night. Our carriage door was thrown open by militia shouting 'Dokumenti' and they tried to throw Chukseyev off the train onto a station platform to be left there in the middle of a snowstorm. Only my travel documents for Chernobyl saved him!

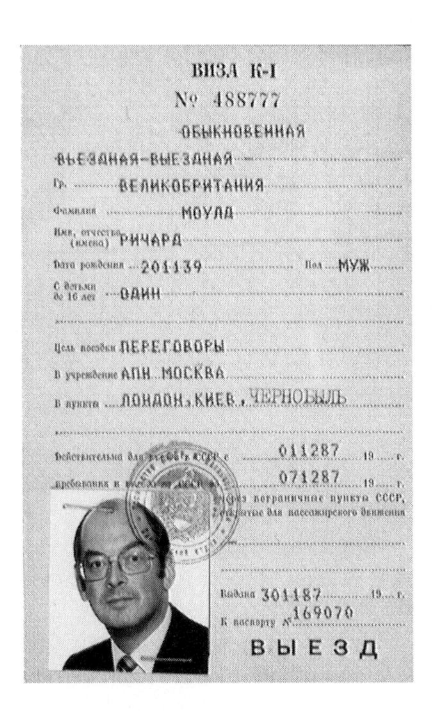

ВИЗА К-I
№ 488777
ОБЫКНОВЕННАЯ
ВЬЕЗДНАЯ-ВЫЕЗДНАЯ —
Гр. ВЕЛИКОБРИТАНИЯ
Фамилия МОУЛД
Имя, отчество (имена) РИЧАРД
Дата рождения 201139 Пол МУЖ
С детьми до 16 лет ОДИН
Цель поездки ПЕРЕГОВОРЫ
В учреждение АПН МОСКВА
В пункты ЛОНДОН,КИЕВ,ЧЕРНОБЫЛЬ
Действительна для ... СССР с 011287 19... г.
пребывания в ... 071287 19... г.
... пограничные пункты СССР, ... для пассажирского движения
Выдана 301187 19... г.
К паспорту № 169070
ВЫЕЗД

Figure 6.11. The Author's Soviet visa for December 1987 granting permission to visit Moscow, Kiev and Chernobyl.

Chapter 7

Sarcophagus

Introduction

The Sarcophagus built, apparently in the minds of the general public but not in actual fact, to outlast the pyramids in Egypt, is now in a situation where it might well collapse, in part if not in full, within the next few years, see section 8.3. This chapter describes its construction and the location of most of the remaining nuclear fuel masses, termed *lava*, including the masses which are known as the *Elephant's foot*, see Plate V. Photographs have already been included of radiation monitoring inside the Sarcophagus, figure 2.8, the measurement recording laboratory, figure 2.9, and in section 8.3 proposals are given for a possible second Sarcophagus.

7.1 Meteorological, geological and seismic conditions of the site

7.1.1 Meteorology

January is the coldest month of the year with an average temperature of −5.6 °C and July is the warmest month with an average of 19.1 °C. The absolute minimum is −44.9 °C and the absolute maximum is 42.2 °C. The normal depth of soil freezing at the site is 110 cm and the annual relative air humidity is 77%. The maximum layer of precipitation is 190 mm. The earliest date when snow cover appears is 6 October and the latest date is 15 December. The earliest date when snow cover disappears is 21 February and the latest is 22 April. The number of days with snow cover is in the range 90–102. The average depth of snow in the open country is 8 cm but the average greatest depth annually is 17 cm with a maximum of 41 cm[1].

The north-west wind predominates in the area during the warm period of the year and during the cold seasons, south-east and north-west winds prevail. The average wind velocity is 4.2 m/s but the maximum has reached

47.3 m/s. The area is classified as tornado-hazardous and the probability of a tornado passing through the western part of the area is 3×10^{-6} per year[1].

7.1.2 Geology

The area is a zone of the Pripyat and Dnieper artesian basin junctions. Of practical interest is a water bearing layer which is used for water supply. The cooling pond, see Plate I and figure 1.2, which is at the south-east of the NPP site, has an area of 22 km^2 and its normal water level is 3.5 m lower than the level of the NPP's foundations. Chalky clays are 10–30 m thick, but there are no absolute water-confining layers. The water bearing complex is directly connected to the waters of the Pripyat river and during the spring floods the Pripyat river ground water level is 1.0–1.5 m higher than that of the flooded land terraces[1].

It was this geology which gave rise to concern that molten fuel would travel through the lower structures and the foundations of Unit No. 4 and contaminate the ground water, spreading radioactivity first to the river Pripyat and then to the Dnieper and its tributaries, of which there are many. This contamination would then have spread to the Kiev reservoir and eventually down to the Black Sea.

7.1.3 Earthquakes

Earthquakes in the adjacent earthquake active regions, as well as local earthquakes in the platform part of the Ukraine, may be dangerous for the Sarcophagus. The magnitude may be up to six or seven on the Richter scale. However, the probability of occurrence is once in 10 000 years[1].

7.2 Cooling slab to prevent contamination of the ground water

Because of the possibility of contamination of the ground water and the occurrence of what the public call a *China Syndrome*, see section 8.1, a special *cooling slab* was built underneath the damaged Unit No. 4. A group of some 400 coal miners from Tula and the Donets basin were drafted for this emergency, and they also dug the underground bunker 600 m from the damaged unit, which acted as a control outpost for the coordination of site operations.

This cooling slab consisted of reinforced concrete incorporating a flat design of heat exchanger but an access tunnel had first to be dug. Because of the conditions, limited working times were set for the miners of three-hour shifts. The first few metres were the most difficult as the tunnellers had to drive to a depth of 6 m in solid sandstone, constantly monitoring radiation levels at all times. The starting phase was first digging a large pit near Unit No. 3. The tunnel, completed in 15 days on 24 June 1986, is encased in reinforced concrete and is 168 m in length and 1.8 m in diameter. The final 5–6 m were worked manually by the miners and the rate of tunnelling was 60 cm of sandstone rock per hour.

Service lines and rails for buggies were laid inside and 13 galleries were dug off the tunnel. In these galleries, riggers assembled what were called *dampers*, which are devices for cooling the foundations off the reactor. Finally, a monolithic reinforced concrete slab was installed underneath the damaged reactor[2,3].

7.3 Construction of the Sarcophagus

The damage to Unit No. 4 has already been described in chapter 3 and shown in the artist's drawing of figure 3.6, the aerial photographs of Plate I and figure 3.4, and the interior view of the Sarcophagus in figure 2.8.

The containment structure built over the damaged Unit No. 4 reactor was originally termed *Sarcophagus* because of the analogy to the coffin-like burial containers in ancient Egypt and Greece. This term was later changed to Shelter, and finally to Ukritiye Encasement[1]. I have retained the original term *Sarcophagus*.

7.3.1 Architect's drawings

Architect's drawings[1] are shown in figures 7.1 and 7.2 with both cross-sections passing through the centre of the reactor hall. In figure 7.1 the materials dropped by the helicopters to put out the fire can be seen above and on either side of the displaced biological shield, indicating that they missed their target.

Within the reactor hall are 27 metal tubes of 1220 mm diameter and 34.5 m long placed on metal girders. The roof over this hall is made of what are termed large-size metal shields and is installed over the tubes. A new roof was also constructed over the turbine hall. The ventilation stack upper section columns were displaced 900–1000 mm in one axis and there are 100–150 mm wide cracks in the lower sections. The stack has had to be stabilized.

Figures 7.1 and 7.2 include the *cascade wall* (also called the *step wall*), the wall with the buttresses, the central reactor hall [U3] and its 2000 tonne biological shield, the southern central water circulating pump [I0ΓUH], the drum separators [bC], the mammoth beam [bM], the turbine hall [M3], the ventilation stack [Δ3], the reactor's emergency cooling system [CAOP] and the protective wall [PC] built between Units No. 4 and 3.

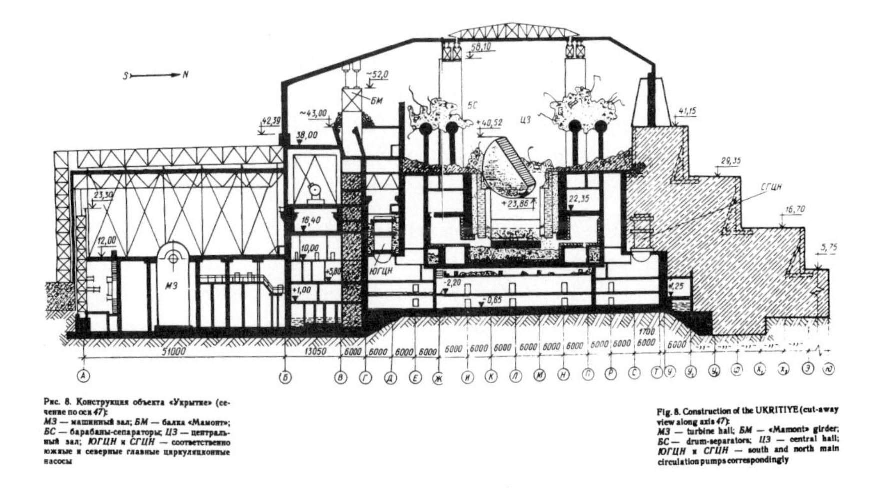

Figure 7.1. Architectural cross-section drawing[1]. (Courtesy: Chernobylinterinform.)

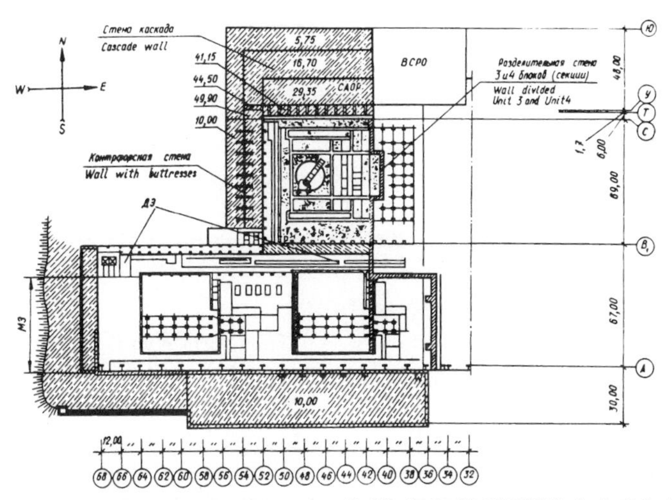

Figure 7.2. Architectural cross-section drawing[1]. (Courtesy: Chernobylinterinform.)

In figure 7.2 the cascade wall is towards the centre-top of the diagram and the reactor is symbolized by a circle. The wall with buttresses is at 90° to the cascade wall and the small rectangle at the bottom of this wall indicates where one enters the Sarcophagus. The turbine hall is at the bottom left of both figures 7.1 and 7.2. The wall separating Units No. 4 and 3 is in the centre of figure 7.2 and extends from the outer wall of the turbine hall.

Figure 7.1 is labelled through axis 47, which is through the central reactor hall and the reactor of Unit No. 4 and shows the maximum height of the Sarcophagus as 58.10 m. Prior to the accident[2] this maximum height was 71.3 m, see figure 3.2. The mammoth girder is in this view as a square cross-section, its upper extremity being at ~52.0 m and its lower extremity at ~43.00 m. The mammoth grider weighs some 160 tonnes, is based on some very questionable foundations, and by 1993 was already rusting so badly that it was estimated that it would soon collapse[4].

The Sarcophagus was completed on 19 November 1986 in a total of only 206 days. Because of the necessity for heavy bearing loads, some of the walls are extremely thick, such as that erected in the turbine hall which is 2.3 m thick ferroconcrete. The step wall was built using 12 m high steps and the wall with buttresses is 50 m high.

7.3.2 Interior views

Some 2000 metal pipes connected the reactor to the basement rooms together with inflammable cables which with their polymer covering present a considerable fire hazard, figure 7.3. Some of the equipment is still recognizable, such as the central water circulating pump at the southern end of the building, figure 7.4. Two such pumps existed, see figure 7.1, but the northern pump was completely destroyed.

Figure 7.5 shows part of the inside of the cascade wall. The two central white splashes of light are from the afternoon sun shining through holes in the walls of the Sarcophagus. The total area of such holes is 1500 m^2 but it is not true that these have been formed by the walls rusting away. These holes, 2, 6 and 10 m^2 in size, were intentionally placed in order to dissipate heat as the Sarcophagus was never planned to be hermetically sealed[5]. Radioactive dust flow through the some of the holes is monitored. Figure 7.6 is another view of the damage, also from the side of the cascade wall, which is partially in view at top left

Figure 7.7 shows the remains of the central reactor hall in which the ceiling and walls were destroyed. At top left is the underside of the roof of the Sarcophagus.

Figure 7.3. Room 009 in Unit No. 4, 1996. (Courtesy: Chernobylinterinform.)

Figure 7.4. The damaged main circulation pump in 1996, IOГUH in figure 7.1. (Courtesy: Chernobylinterinform.)

Figure 7.5. Inside the Sarcophagus near the cascade wall, 1996. (Courtesy: Chernobylinterinform.)

7.3.3 Exterior views

The roof of the Sarcophagus viewed from the outside of the building is seen in figure 7.8 which was taken in October 1986 when the cascade wall was still under construction. The poor photographic quality of figure 7.9, which shows the lowering by cranes of the roof of the Sarcophagus, is due to the fact that because of the high dose rates the photograph had to be taken through lead glass. Afterwards the camera was so badly contaminated that it had to be thrown away[6].

Photography in this early period was extremely hazardous and, for example, when one TASS photographer was filming from a helicopter over the burning reactor, his hat blew off and when he returned to the helicopter base at Chernigov he had to have his head completely shaved because of the contamination[6], figure 8.10.

Figure 7.10 is of the completed Sarcophagus with a view which shows both the cascade wall and the wall with the buttresses. The door at bottom far right is that used for entry into the Sarcophagus. Also at the far right is one end of the exterior of the turbine hall for which figure 7.11 shows the interior in December 1987. The floor of the turbine hall is of a honeycomb design of pressed metal, such as are used on fire-escape stairs

Figure 7.6. Inside the Sarcophagus near the cascade wall, 1996. (Courtesy: Chernobylinterinform.)

outside buildings. This design made it virtually impossible to fully decontaminate and the measure taken, still in existence in 1998, was to cover it with heavy duty plastic which was kept in place by heavy concrete pots containing rubber plants. The protective wall between Units No. 4 and 3, see figure 7.2, is at the far end of the turbine hall in figure 7.11 and is also shown in figure 7.12.

Figure 7.13 is a view of the Sarcophagus from the buttress wall end and far right is the end of the turbine hall which can also be seen in Plate I. The cascade wall is to the left of the buttress wall in this photograph but only a single end of this wall is visible. The smaller building at bottom left is the dust suppression unit, installed in 1990, which is essential to combat what is currently one of the major problems within the Sarcophagus. The door which is two buttresses from the right in figure 7.13, is the entrance used for moving large and heavy equipment into the Sarcophagus.

Figure 7.7. Inside the Sarcophagus in the central reactor hall, 1996. (Courtesy: Chernobylinterinform.)

7.4 Status of the nuclear fuel

7.4.1 Searches for the fuel locations

The radioactivity still within the Sarcophagus amounts to some 96% of the total radioactive content of Unit No. 4 and is estimated to include 180 tonnes of uranium, 400 kg of ^{239}Pu, 170 kg of ^{240}Pu, a significant quantity of transuranic elements and about 70% of the original quantity of caesium radionuclides[7].

At first the scientists could not find the location of the nuclear fuel masses. It had been expected that the majority would be at the base of the reactor hall. It took specialist oil industry engineers 18 months to drill through the heavy concrete walls, as they were slowed down by the need to monitor radiation levels, to reach the reactor hall. When it was found to be empty the next stage in the search was unclear. What they did was to buy a child's toy tank for 15 roubles and strap a camera to it as a remote controlled device, but this did not work either[6].

It was only when it was found that the base of the reactor hall had dropped 4 m in the explosion and consequently the molten fuel had flowed downwards out of the reactor hall that it was eventually located. A typical

Figure 7.8. During the construction of the Sarcophagus, October 1986. (Courtesy: TASS.)

Figure 7.9. Lowering the roof on the Sarcophagus, 1986. (Courtesy: V Zufarov.)

bore hole required when searching rooms, or their remains, for fuel, was a length of some 11 m and a diameter of 112 mm. Dose rates in R/hr were measured along the bore hole as the drill progresses and in one room[7] where the fuel masses were eventually located, the exposure rate went from 1 R/hr up to 4 m depth to 1225 R/hr beyond 11 m.

The scientists tracking the fuel were members of a team called the *Complex Expedition* and typical maximum working times in a room were sometimes less than ten minutes but never usually more than 30 minutes because of the radiation dose rate which could vary widely when a corner was turned in the corridors or rooms and change from 1–5 R/hr to 500 R/hr. The work was carried out for five years, 1986–91.

7.4.2 The lava

Some of the fuel, having melted the sand, concrete and other materials as a result of the fire and the internal heat release, then formed flows of a sort of *lava* which then penetrated into the corridors and rooms of the lower part of the reactor[7]. Lava constituents are given in table 7.1. The amount of fuel varies with samples and is in the range 0.2–18% and the lava melting point is approximately 1200 °C.

Analysis has also shown that unexpected artificially produced materials have been formed in the lava: uranium and zirconium silicates. These have been named *Chernobylite*, Plate V, and are yellow and black in colour.

Figure 7.10. The Sarcophagus, 1989. (Courtesy: TASS.)

Figure 7.11. The turbine hall, December 1987. (Photograph: R F Mould.)

Table 7.1. Lava constituents[7].

Lava segment	Contents (%)					
	SiO_2	U_3O_8	MgO	Al_2O_3	PbO	Fe_2O_3
Slag	60	13	9	12	0	7
Glass	70	8	13	2	0.6	5
Pumice	61	11	12	7	0	4

Eventually the lava will disintegrate into more mobile forms, mainly forming dust and partly soluble compounds.

Plate V shows some of the solidified lava beneath the floor of the shattered central reactor hall. This particular mass of lava is called the *Elephant's Foot* and was discovered only in December 1986. By June 1998 it was beginning to break up from its original ceramic-like format and to crack in parts and to crumble on the surface to dust. The shape of the *Foot* varies because of the chemical reactions occurring inside it. In some parts it looks like glass and it has, because of the heat generated, penetrated through at least 2 m of concrete.

It has many layers, like the bark of a tree[5], and pieces of this bark can be removed by the bullets of a Kalashnikov rifle. This sounds like a very archaic way of obtaining samples for analysis to determine parameters such as percentage weight of nuclear fuel in the lava. However, since remote controlled robots are of no use because the high dose rates affect their electronics, Kalashnikovs are the only possible alternative.

7.4.3 Dust and water problems

Two of the major problems are radioactive dust and water. For example, some 1000–2000 m^3 of radioactive water lies on the floors of many rooms. Currently it is estimated that there are 30 tonnes of dust in the Sarcophagus and this will increase as the lava turns to dust. This can be extremely hazardous as when some machine tools were dropped and a fog of re-suspended dust was formed such that the area had to be evacuated. It took two months to bring this particular fog under control[5].

One of the future concerns is that the displaced biological shield, figures 3.6 and 7.1, which has been given the name *Elena* by the NPP workers, is supported only by crushed and rusting steam pipes. If these fail, and *Elena* collapses to what remains of the floor, there will be an enormous dust storm.

Figure 7.12. The first TASS released photograph of any view inside the Sarcophagus. Taken in September 1986 it shows the construction of the wall between Units No. 3 and 4, which is also seen at the far end of the photograph of the turbine hall in figure 7.11. (Courtesy: TASS.)

Figure 7.13. The Sarcophagus, dust suppression unit on the left, turbine hall on the right, June 1998. (Photograph: R F Mould.)

Figure 7.14. Hazard warning and instruction notice inside the Sarcophagus entrance. Various hazards are indicated, except the obvious one of ionizing radiation. It is perhaps the only building in the world where a very significant radiation hazard exists, but is so well known, that it is superfluous to show a warning. The trefoil symbol which is used internationally as a radiation warning is seen in figure 11.2 and in the *top left* drawing in Plate VI. (Photograph: R F Mould.)

С П Р А В К А

Дана г. Ричарду Ф. Моулду в том, что за время посещения 4-го блока Чернобыльской АЭС «3» июня 1998 года им получена доза облучения 0,01 Бэр.

Начальник ЦРТК
«Объекта «Укрытие» В.А. Балюн

C E R T I F I C A T E

This document was given to Richard F. Mould

It confirms that he (she) has received the radiation dose of 0,01 Rem, attending the 4th disastrous unit of Chernobyl Power Plant on the «3 th» 06 1998.

Head of the Radiation Safety
Division Object «Shelter» Vladimir Baljun

Figure 7.15. On handing back the personal pen-type pocket ionization chamber dosimeter and leaving the Sarcophagus, a visitor is handed a certificate in Russian on one side and English on the reverse, giving the dose received: recorded as 0.01 rem in this instance. The conversion between Sv and rem is as follows: 1 mrem = 0.001 rem = 10 μSv and thus 0.01 rem = 100 μSv which leads me to believe, as the dose rate outside the Sarcophagus was 250 μSv/hour, only 0.01 rem is somewhat of an understatement. A Dutch nuclear engineer who was the only other visitor with me was carrying a pocket dosimeter which kept going off like an alarm clock throughout the visit—much to the consternation of our guide.

Chapter 8

Nuclear Power Past and Future

Introduction

Any history of the Chernobyl accident and its consequences should include at least a short summary on its worldwide effect on nuclear power for electricity production. As will be seen, in spite of many anti-nuclear protests in 1986, nuclear power has not declined in terms of available electricity capacity. The major effect of the catastrophe on nuclear power policy has been more in the improvement of nuclear safety which includes not only design features of nuclear reactors but also the training and education of power plant staff. In terms of Chernobyl NPP it is a different matter and the future of the NPP for the generation of electricity is strictly limited whereas the Sarcophagus currently presents a further disaster in the making.

8.1 Before Chernobyl

The penetration of nuclear power into the electricity market commenced in 1954 with the building of the first nuclear power reactors in the USSR and the USA. These were at Obninsk, 100 km from Moscow, with a 5 MW(e) capacity and at Shippingport, Pennsylvania, with an output of 2.40 MW(e). The Chernobyl RBMK-1000 reactors have an output of 1000 MW(e).

By 1960 there were 17 power reactors in operation with a total electricity capacity of 1200 MW(e) in four countries: France, the United Kingdom, the USSR and the USA. A decade later, in 1970, these figures had increased to 90 units operating in 15 countries with a total capacity of 16 500 MW(e). This expansion continued so that by 1980 there were 253 operating nuclear power plants with 135 000 MW(e) capacity in 22 countries and, in addition, some 230 units with more than 200 000 MW(e) were under construction[1].

The 1970s were the time of the oil price shocks and this gave a boost to the development of nuclear power but it was during this period that the public became increasingly aware, interested and concerned. Some associated it with the atomic and hydrogen bombs and confidence in the nuclear industry was not helped in the Spring of 1979 when Hollywood released the motion picture *The China Syndrome*, starring Jack Lemmon and Jane Fonda, which centred on the discovery of a flaw in the design of a nuclear plant and the efforts of a TV reporter and a nuclear engineer to expose an official cover-up.

Much worse was to come two weeks after the release of *The China Syndrome* when on 28 March 1979 at 4am the Three Mile Island (TMI) accident occurred at the power plant near Harrisburg, Pennsylvania. Twenty years later it is still of public interest and is the subject of an exhibit in the Smithsonian Institution[2] in Washington DC, together with the Love Canal disaster in New York state in 1978 when a housing estate was built over a toxic waste dump and 2500 people had to be evacuated.

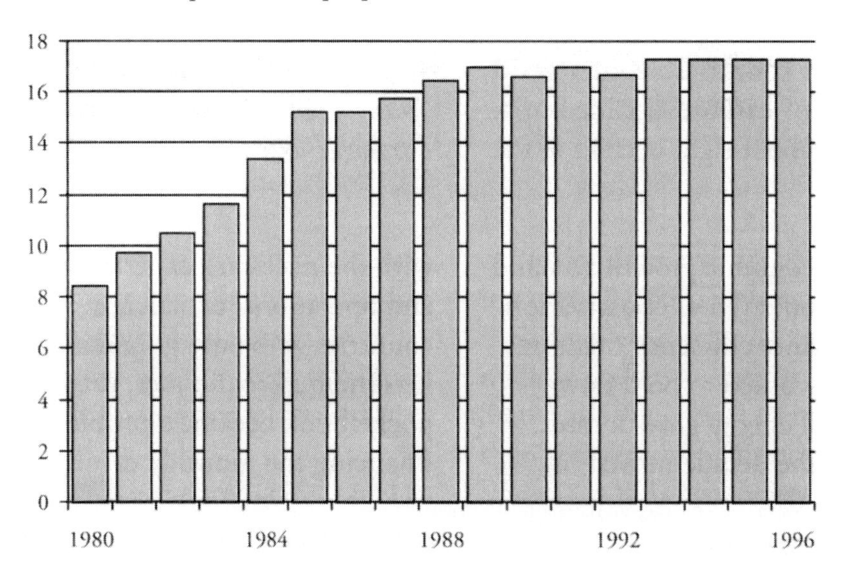

Figure 8.1. Nuclear power growth 1980–96 as a percentage of total electricity generation[4]. (Courtesy: IAEA.)

One of the initial consequences of the TMI accident was that new constructions of power plants declined, although installed nuclear capacity kept increasing as existing plants went on line. TMI also emphasized to the nuclear power industry that many improvements had to be made in the design, construction and operation of nuclear plants with respect to safety and reliability. There were also recommendations following the Chernobyl accident and those of 1986 from INSAG[3] are reproduced in table 8.1. 'Performance pressures' are mentioned in the fourth recommendation and this refers to the need for the experiment commencing on Unit No. 4 at Cher-

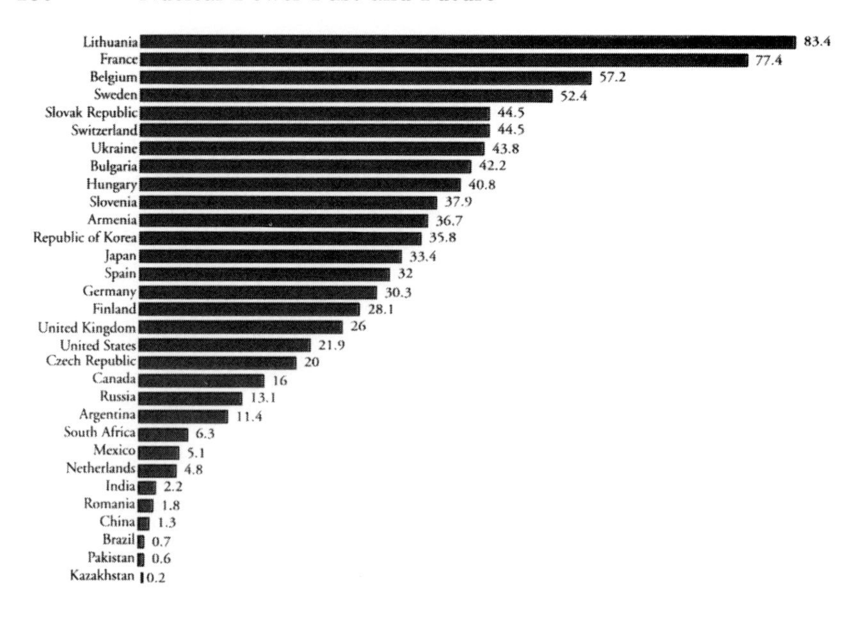

Figure 8.2. Nuclear share of electricity generation, percentage as of January 1997[4]. (Courtesy: IAEA.)

nobyl on 25 April 1986 to be completed as soon as possible, as this was the last opportunity in terms of the deadline to be met for the guarantee of bonuses for the staff. This was the underlying reason for the decisions to override various safety procedures.

At the beginning of 1986 nuclear power was showing a recovery, nuclear power growth had stabilized to about 17% of the world's total electricity generation[4], figure 8.1, for the 31 countries who had invested in nuclear power[4], figure 8.2, and the world had crossed the statistic of 3500 reactor years of operational experience without a single fatal accident involving radiation[1], table 8.2[5a]. More extensive details, for 1945–99, are given in section 8.4, table 8.5[5b].

8.2 After Chernobyl

The Chernobyl accident in the early morning of 26 April 1986 was to change nuclear power policies that year and from the 13 construction starts in 1985 there was only a single start, Ikata-3 in Japan, in 1986 although 23 units were connected to electricity grids in 1986. This compares with 34 electricity grid connections in both 1984 and 1985[1].

Immediately following the accident there were anti-nuclear demonstra-

Table 8.1. Lessons and recommendations submitted in 1986 to the Director-General of the IAEA by the International Nuclear Safety Advisory Group[3].

- Nuclear plant design must be, as far as possible, invulnerable to operator error and to deliberate violation of safety procedures.
- Procedures relating to the operation of the plant must be most carefully prepared with the safety significance of what is intended continuously in view. This is particularly important for cases where unusual operations are intended.
- When special procedures are intended, whereas the initiative and indeed the detailed intent might be in the hands of specialists, the ultimate responsibility for the safety of the operation must lie within the plant management. In such work, an evaluation of the intent from the safety point of view must be provided by staff with a broad understanding of all the implications. It is also important for the technical specialists to be directly involved in the performance of the special work on the plant, though no overriding authority in safety matters is implied by this.
- In the final analysis, reliance on operating staff to follow defined procedures is necessary. To ensure that they do so, an appropriate atmosphere giving the right balance between performance pressures and safety is necessary, in which *quality* checks are made on operational safety practices and tedious and demanding safety practices are seen as a benefit rather than a hindrance.

tions worldwide of which figures 8.3 and 8.4 are only two examples from May 1986: in Switzerland and in Greece. Greece, in particular, was subject to widespread radiation phobia and this over-reaction was clearly seen in the increase in artificially induced abortions in the month of January 1987. The expected number of live births for that month were 9103 whereas the observed number was 7032, which was a 23% fall in the expected level[6]. This was due to conflicting data and rumours of high risk of birth abnormalities which were reported in the media. However, for February and March 1987 the number of observed & expected live births were not significantly different: 7255 & 7645 and 8350 & 8453. The cartoons which appeared in the newspapers and magazines did not help public perceptions with, for example, the equating of the Chernobyl accident with a nuclear bomb and jokes about the contamination of foodstuffs, figure 8.5.

By 1987[1] only one country (Sweden) had a policy, in effect, of phasing out nuclear power; only one (Austria) after starting a nuclear power programme outlawed it; and in the Philippines construction of the first plant was suspended. Although in this case the decision was influenced by the opposition to the then President Marcos and the fact that the power plant site could only be approached in daylight because of the opposition army.

Table 8.2. Fatal radiation accidents in nuclear facilities and non-nuclear research industry, research and medicine (excluding patient-related events) which were reported to the IAEA, 1945–85[5a].

Year	Location	Radiation Source	Fatalities	
			Worker	Public
1945	Los Alamos, USA	Critical assembly	1	
1946	Los Alamos, USA	Critical assembly	1	
1958	Vinca, Yugoslavia	Experimental reactor	1	
1958	Los Alamos, USA	Critical assembly	1	
1961	Switzerland	Tritiated (^{3}H) paint	1	
1962	Mexico City, Mexico	Lost radiography source		4
1963	China	Seed irradiator		2
1964	Germany, Federal Republic	Tritiated (^{3}H) paint	1	
1964	Rhode Island, USA	Uranium recovery plant	1	
1975	Brescia, Italy	Food irradiator	1	
1978	Algeria	Lost radiography source		1
1981	Oklahoma, USA	Industrial radiography	1	
1982	Norway	Instrument sterilizer	1	
1983	Constituyentes, Argentina	Research reactor	1	
1984	Morocco	Lost radiography source		8
Totals			11	15

The influence of the anti-nuclear demonstrations was therefore essentially only a temporary setback to nuclear power generation of electricity. This is in spite of the public reaction and that of the media to the accident in 1986 when initially it was thought that the future of nuclear power for electricity generation would be a significant reduction worldwide, and in the long-term replacement by alternative sources of energy. This trend has not occurred and the nuclear power industry will largely have recovered by the end of the century and overcome the effects of the disasters at Three Mile Island and Chernobyl, although the total electricity generating capacity is estimated to increase more rapidly than its nuclear share.

Data for the end of 1992 are given in table 8.3[7] which shows that there were 424 units in operation with a total electricity capacity of 327 GW(e). It is predicted[8] that these figures will grow and that by the year 2015 the total electricity capacity provided by nuclear power will be in the range 374–571 GW(e), figure 8.6. Although the growth of nuclear power has slowed markedly over the past two decades, steady development continues.

The demand for and the construction of new nuclear power plants in North America is virtually stagnant and currently the USA generates only

Figure 8.3. 150 parents and children demonstrated in Zurich outside the cantonal parliament, 12 May 1986. (Courtesy: G Souchkevitch.)

Table 8.3. Nuclear power reactors in the world[1,7] at the end of 1992.

Region	No. of countries	No. of units in operation	Total electricity capacity (GW(e))	Share of total electric energy supplied (%)	No. of units under construction
W. Europe	9	153	121	36	6
N. America	2	130	114	34	4
Asia	6	70	48	15	21
E. Europe	10	65	44	13	36
Latin America	3	4	2.2	1	5
Africa	1	2	1.8	0.5	0
Worldwide	31	424	327		72

about 20% of its electrical energy with nuclear reactors[2]. It has also been reported[9] in 1999 that Germany is set to abandon nuclear power. However, this remains to be seen as currently Germany has 20 nuclear power stations

Figure 8.4. 10 000 people took part in an anti-nuclear demonstration in Athens on 13 May 1986. (Courtesy: F Dermentzoglou.)

Figure 8.5. French cartoon. (Courtesy: B Asselain.)

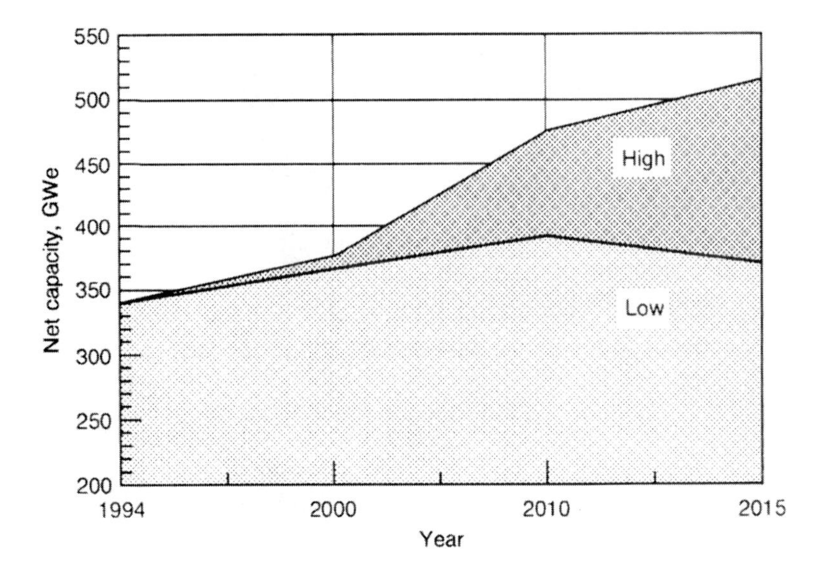

Figure 8.6. Worldwide nuclear power outlook to 2015[8]. (Courtesy: IAEA.)

and these provide almost one-third of the country's electricity needs. Closing these plants over the next five to ten years would cost some 92 billion DM (about US$ 53 billion) and lead to the loss of some 150 000 jobs. However, if Germany is to meet the 1997 Kyoto climate summit agreement for reducing the emissions of carbon dioxide then nuclear power would have to be retained and so the future policy in Germany on nuclear power is unclear.

Nuclear power is moving ahead in France and several countries, in spite of the slowdowns in the USA and the current discussions in Germany, are positively reconsidering the nuclear option. In Asia, new constructions are moving ahead quite strongly in Japan, the Republic of Korea, the Peoples' Republic of China and India. Several of the Eastern European countries depend very strongly on nuclear electricity, and the operation of nuclear power plants is a stabilizing economic, political and social factor, in particular when electricity can be sold to western countries for hard currency. This is true for the Ukraine in its sales of electricity to Austria.

One current move to increase nuclear power electricity production in the Ukraine is for a joint German–French venture by Siemens and Framatome to build two reactors in Rovno and Khmelnitzki, replacing the requirement for the continuation of Chernobyl as a contributor to the national grid. Part of the funding would come from the European Economic Community. There is, however, a protest in Germany by the Green Party and others against this proposal, for example, calling for a boycott of Siemens equipment. This is seen in figure 8.7 where nuclear power is described as the 'black sheep in the family' of Siemens products.

Figure 8.7. Anti-Siemens protest in June 1999 in Heidelberg against building nuclear power stations K2/R4 in the Ukraine. (Courtesy: Internationale Ärzte für die Verhütung des Verhütung des Atomkrieges/Ärzte in sozialer Verantwortung e.V.)

Whatever the future holds, energy is perhaps the key controlling factor for economic growth and development in the 21st century and the practical choices for sources of energy come down to a select few: each with its own consequences. No matter how many economic cost–benefit analyses are undertaken for different sources of electrical power, the nuclear option will continue for the foreseeable future, and it is therefore imperative on national and international organizations that the associated risks, including those of radioactive waste disposal, are minimized as much as is humanly possible.

8.3 Future of the Chernobyl nuclear power station including the Sarcophagus

The future for the continuation of electricity production at the Chernobyl NPP is strictly limited. At the end of 1998 there was only a single reactor in operation*, Unit No. 3, and the 1995 *Memorandum of Understanding* on the closure of the NPP signed by the G7 countries and the European Union appeared to herald the closure by the year 2000. However, life is not so simple. The Ukraine objects to the separation of the two issues.

- Closure of the Chernobyl NPP.
- Completion of two new nuclear units at Khmelnitsky and Rovno.

It also considers the second issue the priority, and the Prime Minister of Ukraine, Yuri Marchuk, made the following statement[10] in 1996.

> We should like to confirm our intention to close the Chernobyl NPP by the year 2000, as stated in the *Memorandum of Understanding* between the Government of the Ukraine and the Governments of the G7 countries and the European Commission. However, without financial assistance from the world community, Ukraine will not be able to go through this on account of its difficult economic situation. It is to be noted that, after having opted to become a non-nuclear weapon state, the Ukraine was promised financial support by several countries, but to date has not received enough aid to resolve this problem.

Marchuk continued[10] to state that

> In all more than US$2500 million has been invested in major construction work and that the total expenditure on eliminating the consequences of the accident over the period 1992–96 alone, paid for out of the Ukranian national budget, exceeds US$3000 million. In the budget for 1996, more than US$600 million has been allocated and such sums are likely to be required for many years ahead.

Financing agreed in the *Memorandum of Understanding* is US$350 million for the following three main elements.

- Provision of an interim spent fuel facility to permit safe storage of spent fuel assemblies.
- Provision of a liquid radioactive waste treatment facility to immobilize the backlog of operational waste stored at Chernobyl.

* Unit No. 2 was shut down in October 1991 when the turbine flamed and destroyed part of the roof and a wall. Unit No. 1 was shut down in December 1997.

- Safety improvements to Unit No. 3 for implementation of short-term and operational safety improvements prior to closure.

However, because of the essential need in the Ukraine to first replace the Chernobyl generating capacity and also because of the dire economic situation in the republic, the Chernobyl power station may not close in the immediate future†.

Funding is not the only problem, although media claims that the nuclear fuel masses in the Sarcophagus might become critical and cause a Hiroshima-type nuclear explosion can be ignored. However, part of the Sarcophagus really is in danger of collapse (a 70% probability by the year 2003 according to Evgenii Velikhov[14] (Velikhov was the chief scientist involved in the clean-up operations and scientific advisor to Mikhail Gorbachev)) bringing with it a significant radioactive dust hazard and if this occurs it will be a major environmental problem as well as a health problem to those within the 30 km zone at the time of the collapse and perhaps also to those further afield. The latter qualification is necessary because it is not known how much activity will be released in the dust and what the strength and direction of the wind will be if this occurs before measures are taken to ensure the safety of the Sarcophagus.

The 1991 TV programme *Suicide Mission to Chernobyl*[15] describes the three future options for the Sarcophagus, detailing advantages and disadvantages, table 8.4, and emphasizes the current worst situation scenario of a radioactive dust accident as there is estimated to be 30 tonnes of dust. Even a *minor* accident with dust can cause enormous problems, as illustrated in the case already mentioned when some machine tools were accidentally dropped and a *fog* of re-suspended dust particles was formed. It took two months to bring this particular dust situation under control.

† This is linked to the economic assessment of proposals for the building of the Khmelnitsky 2 and Rovno 4 (often termed K2/R4) nuclear reactors in the Ukraine. In February 1997 a report was commissioned by the European Bank for Reconstruction and Development (EBRD), the European Commission and the US Agency for International Development from an international panel of experts chaired by Professor John Surrey of the Science Policy Research Unit of the University of Sussex, United Kingdom[11]. The main question which was addressed by this panel was 'Whether completing K2/R4 is economic and whether these two 1000 MW reactors form part of a *least cost* plan for the development of Ukraine's energy sector'. The conclusion was 'that K2/R4 are not economic and completing these reactors would not represent the most productive use of US$1 billion or more of EBRD/EU funds at this time' (the December 1995 *Memorandum of Understanding* had promised US$1.8 billion of western assistance for energy projects in the Ukraine). Following the Surrey report, the US consultants Stone and Webster submitted their own report[12] to the EBRD in May 1998. The Stone and Webster conclusions were the direct opposite of those of Surrey, but the approaches and assumptions are not identical. A critique of Stone and Webster was funded by the Austrian Energy Agency[13] following the EBRD giving provisional approval for the K2/R4 project to go ahead. This is the situation at the time of writing.

Table 8.4. Future possibilities for solving the current situation with the Sarcophagus[15].

Solution	Problem
• Permanently entomb in concrete	Very difficult to monitor the nuclear fuel masses.
• Cover with sand	Would form a heat trap and the fuel might overheat.
• Second Sarcophagus to last for a few centuries	Expense.

The second Sarcophagus was the subject of an international competition[16] in 1992 for the best long-term solution. Figure 8.8 shows the cover of this tender document which in its introduction describes the problem of the Sarcophagus (now termed Ukritiye, i.e. encasement) as follows.

> It cannot be considered as a storage for spent nuclear fuels or radioactive wastes since it does not comply with the national and international standards on nuclear and radiation safety of the regular nuclear power stations. At present the Ukritiye is a temporary system which localizes nuclear fuels and radioactive materials. However, in the future it ought to be converted into an ecologically safe system which may include a full removal of nuclear fuels and their burial in accordance with operating international standards and regulations.

The results of this competition were announced in 1994 following detailed evaluation of six of 24 applications. It was won by a consortium, Alliance, led by Campenon Barnard of France for the construction of a Supersarcophagus. The design work is estimated to cost US$20–30 million, the construction to take five years and cost US$300 million and the final disposal of the radioactive waste to take some 30 years[17].

It is also important to be aware that following closure of the power station for electricity production for the Ukranian national grid, all the associated problems will not just vanish. There will still remain much work to be accomplished, not least the building of the second Sarcophagus[16] and the continuation of environmental work in the 30 km zone.

The non-implementation of the *Memoranda of Understanding*, lack of funding and even in some cases the probability in the future of not being able to pay some of the staff at Chernobyl, poor salaries (the average at the NPP in late 1998 was US$200 per month which even so is double the monthly salary of many staff at Kiev University Institutes[18]) and the continual devaluation of the Ukranian currency against the US$, are precursors

МИНЧЕРНОБЫЛЬ УКРАИНЫ

АКАДЕМИЯ НАУК УКРАИНЫ

ОПИСАНИЕ ОБЪЕКТА «УКРЫТИЕ» И ТРЕБОВАНИЯ К ЕГО ПРЕОБРАЗОВАНИЮ

Исходные данные для выполнения конкурсной работы на проект и технические решения по преобразованию объекта «Укрытие» Чернобыльской АЭС в экологически безопасную систему

Figure 8.8. Competition tender document for a second Sarcophagus[16]. (Courtesy: Chernobylinterinform.)

of a further environmental and human disaster at Chernobyl and its immediate surroundings. Nor does bureaucracy help: as an example[15] work stopped for the scientists in the Sarcophagus for an entire two months due to bureaucratic paperwork and signature authorization for requirements for adequate clothing, even socks.

It is also important that funding, when it is available, should be properly directed but, according to the German Federal Minister for the Environment, Nature Conservation and Nuclear Safety, A Merkel, there is some doubt about this[19]:

> The assistance provided by the West should be looked at critically, as to whether sufficient help was given or whether it sometimes missed the target. Western consultancies were paid quite considerable sums for their activities. There is nothing wrong with this. But I cannot understand that there was not enough money to provide children suffering from thyroid cancer with the medicine they need.

Also, many of the workers at the Chernobyl NPP, in Chernobyl town and its surroundings, consider that the world has largely forgotten the catastrophe, and this opinion was voiced in 1996 by the President of Belarus, A G Lukashenko[20]: 'For those who did not face directly the radiation disaster it may seem that the problem of Chernobyl has lost its intensity and topicality'.

8.4 Major radiation accidents 1945–99

Table 8.2[5a] is only a brief summary of fatal radiation accidents whereas table 8.5[5b] presents more detailed information, quoting not only the fatalities but also the number of overexposures. These significant exposures are defined as follows: > 0.25 Sv to the whole body, blood forming organs or other critical organs, ~ 6 Gy to the skin locally, ~ 0.75 Gy to other tissues or organs from an external source, or exceeding half the annual limit on intake. Mixed radiation (MR in table 8.5) refers to various types of radiation with different LET values such as neutrons and gamma rays, or gamma rays and beta rays.

It should be noted that table 8.5 contains data for accidents not only in nuclear facilities but also in non-nuclear research industry, and in medical research and radiation treatment for cancer. For example, that in 1961 in Plymouth, United Kingdom, was a radiation therapy accident where a metal filter was wrongly left out of the machine during treatment delivery and the 11 cases overexposed were patients. It is also noted that overexposures of cancer patients can not only occur due to machine faults caused by human error, but can also occur due to incorrect medical prescriptions. However, assessments of overexposure in these instances is in the province of a Court of Law with expert witnesses assessing what was clinically acceptable in terms of dose–time fractionation and dose distribution planning at the time when the treatment was given[21].

It should also be noted that table 8.5 refers only to those accidents reported to the IAEA and therefore does not include all radiation accidents that have occurred during this period from 1945 to September 1999. For example, the ^{60}Co radiotherapy accident at the Royal Devon and Exeter Hospital, United Kingdom, is not included by the IAEA. This occurred in 1988 when 207 cancer patients received significant overdoses. The error was

Table 8.5. Major radiation accidents reported to the IAEA, 1945 to 1998, for which there were one or more fatalities and/or two or more overexposures. For accidents reported to the IAEA which resulted in one overexposure but no deaths, see the original IAEA reference[5b]. The number of deaths are not always consistent between the two references sometimes being given in one[5a] but not[5b] in the other. When the former is used it is referenced. See also references[5c–5k] for the accident reports published by the IAEA which include the 1993 accident in Hanoi[5j] and the 1993 accident in Tomsk[5k].

Year	Place	Source and dose or activity intake	Number over-exposed (Number of deaths)
1945/46	Los Alamos, USA	Criticality: up to 13 Gy MR	10 (2)
1952	Argonne, USA	Criticality: 0.1–1.6 Gy MR	3
1953	USSR	Experimental reactor: 3.0–4.5 Gy MR	2
1958	Oak Ridge, USA	Criticality: 0.7–3.7 Gy MR	7
1958	Vinca, Yugoslavia	Experimental reactor: 2.1–4.4 Gy MR	8 (1[5a])
1958	Los Alamos, USA	Criticality: 0.35–45 Gy MR	3 (1)
1960	Lockport, USA	X-rays: to 12 Gy non-uniformly	6
1960	USSR	^{137}Cs: 15 Gy, suicide case	1 (1)
1960	USSR	Ingestion of RaBr: 74 MBq (death after 4 years)	1 (1)
1961	USSR¶	Submarine accident: 1.0–50.0 Gy	>30 (8)
1961	Miamisburg, USA	^{238}Pu	2
1961	Miamisburg, USA	^{210}Po	4
1961	Switzerland	^{3}H: 3 Gy	3 (1)
1961	Idaho Falls, USA	Explosion in reactor: up to 3.5 Gy	7 (3)
1961	Plymouth, UK	X-rays: local overdosage	11
1962	Richland, USA	Criticality	2
1962	Hanford, USA	Criticality: 0.2–1.1 Gy MR	3
1962	Mexico City, Mexico	^{60}Co capsule: 9.9–52 Sv	5 (4)
1963	China	^{60}Co: 0.2–80 Gy	6 (2)
1963	Saclay, France	Electron beam	2
1964	Germany (FRG)	^{3}H: 10 Gy	4 (1)
1964	Rhode Island, USA	Criticality: 0.3–46 Gy MR	4 (1)
1966	Portland, USA	^{32}P	4
1966	Pennsylvania, USA	^{198}Au	1 (1)
1966	China	Contaminated zone: 2–3 Gy	2
1966	USSR	Experimental reactor: 3.0–7.0 Gy	5
1967	Pittsburgh, USA	Accelerator: 1–6 Gy	3
1967	USSR	Diagnostic x-rays: 50 Gy to head (death after 7 years)	1 (1)
1968	Burbank, USA	^{239}Pu	2
1968	Wisconsin, USA	^{198}Au	1 (1)

Table 8.5. (Continued)

Year	Place	Source and dose or activity intake	Number over-exposed (Number of deaths)
1968	Chicago, USA	^{198}Au: 4–5 Gy to bone marrow	1 (1)
1968	USSR	Experimental reactor: 1.0–1.5 Gy	4
1970	Australia	X-rays: 4–4.5 Gy local dose	2
1971	Japan	^{192}Ir: 0.2–1.5 Gy	4
1971	USSR	Experimental reactor: 7.8 Sv, 8.1 Sv	2
1971	USSR	Experimental reactor: 3.0 Gy	3
1972	China	^{60}Co: 0.4–5.0 Gy	20
1972	Bulgaria§	^{137}Cs capsules: >200 Gy suicide case	1 (1)
1974	Illinois, USA	Spectrometer: 2.4–48 Gy local dose	3
1975	Bescia, Italy	^{60}Co: 10 Gy	1 (1[5a])
1975	Columbus, USA	^{60}Co: 11–14 Gy local dose	6
1977	Peru	^{192}Ir: 0.9–2.0 Gy total, 160 Gy to hand	3
1978	Algeria	^{192}Ir: up to 13 Gy	7 (1[5a])
1979	California, USA	^{192}Ir: up to 1 Gy	5
1981	Saintes, France	^{60}Co medical facility: >25 Gy	3
1981	Oklahoma, USA	^{192}Ir	1 (1[5a])
1982	Norway	^{60}Co: 22 Gy	1 (1)
1983	Argentina	Criticality: 43 Gy MR	1 (1)
1983	Juarez, Mexico‖	^{60}Co: 0.25–5.0 Sv protracted exposure	10
1984	Morocco	^{192}Ir	11 (8)
1984	Peru	X-rays: 5–40 Gy local dose	6
1985	China	Electron accelerator	2
1985	China	^{197}Au treatment mistake: internal dose	2 (1)
1985	China	^{137}Cs: 8–10 Sv	3
1985	Brazil	Radiography source: 160 Sv local dose	2
1985/86	USA	Accelerator	3 (2)
1986	China	^{60}Co: 2–3 Gy	2
1986	Chernobyl, USSR	RBMK NPP: 1–16 Gy MR	134 (28)
1987	Goiânia, Brazil[5a]	^{137}Cs: up to 7 Gy MR	50 (4)
1989	El Salvador[5c]	^{60}Co irradiation facility: 3–8 Gy	3 (1)
1990	Soreq, Israel[5d]	^{60}Co irradiation facility: >12 Gy	1 (1)
1990	Saragosa, Spain	Radiotherapy accelerator	27 (11)
1991	Nesvizh, Belarus[5e]	^{60}Co irradiation facility: 10 Gy	1 (1)
1992	China	^{60}Co: >0.25–10 Gy local dose	8 (3)
1992	USA	^{192}Ir brachytherapy: > 1000 Gy	1 (1)
1994	Tammiku, Estonia[5f]	^{137}Cs waste repository: 1 830 Gy thigh, 4 Gy whole body	3 (1)
1996	Costa Rica[5g]	^{60}Co radiotherapy: 60% overdose	115 (13)
1998	Turkey[5h]	^{60}Co: various up to 3 Gy whole body	10

Table 8.5. (Continued)

¶ This Russian submarine was the K-19, the *Hiroshima*, and in July 1961 the hermetic seal to the reactor was breached, causing the accident. In 1972 a fire caused the *Hiroshima* to sink and 28 lives were lost[28].

§ A capsule more often associated with Bulgaria is that containing the poison ricin and was used in London in 1978 to assassinate Georgi Markov in the so-called Poisoned Umbrella Murder. The technology was attributed by two Scotland Yard detectives who visited Sofia for a week[23], to that of a ^{198}Gold implantation gun designed for the treatment of cancer[24] by implanting small radioactive ^{198}Au grains into the tumours, such as malignant nodes in the neck. However, it turned out that Scotland Yard were not correct because the gun was of such a size that it could not have been easily hidden within an umbrella. It takes a magazine of 14 grains not one, but more importantly the diameter of the capsule in Markov's leg was larger than the diameter of the gun barrel[25]. The most likely technology used was that of the simple design of instrument used for implanting ^{222}Radon seeds, which were the forerunner of ^{198}Au grains. The design is based on a hollow tube with a sharp end to penetrate tissue and a plunger to force the ^{222}Rn seed into the tissue[26]. Following publication of this solution[25] a BBC TV *Panorama* programme was set-up for Scotland Yard to publicly comment. However, at the last minute the Ministry of Defence placed a censorship D Notice on the programme so that it could not take place: the implication is that the ^{222}Radon seed inducer was the correct technology.

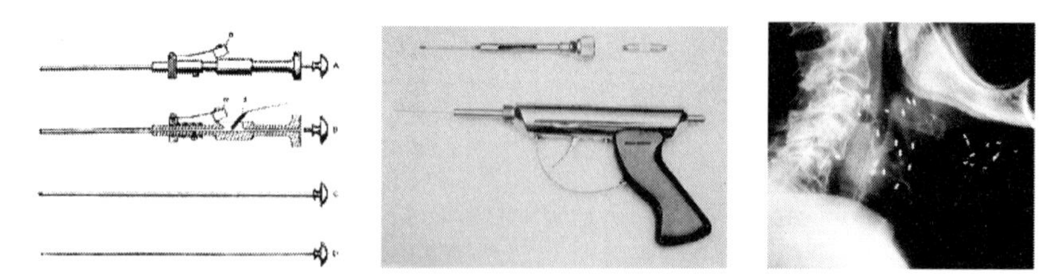

Figure 8.9. Cancer treatment technology using small radioactive sources: (*left*) ^{222}Radon in the 1920s–1950s and (*centre and right*) ^{198}Gold in the 1960s–1980s.

‖ This radiation accident was detected in a most bizarre manner[27]. A radiotherapy machine in a hospital in Lubbock, Texas, with a source consisting of 7000 tiny pellets of ^{60}Co which in 1969 had an activity of 3000 Ci, was sold in 1977 and shipped to the Centro

Medico in Juarez, Mexico. It was never installed and in 1983 someone had taken the decision to dismantle it and it was then stolen and ended up in the Junke Fenix scrap yard. This sequence of events only became known by chance in January 1984 when a truck loaded with steel rods from scrap took a wrong turning, passed the Los Alamos laboratory in New Mexico, and set off an alarm from a radiation sensor in the road. It was then discovered that some 5000 tons of reinforcing rods and some 18 000 table legs had left Mexico. Also, at least 12 children had played on the highly contaminated truck before it was removed to safety, still containing some of the ^{60}Co pellets.

due to a wrong calibration of the output of the ^{60}Co machine. The physicist who was responsible measured the output for 40 seconds but then carelessly forgot to correct the figure by a factor of 60/40 to obtain the output dose per minute. The error therefore implied that the gamma-ray output was lower than it was in reality and cancer patients therefore remained under the ^{60}Co machine for a longer time than was necessary to deliver the prescribed treatment dose.

The types of accident which have occurred during this period of just over 50 years are very varied and are by no means limited to reactor accidents and, although any radiation accident is unacceptable, it has to be recognized that nuclear power for electrical energy production will continue for the foreseeable future. It is predicted that with the expanding world population, worldwide energy consumption will increase by 50% by the year 2020 and could double by the mid-21st century[22]. Currently a total of some 430 nuclear power reactors, half in Europe, operate in 31 countries and produce 16% of the world's electricity. Safety improvements are therefore essential, including improved training and education of NPP staff. With regard to the Chernobyl type RBMK reactors some 15 plants are still in use and although now better equipped with safety upgrades, these still fall short of current standards and must be phased out as soon as alternative energy supplies can be funded and installed[22].

Figure 8.10. April 1986 photograph of the damage to Unit No. 4, taken from a helicopter by the TASS photographer Valery Zufarov whilst the burning reactor was still sending out clouds of smoke. When the helicopter landed back in Chernigov, Zufarov had to have his head shaved because of the contamination. He died in the mid-1990's of what was described as a 'blood disease'. (Photograph: V Zufarov.)

Chapter 9

Dose Measurement and Estimation Methods

Introduction

Under normal circumstances radiation workers carry personal dosimeters and are monitored at regular intervals. When the Chernobyl accident occurred, radiation dose monitors went off scale, there was obviously no infrastructure to assess doses using personal dosimeters: small ionization chambers, film badges or TLD. Doses had therefore in the main to be reconstructed using some form of retrospective measurement for the estimates.

Such retrospective estimation with or without statistical models is extremely very difficult. For example the atomic bomb dosimetry system T65D (tentative 1965 dosimetry[1]) which was used as the most accurate method of estimating the doses received by individual survivors has now been replaced the DS86 system (dosimetry system 1986*). Thus dose estimation was still being improved upon some 40 years after the atomic bombings: so we cannot really expect very accurate estimates of individual doses from the Chernobyl accident only some 15 years after the event.

9.1 Military personnel doses

One of the first summaries on personnel radiation doses has finally been published[2] in 1995 although it was compiled on 11 November 1986, ta-

* The T65D dose estimation system was based on experimental data obtained for Nagasaki-type atomic bombs in Nevada, USA. It was devised in 1965 as a formula that incorporated various parameters such as distance from the hypocentre, and transmission factors for shielding materials. DS86 was devised to permit more detailed calculations than T65D. It is based on elementary physical processes, and enables computer coding of the different processes involved from the time of emission until arrival at various human organs[1].

ble 9.1. Presumably this is because it was previously censored by the USSR and then by Russia because it relates only to servicemen: defined as chemical and engineering troops, civil defence and military helicopter crews.

Six months earlier[2], on 2 May 1986, Colonel-General Vladimir Pikalov the commander of the chemical troops, reported the doses received in Chernobyl by a cohort of 671 servicemen (the actual number involved was much higher) on 1 May: 61% received up to 10 R, 6.7% up to 20 R, 16% up to 30 R, 4.4% up to 50 R and 11% over 50 R. An approximate numerical equality is 1 R = 1 cGy.

The permitted emergency dose in the USSR was only 25 cGy and therefore this limit had been exceeded by some one-third of this group of 671. It is not stated, but it is reasonable to assume that the military were better equipped with dosimeters than were other groups of personnel, but even so it must be assumed that these early doses can only be considered to be estimates: but they are all that are available from this early period.

Table 9.1. Radiation doses to servicemen as of 11 November 1986. Data from the Central Military Medical Board[2] (R denotes the roentgen unit of exposure). In addition to the population of 18 614 still remaining in the area after 11 November 1986 there are also an additional 23 583 of whom 19 352 were reservists.

Group	No. of servicemen	Servicemen subgroup				Percentage
		Generals	Officers	Warrant officers	Privates & sergeants	
Total exposed	66 752	50	8378	2570	55 754	
Those who left the area	48 141	37	5883	1888	40 333	100
Dose						
<25 R	46 076	17	5195	1778	39 086	95.7
25–50 R	2041	19	674	106	1242	4.24
>50 R	21	1	14	4	2	0.04
Those remaining in the area	18 614	13	2495	682	15 424	100
Dose						
1–10 R	13 018	11	1732	546	10 724	69.9
11–15 R	2629	2	321	71	2235	14.1
16–20 R	1704	0	243	36	1515	9.2
21–25 R	1173	0	199	29	945	6.3

9.2 Group dose and itinerary dose

The major problems encountered in trying to obtain dose measurements were fourfold.

- The enormous scale of the accident.
- Lack of a sufficient number of available measurement techniques.
- Organizational difficulties.
- Some measurements turned out to be insufficiently reliable.

Those dosimeters which were available were only valid within the dose range 0.5–20 mGy which was useless during the accident[3]. Alternatives therefore had to be used to estimate doses and these were defined as a *group dose* and an *itinerary dose*, table 9.2.

Table 9.2. Dose estimation methods for liquidators classified into three groups according to reliability[3].

- Exposure or absorbed dose recorded by an individual dosimeter: the maximum error is about 50%.
- Group dose assigned to the members of a group performing an operation in the zone, based on the readings of an individual dosimeter held by one member of the group: the maximum dose error in the group can be as high as 300%.
- Itinerary dose, which was estimated from exposure dose rate in the workplace and the duration of stay of the group there: the maximum error in the group can be as high as 500%.

9.3 Thyroid and whole-body dosimetry

The WHO IPHECA dosimetry methods[4] were divided into two categories: dosimetry of the thyroid gland and whole-body dosimetry. For the latter there were insufficient numbers of whole-body counters available, and those that were used were not all of the same design: in some the measurement geometry was with the person in a chair and in others it was not.

Numerous specialist teams took direct measurements in May–June 1986 of the radioactive ^{131}I content in the thyroid gland and the results were coordinated at several different institutes in the USSR. There were, though, some differences in the approaches used by the three republics, Belarus, Russia and Ukraine.

In Belarus, for example, the ^{131}I content in the thyroid was calculated using a formula[4] which was based on the exposure rate measurement near the surface of the neck. For the 130 000 persons who were measured the data were divided into three groups according to the dose determination error. For 10 000 there was an error of <50%, for 40 000 the mean error was <300% and for the remainder the error exceeded 300% and was often some 500%.

In Russia some 31 000 residents were monitored using portable scintillation detectors. The results were correlated, when possible, with those from whole-body counting and with spectrometry. In all, though, the thyroid doses were reconstructed by mathematically modelling for some 170 000 people including 37 000 children. Similar methods were also used in the Ukraine.

In spite of such large population numbers, there are many sources of uncertainty in retrospective modelling for the reconstruction of thyroid doses[4,5]. For example, not all of the measurements were made with an instrument which could be used with a known energy range measurement window; measurement geometry could not be standardized for this large population of measurements and there were also errors in detector positioning.

Several different approaches have been used in modelling the reconstruction of thyroid dose and table 9.3 summarizes four of these approaches[5].

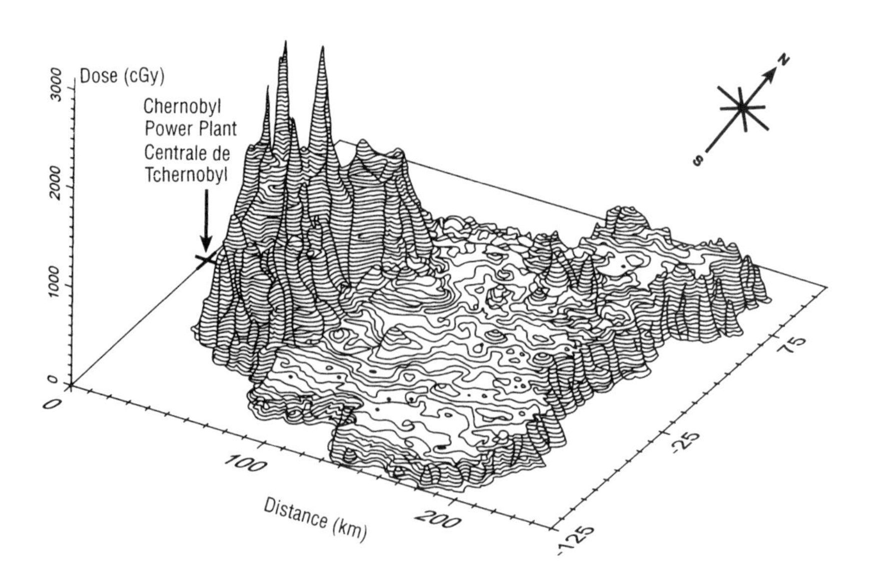

Figure 9.1. Average distribution of thyroid doses in children born in 1986[6]. (Courtesy: WHO.)

Table 9.3. Reconstruction of thyroid doses[5].

Development centres	Assumptions
Jülich Research Centre	• A constant ratio of ^{131}I and ^{137}Cs with the long-lived radionuclide ^{129}I. • The concentration of ^{129}I can be used to assess thyroid dose.
IRH St Petersburg	• Linear relationship between thyroid dose derived from ^{131}I thyroid activity and ^{137}Cs contamination, air kerma rate and mean ^{131}I concentration in milk assumed when evaluating average dose. • Method based on an observed correlation (for Bryansk, Russia) between the total ^{131}I content in the thyroid and the whole-body ^{137}Cs content measured during the first few months after the accident. • Data on individual milk consumption during the first weeks after the accident were also taken into account.
IBPh Moscow, MRRC Obninsk and IRM Minsk	• Method uses the correlation between the mean dose calculation based on ^{131}I thyroid activity measurements and ^{137}Cs contamination in soil. • Available ^{131}I soil contamination data were taken into account. • A semi-empirical method infers ^{131}I soil contamination from ^{129}I measurements when those for ^{131}I are not available.
SCRM Kiev and GSF Munich	• The model is for evacuees from Pripyat based on measurement of ^{131}I activity and questionnaire data.

Examples of the results[4,6] of one of the thyroid dosimetry models are shown in figures 9.1 and 9.2 for Chernigov oblast in the Ukraine. The doses are shown at distances from the NPP of up to 250 km over an area of 250 km $\times$ 250 km. It is quite clear that the doses to children born in 1986, figure 9.1, are higher than those of adults, defined as of age greater than 18 years, figure 9.2.

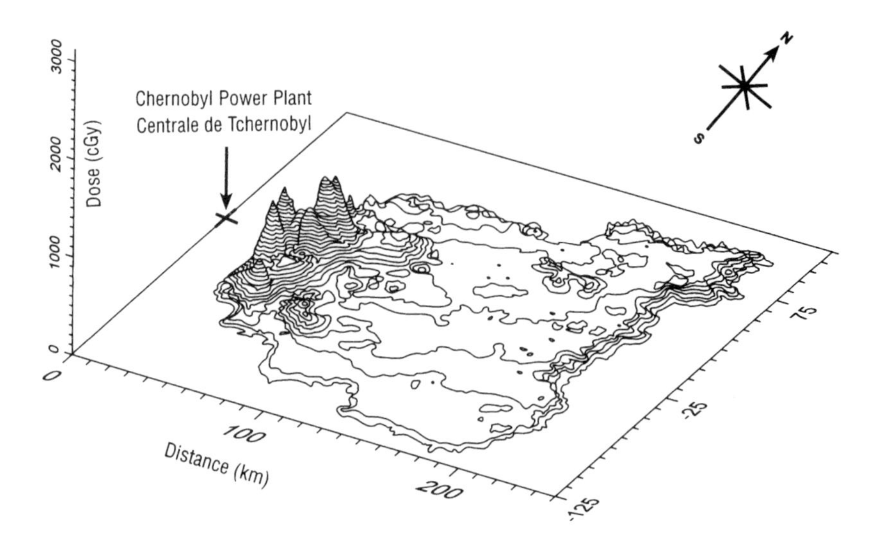

Figure 9.2. Average distribution of thyroid doses in adults[6]. (Courtesy: WHO.)

9.4 Electron paramagnetic resonance dosimetry with tooth enamel

9.4.1 Measurement of Chernobyl doses

Electron paramagnetic resonance (EPR) which is also termed electron spin resonance (ESR) has been used[4–6] for Chernobyl accident dosimetry because human tooth enamel is a natural detector of accumulated dose. Radiation-induced paramagnetic centres which can be recorded by the EPR method arise in the enamel following exposure to ionizing radiation. Some of these centres are unstable and disappear within a few days. The remaining radiation-induced centres are stable and persist far longer than the human lifespan. The radiation dose received by the tooth enamel is determined from the EPR signal strength from the stable centres.

One method[4] used to prepare Chernobyl samples is to extract a tooth and then mechanically separate the crown from the root. The dentine is then removed with a hard alloy dental drill and the enamel removed from it. This enamel is then crushed into pieces measuring 1–2 mm. The extent of removal of the dentine from the enamel is verified from the strength of the background signal in the EPR spectrum due to the radicals of the organic matrix of the enamel and the remains of the dentine.

The samples are then kept in an alcohol–ether mixture for several days for degreasing and drying. Following this processing the amplitude of the background signal is reduced. The ground-up enamel is weighed (the optimum weight is about 100 mg) on an analytical balance and placed in quartz ampoules for ESR spectroscopy. The errors with this technique are stated[4] to be some 20–50 mGy for small doses of 100–200 mGy and with larger doses 10%. Calibrations are made with samples of known activities of ^{60}Co and ^{137}Cs.

The lowest measurable absorbed dose in tooth enamel was found to be limited not by the sensitivity of the EPR spectrometer but by the existence of background EPR signals in the sample. The background spectra in teeth from different donors, including children and adults below the age of 25 years were found to be identical in line shape and asymmetric rather than symmetric. A tendency for dose over-estimation was found when symmetric models of the background were used. It was also observed that teeth with caries were more sensitive to radiation than healthy teeth and that the increase in sensitivity was dose-dependent[5]. Results from teeth sampled from adult residents living in contaminated territories in Belarus are given in table 9.4 for doses accumulated over age and nine years exposure to contamination.

Table 9.4. EPR results from Belarus residents[5].

Contamination (TBq/km^2)	No. of samples	Mean values of total absorbed dose (mGy)
0.6–1.5	44	129
0.2–0.6	14	92
0.04–0.2	12	69
<0.04	31	43

Several different procedures have been used and unfortunately the accuracy and precision differ significantly amongst them, although almost all could reconstruct doses above 500 mGy to ±25%, and the potential is present with EPR to measure doses in the range 0.1–1 Gy (i.e. 100–1000 mGy)[5]. Work is continuing with EPR dosimetry and more than 300 teeth of liquidators have been collected with preliminary results[4] showing that 15% received doses apparently higher than 20 cGy.

9.4.2 Measurement of doses from other accidents

To date EPR spectroscopy of tooth enamel has been used widely for accidents other than Chernobyl and for some ten years has been studied by Romanyukha and others at the Institute of Metals in Ekaterinburg in the Urals, particularly for residents in the Techa river area[7–9] in the eastern Urals, see section 4.9, including the Mayak facility for which the radionuclide release data are given[10] in figure 9.3a, revealing a peak release in

1950 of some 4000 Ci/day. An EPR dose reconstruction study[9] for the middle and lower Techa river population revealed ultrahigh doses in tooth enamel for individuals born between 1945 and 1949, figure 9.3b. These can be explained by the ultrahigh local ^{90}Sr concentration in tooth enamel for this particular age group. This study also found that tooth position (since teeth develop at different rates) as well as age group are important when analysing results from ^{90}Sr internal exposure: the most sensitive tooth being the first molar. A tooth from this position can give an additional enhancement factor of 4–6. Thus selection by age group and tooth position at sample collection can improve the detection and measurement of former ^{90}Sr releases by a factor of 200.

Although the major test site for nuclear weapons testing in the USSR was at Semipalatinsk in Kazakhstan, see section 4.8, testing also occurred in the southern Urals at Totskoye which is in Orenburg oblast[11]. On 14 September 1954 a 40 kilotonne nuclear device was detonated at a height of 350 m above ground level. The aim of this test was to determine how well the army could operate during nuclear warfare. In all, 45 000 soldiers returned to the area after the explosion, the first units after only 45 minutes and the remainder were in place within two to three hours. In addition, some 60 000 of the local population were exposed as there was a strong wind which focused the fallout along a narrow path of some 210 km, which included several villages. EPR dosimetry with tooth enamel has been used for dose reconstruction and showed doses received of up to 3 Gy. This conflicts with the official estimates of the doses received by the military and civilians of 0.08–0.7 Sv. It was also found that there was a strong dependence of the dose on the distance between the tooth donor's location and the explosion site. In addition, these authors concluded that it is very probable that when the EPR method is used in combination with the fluorescent *in situ* hybridization (FISH) technique, see page 168, it will provide more reliable dosimetric information than is currently available[11]. This would have a significant effect on the dose reconstruction of those irradiated by the Chernobyl accident.

This Ekaterinburg work on EPR is now being extended at the US National Institute of Standards and Technology (NIST) in Gaithersburg[9,12,13]. Figure 9.4 illustrates the additive dose method for reconstruction which is based on the re-irradiation (d_1 to d_5) of a tooth sample to obtain a sample-specific dose response curve which is used to back-extrapolate to the absorbed dose value, the accident dose[12]. The attributes of EPR tooth dosimetry method are given in table 9.5, a summary of the Ekaterinburg procedure for sample preparation in table 9.6, and some cautionary notes[12] on technique in table 9.7. These details are included because the future of low dose reconstruction for Chernobyl populations such as evacuees may well lie with this EPR method.

The NIST group have also used EPR accident dosimetry reconstruc-

Table 9.5. Attributes of the EPR tooth dosimetry method[12].

- Dose dependence ranges from 100 mGy (but this can now be reduced to 20 mGy[16]) to >10 kGy.
- Although interfering signals (e.g. produced by UV) can occur, the EPR signal is specific to ionizing radiation.
- Rapid estimates are possible, with confirmation available within a few days.
- The lifetime of the radiation-induced EPR signal far exceeds the human lifespan.
- Determination of partial-body exposure is possible.
- The method is applicable to fractionated and chronic exposures.
- All of the different radiation qualities are covered by the method.
- Absorbed doses due to internal emitters are measurable.
- The measurement method is capable of being transferred from experts to technical staff.

Table 9.6. Summary of the preparation procedure for tooth sample preparation[11].

- Cutting off the root from the crown.
- Crushing the crown into 3–4 pieces with an agate mortar and pestle.
- First ultrasonic treatment with a 30% NaOH aqueous solution for 60 minutes to soften dentin.
- Washing sample with distilled water 5–10 times by adding water, shaking and decanting.
- Removing soft dentin and parts of tooth enamel coloured due to a disease with a dental drill.
- Second ultrasonic treatment with a fresh 30% NaOH solution for 120 minutes.
- Final washing of the samples with distilled water five times for 30 minutes each in an ultrasonic bath with fresh water replacement each time.
- Overnight drying in a dessicator.
- Final sample crushing in a mortar with a pestle followed by sample sieving to grain sizes of 0.3–0.6 mm.

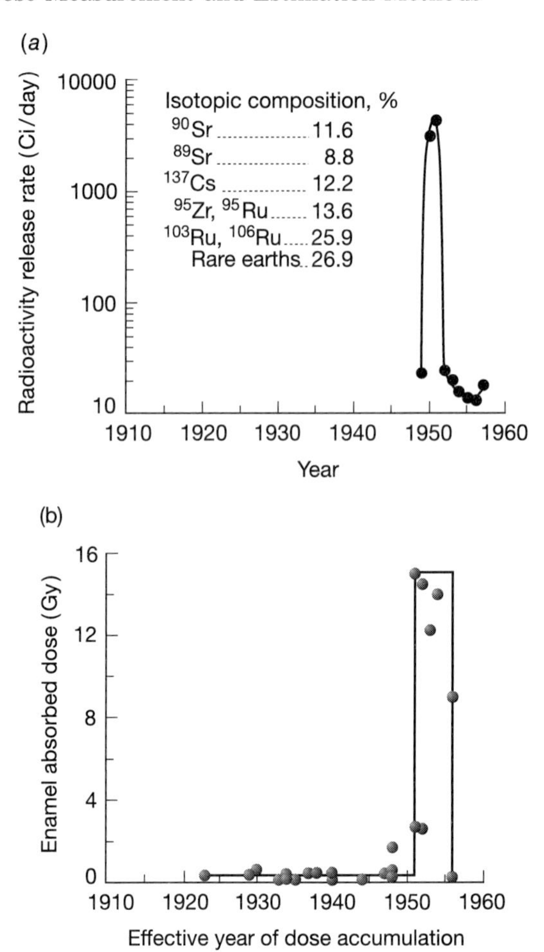

Figure 9.3. (*a*) Mayak release data and (*b*) tooth enamel dose measured by EPR *versus* year of onset of dose accumulation. (Courtesy: A A Romanyukha.)

tion using bone rather than teeth. These studies[14,15] were related to an accidental overexposure at an industrial accelerator facility in Maryland in 1991, in which two radiation-damaged fingers of the victim were amputated. Dose estimates assigned to three bones measured were 55, 79 and 108 Gy; but the value of the EPR method for Chernobyl, if the work of NIST can be extended to other laboratories such as those in the Ukraine, is the ability for dose reconstruction down to lower levels[16] of some 20 mGy.

A further example of the use of EPR bone dosimetry followed the radiation accident which occurred at a ^{60}Co industrial irradiation facility

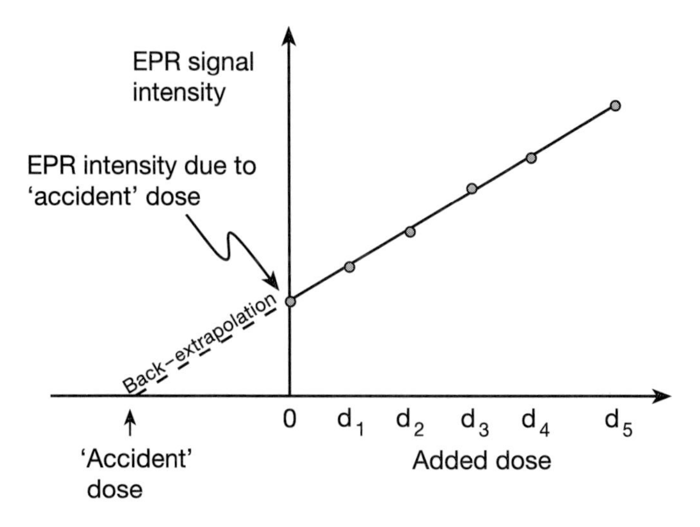

Figure 9.4. EPR dose reconstruction. (Courtesy: M F Desrosiers and A A Romanyukha.)

Table 9.7. Cautionary notes for the use of EPR tooth enamel dosimetry[12].

- When using dental drills and saws to mechanically separate tooth enamel from dentin, one is strongly advised not to overheat the sample since this can lead to the appearance of an interfering EPR signal. Separation of enamel from dentin can also be achieved by the gravitation method based on the differences in their densities.
- The application of UV light to better differentiate between enamel and dentin is also dangerous because UV light produces an EPR signal in the hydroxyapatite $[Ca_{10}(PO_4)_6OH_2]$. The UV-induced signal has EPR parameters very similar to the radiation-induced signal.
- Since the organic fraction of tooth enamel is bound to hydroxyapatite very tightly, extensive ultrasonic treatment with a concentrated NaOH solution should be applied to remove the organic component.
- To minimize effects that arise from the orientation dependence of EPR spectral intensity on the externally applied magnetic field, tooth enamel should be crushed to small grains (0.3–0.5 mm). The precision of the EPR measurement will also be improved if the range of grain size is kept to a minimum.

in El Salvador in 1989[17]. Sections of bone fragments from the amputated lower legs of two victims were obtained and doses for one patient were measured[13] as 11, 32 and 69 Gy and for the second patient as 7 and 25 Gy,

The availability of bone tissue is rare† but is relevant to Chernobyl if such tissues exist for EPR dosimetry.

It is also noted that EPR spectroscopy dosimetry has been used for studying the dose received by atomic bomb survivors[19] but to date, as with Chernobyl, limited data are available.

9.5 Dosimetry based on chromosome aberrations

9.5.1 Introduction

Ionizing radiation is well known to produce chromosome aberrations in living cells and figure 9.5 illustrates the chromosome pattern[20] of a surviving fireman who reached the NPP shortly after the fire brigade team who died. The arrows indicate the aberrations.

As well as physical dosimetry, which is not always possible in emergency situations, cytogenetic methods can also be used for radiation dose reconstruction and this has been shown for atomic bomb survivors[21], figure 9.6.

The use of peripheral blood lymphocytes in such biological dosimetry has several advantages. These include easy sampling, a standard investigation protocol, low and a relatively stable level of spontaneous damage to chromosomes. Also, chromosome damage induced *in vivo* can be compared with *in vitro* dose response curves.

Stable chromosomal aberrations (inversions and translocations) do not disturb cell proliferation. Thus their frequencies are expected to remain

† Because of the extremely long lifetime of 10^6 years of the radiation-induced EPR signal, EPR methods are also used for archaeological dating[18] as it can cover far beyond the limit of radiocarbon ^{14}C dating techniques. Another use to which it might be applied if the lower dose measuring limit is truly 20 mSv, an application which, to my knowledge, has not previously been suggested, is for determining radiation doses received by military veterans in Desert Storm in the 1991 Gulf War in Iraq, particularly those soldiers whose tanks were hit by *friendly fire* armour-piercing rounds which contained a solid depleted uranium (DU) core. DU is a pyrophoric material and will ignite when a DU penetrator core impacts on a hard surface and toxic and radioactive particles produced by the DU combustion can be ingested and inhaled within the confines of the tank. DU is the leftover after *extraction* of ^{235}U from natural uranium which is 99.284% ^{238}U, 0.711% ^{235}U and 0.005% ^{234}U. Enriched ^{235}U ranges from 3% for nuclear energy purposes to 97.3% for nuclear weapons and the order of levels of radioactivity for the three uranium mixtures are $\sim$50 μCi/g for enriched, 0.7 μCi/g for natural and 0.33 μCi/g for DU. It is therefore extremely unlikely that Gulf War syndrome has radioactivity of DU as a major factor of influence. However, for the due process of law in the Courts of the USA and the United Kingdom where at present Gulf War veterans are taking legal action, DU must first be *ruled in*, and then presumably *ruled out* if the doses are so low. The problem is that, to date, no method exists of determining the radiation dose from DU as individual dose estimates and global population doses are too inaccurate. This may perhaps be solved for the legal process as well as for the peace of mind of the veterans if EPR tooth enamel dose estimates can be made such that they can be classed as $<$20 mSv or above this level be estimated as a dose value with an associated standard error.

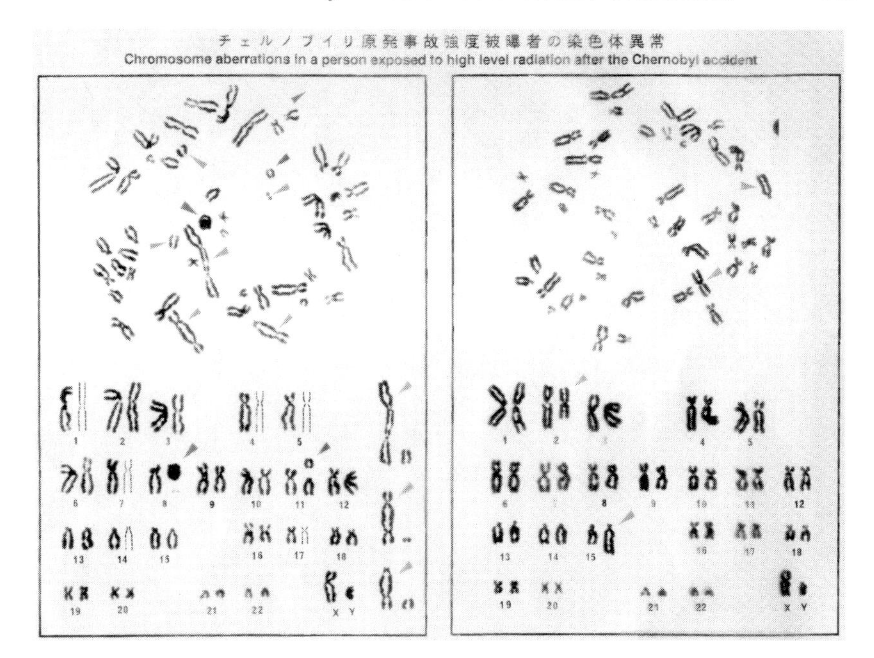

Figure 9.5. Chromosomal aberrations in a fireman who survived Chernobyl[20]. (Courtesy: A Awa.)

constant in the cell clones derived from lymphoid progenitors. This type of chromosomal damage can be detected a long time after radiation exposure, but the difficulty, to date, in detecting stable aberrations by conventional methods has limited its use as a radiation exposure marker.

9.5.2 Terminology

Chromosome

One of several dark staining and more or less rod-shaped bodies which appear in the nucleus of a cell at the time of cell division. They contain the genes, or hereditary factors, and are constant in number in each species. They are composed of deoxyribonucleic acid (DNA) and proteins. The normal number in man is 46.

Cytogenetics

This is the branch of genetics devoted to the study of chromosomes and *clinical cytogenetics* is the scientific study of the relationship between chromosomal aberrations and pathological conditions.

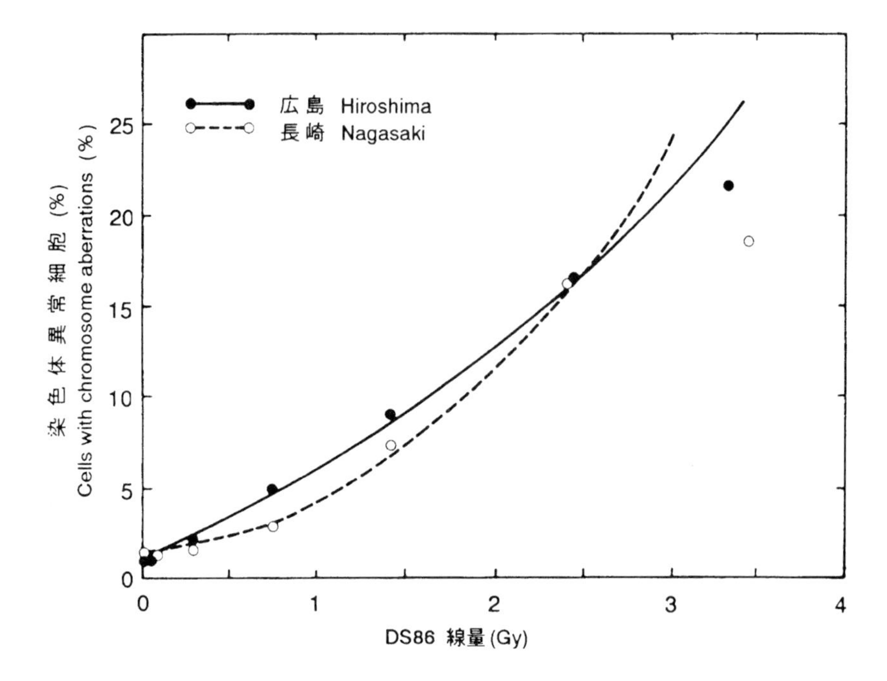

Figure 9.6. Dose-response relationships for chromosome aberration frequencies *versus* estimated doses assigned to individual atomic bomb survivors[21]. (Courtesy: RERF.)

Leucocyte

Blood is composed of a fluid called plasma, in which are suspended red corpuscles, white cells and platelets. Leucocyte is a term for a white blood cell. In normal blood there are approximately 8000 per mm^3. *Leukopenia* is a deficiency of the number of leucocytes in the blood. *Lymphocyte* is a type of leucocyte which has a clear cytoplasm.

Translocation

The transfer of genetic material from one chromosome to another, non-homologous chromosome. An exchange of genetic material between two chromosomes is referred to as a *reciprocal translocation.* Figure 9.7 (left) is a cell showing two abnormal chromosomes, indicated by the arrows, from a 63-year-old male survivor in Hiroshima with an estimated dose of 1.14 Gy. Figure 9.7 (right) is an alignment of the metaphase chromosomes from the left. Aberrant chromosomes are produced by an exchange between broken segments of the no. 2 chromosome and the no. 14 chromosome, see arrows.

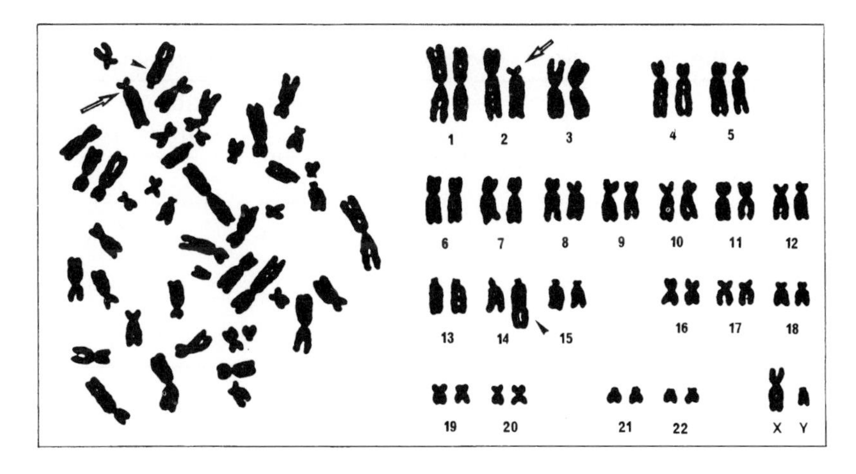

Figure 9.7. Chromosomes of a Hiroshima atomic bomb survivor which show a reciprocal translation[21]. (Courtesy: RERF, Hiroshima.)

This is an example of a reciprocal translocation[20].

9.5.3 Studies in atomic bomb survivors

In a somatic chromosome study[20], cytogenetic data on 1200 survivors in Hiroshima and Nagasaki have shown that the lymphocytes with radiation-induced chromosome aberrations induced in 1945 have persisted for many decades in the peripheral blood of atomic bomb survivors. The rates of cells with such chromosome aberrations increases with increasing radiation doses, as already seen in figure 9.6.

The function and role of lymphocytes with chromosome aberrations in the body have remained unclear and it is necessary to consider whether these aberrant cells may be predisposed to malignant changes in the somatic cells.

In another cytogenetic study[20] the frequency of chromosome mutations induced in parental germ cells was assayed by measuring the frequency of children with chromosome abnormalities. Table 9.8 shows the results of a study on a total of 16 000 participants: 8000 proximally exposed (within 2000 m from the hypocentre) and 8000 distally exposed (2500 m or more from the hypocentre). An increase in the genetic effects due to the atomic bomb radiation exposure has not been demonstrated.

Studies with such large population numbers are beyond the capabilities of Chernobyl investigators but it is interesting to record the comment by Guscova[22] that not all patients examined experienced radiation sickness and in the group that did not, only one-half showed signs of chromosome

Table 9.8. Frequencies of children with chromosome abnormalities[20].

Variable	Children of proximally exposed parents	Children of distally exposed parents
No. examined	8322	7976
Abnormalities		
Sex chromosomes	19 (0.23%)	24 (0.30%)
Structural rearrangements	23 (0.28%)	27 (0.34%)
Trisomics	1 (0.01%)	0
Total abnormalities	43 (0.52%)	51 (0.64%)

aberrations. The dose maximum for these patients was estimated to be in the range 0.2–0.8 Gy.

9.5.4 Fluorescence *in situ* hybridization

Recent advances in chromosome painting using fluorescent *in situ* hybridization (FISH) facilitate the detection of translocations. By this method one can identify not only unstable rearrangements of chromosomes, but also stable aberrations (symmetrical chromosome translocations). *In vitro* dose–response curves have been established for the frequency of symmetrical chromosome translocations after exposing lymphocyte culture to gamma rays. However, this FISH method is only in its early stages as a useful tool for determining radiation doses received by the Chernobyl liquidators[4].

Nevertheless, some results have been obtained in joint studies involving the European Commission[23] and laboratories at the University of Leiden in the Netherlands and at the Institute for Strahlenbiology, Neuherberg in Germany, and three in Minsk, Kiev and Kharkov. The criteria for aberration scoring have been described by Bauchinger *et al*[24]. In brief[23], two-colour derivative chromosomes (yellow/red or yellow/blue) with one centromeric signal were classified either as complete or incomplete symmetrical translocations or insertions. The respective translocation process was evaluated as a single event. Two-colour chromosomes with two or more centromeric signals were classified as dicentrics or multicentrics. Ideally the dose–response data should be fitted by relating the number of aberrations observed to the number of cells analysed. The optimum model is linear-quadratic.

To date the number of liquidators studied is small, and the control groups total 11 adults by Neuherberg and four by Leiden. Results for the controls were respectively 16 translocations in 11 345 cells and one translocation in 1850 cells. There appears to be a slight trend with the

Neuherberg results whereby the translocation yield is higher for the 100–200 mSv group than for the controls and for 0–100 mSv. However, it is less for >200 mSv. The Leiden results show a progressive increase with dose. However, because the number of liquidators analysed is small the standard errors for the results are large and statistical significance between the different dose groups cannot, as yet, be demonstrated.

КОЛОКОЛ ЧЕРНОБЫЛЯ

BELL OF CHERNOBYL

Figure 9.8. 2 Gryvnya Ukranian coin, minted on the tenth anniversary of the accident. The Bell is a symbol of the catastrophe and implies a *Warning* Bell. On 3 March 1987 the Soviet Central Documentary Film Studio premiered at the Oktyabr Cinema in Moscow the film *The Chernobyl Warning Bell*. The rates of exchange between the Gryvnya and the US$ have progressively worsened from 2 Grv = 1 US$ (June 1998) to 3 Grv = 1 US$ (September 1998) to 4 Grv = 1 US$ (March 1999) and the Grv is still falling.

Chapter 10

Population Doses

Introduction

An explanation has already been given in section 1.1.1 for the various dose terms including dose equivalent and collective effective dose equivalent, which are used when population doses are stated. The ICRP recommended dose limits[1] have been given in section 1.1.2, and for exposures above natural background are 1 mSv for the public and 5 mSv for occupational workers, such as power plant workers radiologists and medical physicists and physicians working with x-rays in the fields of diagnostic radiology, radiation oncology and nuclear medicine. In this chapter the effective doses will be stated in either mSv or in μSv where 1000 μSv = 1 mSv and 1000 mSv = 1 Sv, and the collective effective doses in man-Sv.

Using chest x-rays which are a common radiological diagnostic tool to put into perspective in layman's terms 'What is 1 mSv' the WHO[2] equates it to 'about 50 general chest x-rays' and states that on average, people are naturally exposed to about 1 mSv per year from background radiation, which comes from traces of radioactive isotopes which occur in natural and man-made materials.

In order to estimate population doses, information has to be available for many of the individuals who comprise the population under study. For the Chernobyl accident, as seen in chapter 9, radiation doses have had to be estimated in various ways because so few accurate individual measurements were available in Ukraine, Belarus and Russia. This, in turn, will make the population dose estimates in these three republics also subject to uncertainty. However, for countries more distant to the site of the accident the accuracy of population doses will be improved when compared, for example, to that for the liquidators.

10.1 Natural and man-made background radiation

10.1.1 Natural background radiation

The human race and all organisms have been exposed to ionizing radiation from natural sources since the origin of life. This radiation is called natural background radiation. Various types of primary cosmic radiation irradiate the earth from the sun and the Milky Way. Most are energetic protons which interact with nitrogen and oxygen nucleii in the earth's atmosphere and produce secondary cosmic radiation: energetic nucleii, neutrons, mesons, electron, gamma rays, etc.

Gamma rays are emitted from such naturally occurring radionuclides in the ground such as those of uranium and thorium and the potassium nuclide ^{40}K, which enter the body *via* food and by breathing air. ^{40}K is, for the most part, contained in muscle whereas radon, which is a gas and part of the uranium decay series as the daughter product of radium, itself decays into daughter nucleii which may be trapped in the lungs. It is the radon daughters which make up a large part of the internal dose[3], table 10.1, and they were the cause of the lung cancers in uranium miners*.

The global average annual effective dose due to natural background radiation is 2.4 mSv with considerable geographical variation, but based on this figure of 2.4 mSv over a *standard lifetime* of 70 years an individual will accrue 170 mSv.

Table 10.1 gives, after UNSCEAR[3], typical annual effective doses in adults from the principal natural sources. The table also includes what UNSCEAR terms *elevated values*. These values are representative of large regions and even higher values occur locally. This typical annual effective dose of 2.4 mSv from natural sources results in an annual collective dose to the world population of 5.3 billion people of about 13 million man-Sv.

It has already been mentioned[2] that there is considerable variation in effective doses for different geographical location. Lifetime doses are of a few hundred mSv for the average natural background and the variation is about 400 mSv in western Europe[5], figure 10.1.

* The earliest known cases of cancer induced by radioactivity occurred well before Becquerel's discovery of this phenomenon (1896) and the discovery of radium by the Curies (1898). In the ore mountains, the Erzgebirge, which separate Saxony from Bohemia there are the uranium mines of Joachimsthal, now Jáchymov, and Schneeberg. As long ago as the 16th century, Paracelsus (1493–1541) described the occupational diseases of the ore miners and included a disease which killed in the prime of life. Three centuries later this locally known *mountain sickness* (*Bergkrankenheit*) was identified as cancer of the lung and attributed to working in an environment high in atmospheric radon. It was the mines of Joachimsthal which provided Marie Curie with the pitchblende ore from which she separated polonium and radium and which, much later in the Stalin era, were used as a gulag for political prisoners. The mines are now closed.

Table 10.1. Annual effective doses to adults from natural sources[3]. A breakdown of the cosmic ray contribution by external and internal dose is[4], respectively, 0.38 and 0.012 mSv. A breakdown of doses from terrestrial radionuclides is given[4] in table 10.2.

Source of exposure	Annual effective dose (mSv)	
	Typical	Elevated
Cosmic rays	0.39	2.0
Terrestrial gamma rays	0.46	4.3
Radionuclides in the body, except radon	0.23	0.6
Radon and its decay products	1.3	10
Total	2.4	—

Table 10.2. Annual effective doses from terrestrial radionuclides[4]. The total dose from this source is 1.97 mSv.

Terrestrial radionuclide source of exposure	Annual effective dose (mSv)		
	External	Internal	Total
^{40}K	0.13	0.17	0.3
^{238}U series: $^{238}\text{U} \rightarrow {}^{234}\text{U} \rightarrow {}^{230}\text{Th}$	0.14	0.001	0.141
^{226}Ra	—	0.004	0.004
$^{222}\text{Rn} \rightarrow {}^{214}\text{Po}$	—	1.2	1.2
$^{210}\text{Pb} \rightarrow {}^{210}\text{Po}$	—	0.05	0.05
^{232}Th series	0.19	0.08	0.27

10.1.2 Medical irradiation

Medical irradiation will be either therapeutic, mainly as cancer treatment, or diagnostic but doses from diagnostic investigations are several orders of magnitude less than those from therapy. Numbers of cancer treatments are also far fewer than diagnostic investigations. The worldwide annual collective effective dose has been estimated as 1.8 million man-Sv for diagnostic medical exposures and 1.5 million man-Sv for from therapy[3]. As an example of national figures for diagnostic studies table 10.3 lists annual dose equivalents using Japan as the example[6].

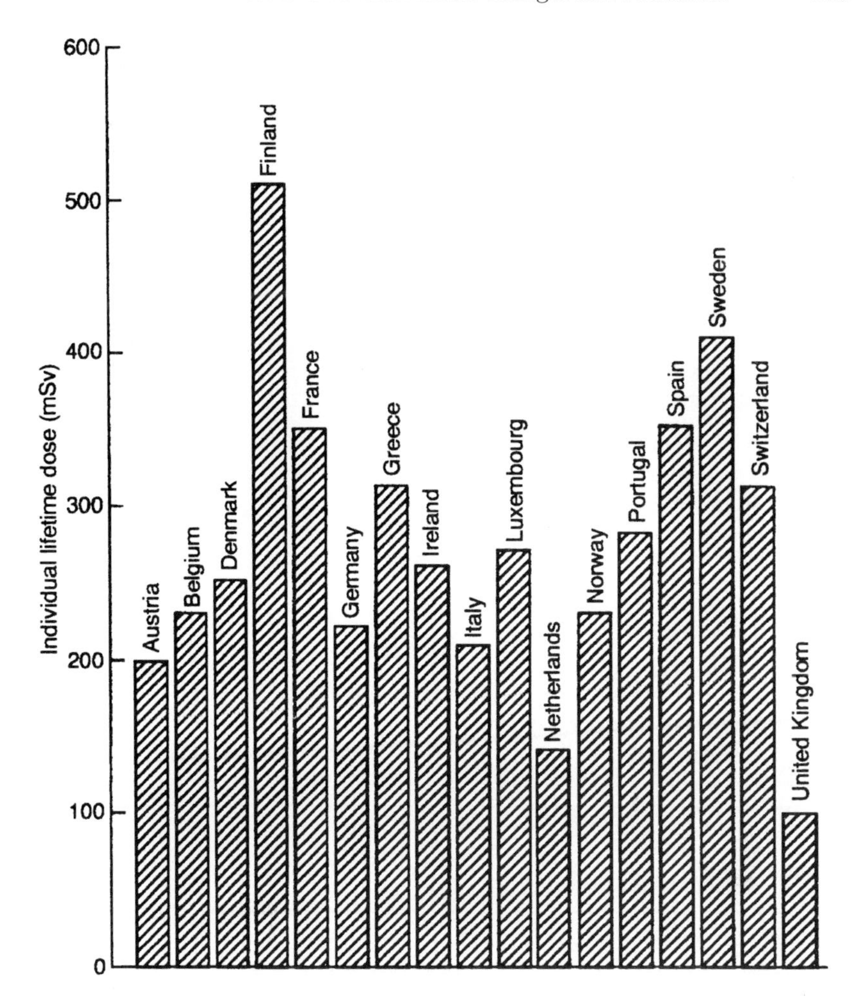

Figure 10.1. Lifetime dose from natural background radiation for selected countries[5]. (Courtesy: IAEA.)

10.1.3 Long-term committed doses from man-made sources

The collective effective doses from nuclear testing, nuclear weapons fabrication, nuclear accidents including Chernobyl, and radioisotope production are given in table 10.4. The figure of 600 000 man-Sv for Chernobyl is an UNSCEAR estimate and of this amount 40% is expected to be received in the territory of the former USSR, 57% in the rest of Europe and 3% in other countries of the northern hemisphere[3,7].

Table 10.3. Annual dose equivalents due to medical examination, Japan[6].

Examination	Annual dose equivalent (mSv)	No. of examinations ($\times 10^4$)
Radiography		
Brain	0.05	6.8
Chest	0.06	53.9
Stomach	2.7	16.7
Barium in bowels	4.3	1.8
Lumber vertebra	1.5	6.3
Bladder	1.9	0.16
Groin	0.32	2.5
CT scanning		
Brain	0.49	1.23
Chest	6.8	0.0037
Stomach	3.8	0.16
Fluoroscopy		
Brain	0.27	0.099
Chest	2.4	1.05
Stomach	2.8	16.0
Barium in bowels	4.2	1.7
Lumber vertebra	3.7	0.50
Bladder	3.0	0.12
Groin	2.3	0.015
Dental x-rays		
Oral	0.029	84.5
Panoramic	0.043	11.2

10.2 First year dose estimates in European countries other than the USSR

Population exposures have been calculated for UNSCEAR for all 34 countries for which measurements are available. These are given as broad averages, for thyroid dose equivalents for adults and infants, primarily from ^{131}I, and for effective dose equivalents from all radionuclides, for rural and urban areas[7].

The country-wide average thyroid dose equivalents for infants are in the range 1–25 mSv with the highest doses in Bulgaria, Greece and Romania. For adults the range is up to 5.5 mSv with the highest values in Yugoslavia and Greece. Adult thyroid doses were usually smaller by a factor of five than infant thyroid doses in the same country in western and

Table 10.4. Long-term committed doses from man-made sources[4].

Source of exposure	Main radionuclides	Collective effective dose (man-Sv)
Atmospheric nuclear testing	^{14}C, ^{137}Cs, ^{90}Sr, ^{95}Zr	30 000 000
Chernobyl accident	^{137}Cs, ^{134}Cs, ^{131}I	600 000
Nuclear power production	^{14}C, ^{222}Rn	400 000
Radionuclide production and use	^{14}C	80 000
Nuclear weapons fabrication	^{137}Cs, ^{106}Ru, ^{95}Zr	60 000
Kyshtym accident	^{144}Ce, ^{95}Zr, ^{90}Sr	2500
Satellite re-entries	^{238}Pu, ^{239}Pu, ^{137}Cs	2100
Windscale accident	^{131}I, ^{210}Po, ^{137}Cs	2000
Other accidents	^{137}Cs, ^{133}Xe, ^{60}Co, ^{192}Ir	300
Underground nuclear testing	^{131}I	200

central Europe. The differences were smaller in northern Europe where the milk was less contaminated because the cows had not been on pasture[7].

The average first year effective doses range from very low to just above 0.7 mSv and some examples are shown[8] in figure 10.2 for selected countries, including those with the highest effective doses: Bulgaria, Austria, Greece and Romania[7]. See figure 4.4 for a comparison of the contribution from the Chernobyl accident to the first year effective dose, for the countries in figure 10.2, to natural background exposure[8].

The measurements for the 34 countries were used to provide a pattern of transfer of radionuclides in air, deposition and diet to dose, which was then used to evaluate doses in all other countries of the northern hemisphere. Such transfer factors were obtained for all pathways and for the major radionuclides contributing to the dose, ^{131}I, ^{134}Cs and ^{137}Cs, as well as some other nuclides. One of the results from these studies was that it was estimated that the lifetime external exposure is approximately eight times the first year external exposure[8].

10.3 Collective effective doses for populations in Ukraine, Belarus and Russia

10.3.1 Residents in contaminated territories

This population refers to those persons living in the strictly controlled zones (SCZs) which surround the 30 km exclusion zones from which the initial evacuation took place, see table 6.2. The average annual effective dose was

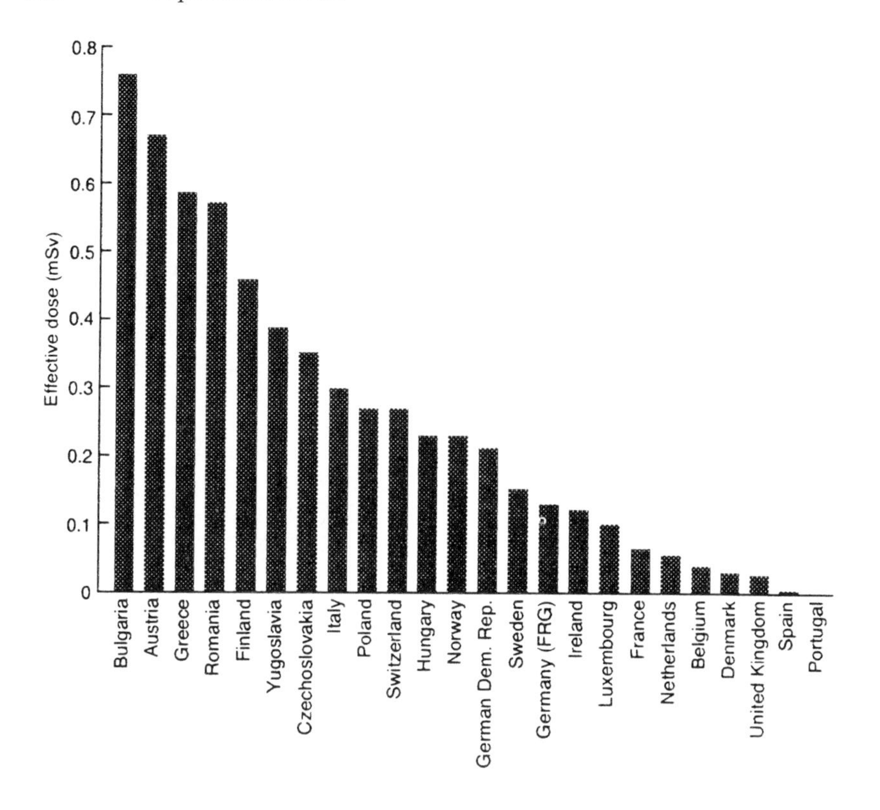

Figure 10.2. First year effective dose in European countries from the Chernobyl accident[8]. (Courtesy: IAEA.)

40 mSv in the year following the accident but had fallen to less than 10 mSv in the following years to 1989[3].

For the individual lifetime doses of those living in contaminated areas it has been estimated[5] that one-third of the dose will be received in the three periods, 1986, 1987–95 and 1996–2056, although this is open to discussion. For example, it has also been stated[9] for the Ukraine that 20% will be received in 1986, 70% in the years 1986–97 and thus only 10% during the next 50–60 years. It has also been estimated[5] for all three republics that the ratio of internal to external doses is 0.3–1.4.

The lifetime doses related to different contamination levels is given in table 10.5, where the doses have been normalized to a surface contamination density assuming that no countermeasures were applied[5]. For a comparison with the lifetime doses from natural background for various countries in Europe, see figure 10.1.

For a contamination level of 1480 kBq/m^2 the individual lifetime doses have also been estimated[10] for rural and urban populations as 80 and 50 mSv respectively, and that for the entire SCZ, assuming that countermeasures are in existence (see chapter 12), the lifetime dose will be in the range 10–30 mSv but that without these countermeasures it could be as high as 200 mSv.

Table 10.5. Individual lifetime doses as a function of contamination level[5] (to convert units of kBq/m^2 to Ci/km^2 divide by 37).

Contamination level (kBq/m^2)	Lifetime dose (mSv)
185–555	5–20
555–1480	20–50
1480–2960	50–100

Maximum permissible dose limits approved by the Soviet Government Commission for Accidental Whole-Body Radiation Doses were introduced into State regulations by the Ministry of Health and in May 1986 were initially set at 100 mSv for the first year after the accident. Then as the hazard from ^{131}I declined considerably within a few months, the limit was reduced to 30 mSv for the second year post-accident and in 1988 reduced further to 25 mSv for the years 1988 and 1989. Thus the total body burden for the four years following the accident was <180 mSv. However, it was virtually impossible for the regulations to be fully complied with among such large population groups as the residents of the SCZs and the evacuees and therefore some individuals received higher doses. In October 1988 the USSR NCRP recommended 350 mSv as the lifetime dose limit for external and internal irradiation from the Chernobyl accident[2].

Estimates of the collective effective doses are given in table 10.6 for the period 1986–2056. Two-thirds of the dose is received between the years 1986 and 1995 and one-third over the period 1996–2056[11].

Table 10.6. Collective effective doses[11] for persons living in SCZs.

Contamination level (kBq/m^2)	Lifetime dose (man-Sv)	Population size
37–555	52 500–150 000	6800 000
>555	15 000–30 000	270 000

10.3.2 Evacuees

Data for the evacuees are more difficult to obtain because they have been relocated throughout the former USSR and the data registries, first for the entire USSR for 1986–91 and then individually for the all independent republics, have concentrated on the liquidator population and prediction modelling for the incidence of thyroid cancer and leukaemia. The numbers in the various population groups for the State registries of the Ukraine and of Russia are given in table 10.7.

Table 10.7. Population groups in the States registries of the Ukraine[12] and of Russia[13] as of 1995.

Population	Number of registrations	
	Ukraine	Russia
Liquidators	168 758	152 325
Evacuees	61 452	12 899
Residents of SCZs	155 047	251 246¶
Children born to irradiated persons	23 037	
Children born to liquidators of 1986–87		18 816

¶ Classified as living or have lived in SCZs.

For the 115 000 evacuees from the 30 km exclusion zone the external doses received by most were less than 250 mSv but a few in the most contaminated areas might have received doses of some 300–400 mSv. The collective dose[3] for the evacuees from external irradiation is estimated to be 16 000 man-Sv. However, the individual doses to thyroid were higher by a factor of ten and the estimated collective thyroid dose is 400 000 man-Sv.

The numbers in each population group, see table 10.7, has not been published for the Belarus State registry but a cohort of 1300 evacuees from the 30 km zone[14], were examined in clinics in Minsk during May–June 1986 and it was found that their doses were in the range 1–850 mSv with the majority, 78.5%, in the range 5–50 mSv. Only 12.4% received doses in excess of 100 mSv. The collective effective doses for the entire population of Belarus, including evacuees, liquidators and those living in SCZs, is given[14] in table 10.8. This table also presents data for the Gomel region of Belarus which because of the rainfall in this region when the radioactive plume was passing, made the region the most heavily contaminated in Belarus. This is reflected in the incidence of thyroid cancers in children and adolescents.

Table 10.8. Collective effective doses for the republic of Belarus[14].

Dose	Population	Collective effective dose (man-Sv)	
		All Belarus	Gomel region only
External	Rural	9407	5393
	Urban	7218	4882
Internal	Rural	3404	1956
	Urban	1910	1244
Thyroid	Total population	466 475	315 748
	Children 0–6 y	117 016	80 797

10.3.3 Liquidators

The average dose from external exposure is estimated to be 120 mSv which for the 247 000 considered by UNSCEAR[3] to be liquidators gives a collective dose of 30 000 man-Sv.

The number of liquidators in the State registries of the Ukraine and Russia are given in table 10.7 as 168 758 and 152 325 respectively[12,13]. However, these numbers vary in different publications and as well as this factor, it should be noted that only a percentage of those registered have known doses. Table 10.9 presents data[11] on the percentage of known doses for liquidators for the three republics.

Table 10.9. Distribution of registered doses in the population of liquidators[11].

Country	No. of liquidators	No. with known doses (%)
Belarus 1986–89	63 000	13.8
Ukraine 1986–87	15 700¶	51.9
Russia 1986–89	148 000	63.2

¶ This was the study sample from a population of 102 000.

The radiation doses to servicemen have already been given in table 9.1 and showed[15] that only 21 of a total of 66 752 received doses greater than 50 R whereas 2041 received doses in the range 25–50 R, and that the population within the lowest dose range quoted of 1–10 R were 13 018, almost 20% of the total. Numerically 1 R $\sim$ 1 cGy $\sim$ 1 cSv = 10 mSv and therefore 50 R $\sim$ 500 mSv.

The highest doses were received by those who worked near the NPP in 1986–87 and of the liquidators from Belarus[14], 30% received 50–100 mSv,

47% received 100–250 mSv. The collective dose for 226 242 liquidators working in 1986–87 has been estimated[16] to be 14 070 man-Gy with the subdivision for the years 1986 and 1987, relating to 116 641 and 109 601 liquidators, being 8970 and 5100 man-Gy respectively. The average doses received for 1986 and 1987 were 77 and 47 mGy.

Figure 10.3. This engraving was published in the Soviet magazine *Youth (Unost)* issue of June 1987 for an article by Yuri Scherbak entitled *Chernobyl Documentary Story*. It depicts a doctor, a Geiger counter operator, workers' apartments in Pripyat, the roof of the reactor hall of Unit No. 4 and a helicopter *bombing mission* to put out the inferno. (Courtesy: Unost.)

Chapter 11

Contamination in Farming, Milk, Wild Animals and Fish

Introduction

Entry into the food chain of radioactive material such as the relatively short-lived ^{131}I or the long-lived ^{137}Cs were the immediate concerns for most of those who had possibly been affected. Milk, vegetables and fruit were the initial major worries of populations. Data for the contamination of milk by ^{131}I which occurred in the early post-accident period as well as data for longer term ^{137}Cs contamination in fish and animals are given in this chapter. In addition, data are included on some animals and aquatic organisms which are not used for food.

11.1 Exposure pathways

The pathways for human exposure were fivefold, table 11.1, and of these the ingestion of contaminated foods and irradiation from deposited radionuclides were most important in the long-term for most of the populations at risk. The exceptions were the population of liquidators and NPP staff who suffered from acute radiation syndrome following external irradiation and inhalation doses.

Table 11.1. Exposure pathways.

- External irradiation by the radioactive cloud.
- Inhalation of radioactive material in the cloud.
- Beta radiation contamination of the skin.
- External irradiation by material deposited on the ground.
- Ingestion of contaminated foods.

Irradiation from deposited radionuclides was in the long-term primarily due to ^{137}Cs. However, during the first month the most important pathway was dietary ingestion due to ^{131}I in milk and leafy vegetables, and only after that due to ^{137}Cs and ^{134}Cs in foods.

Doses from external irradiation by the radioactive cloud and inhalation of radionuclides in the air depended on the wind and rainfall patterns of the first few days. This was particularly true of parts of Bavaria in the Federal Republic of Germany where there was coincidentally a very heavy thunderstorm on the afternoon of 30 April 1986 at the time the Chernobyl cloud was passing, table 11.2.

Table 11.2. Washout contamination rate measurements of grass[1].

Date	Contamination rate (μSv/hr)
Prior to 30 April 1986	0.08
In a few minutes on 30 April 1986	1.0
At the end of 1986	0.12

It should, however, be noted[2] that all soil for agriculture anywhere contains radionuclides to a greater or lesser extent. Typical soils contain approximately 300 kBq/m^2 of ^{40}K to a depth of 20 cm. This radionuclide and others are then taken up by crops and transferred to food, leading to a concentration in food and feed in the range 50–500 Bq/kg.

The radionuclide contaminants of most significance in agriculture are those which are relatively highly taken up by crops, have high rates of transfer to animal products such as milk and meat, and have relatively long radionuclide half-lives. However, the ecological pathways leading to crop contamination are complex and depend on factors such as soil type, cropping system, climate and season[2].

Ingestion of ^{131}I was *via* milk from cows which had eaten contaminated grass or fodder or contaminated leafy green vegetables. The pathways of ^{137}Cs and ^{134}Cs into growing plants were either by deposition on leaves or by uptake from soil through roots. In addition, with ^{137}Cs special considerations were given to game meat such as deer and rabbit and, in particular, reindeer, where the ^{137}Cs concentration can be high because of their diet of lichens. ^{137}Cs activity in fish was also found in freshwater lakes in Belarus and Ukraine to sometimes have risen significantly, although this was not noted in seawater or in estuary fisheries.

Once these radioactive isotopes are ingested, the ^{131}I locates in the thyroid and the ^{137}Cs is almost completely absorbed by the gastrointestinal tract. ^{90}Sr, which in the Chernobyl fallout was only 1% of that from ^{137}Cs, is also absorbed in the gastrointestinal tract.

It has been predicted[3] that the effective dose for the period 1996–2056 will be two to three times lower than that for 1986–95. This is due to the food intake being estimated as representing 60% of the total dose for 1986–95 but that for 1996–2056, only a 20% contribution is predicted from internal dose with 80% being received from external dose. Most of the dose is from ^{134}Cs and ^{137}Cs with a minor contribution from ^{90}Sr and a minimal contribution from ^{239}Pu.

Countermeasures are relatively inefficient in reducing external exposure but can be very efficient in reducing the uptake of radioactive material. In the long-term the appropriate application of agricultural countermeasures can effectively reduce the uptake of caesium into food[4]. Which countermeasures are most appropriate strongly depends on local conditions such as soil type.

Some food products derived from animals that graze in semi-natural pastures, forests and mountain areas, and wild foods, such as game, berries and mushrooms, will continue to show higher levels of ^{137}Cs than food products derived from managed agricultural land[4].

11.2 Evaluation of dose from ingestion of foods

The UNSCEAR 1988 report[5] describes the computational methods for all pathways (external and internal) and these have been summarized in 1996 by Bennett[6]. For most cases this involves multiplication of first year integrated deposition or concentration in foods by suitable dose factors, such as dose per unit intake, and by intake rates, such as food consumption rates and breathing rates.

The projected dose from ingestion of food was estimated from the transfer relationship from deposition to diet, P_{23}. The model for the transfer function previously derived and used in UNSCEAR assessments is

$$P_{23} = b_1 + b_2 + b_3 e^{-\lambda t}$$

where b_1 is the component of first year transfer, b_2 is second year transfer and $b_3 e^{-\lambda t}$ is the subsequent transfer accounting for both environmental loss and radioactive decay.

11.3 ^{131}I contamination in milk

The World Health Organization sponsored meetings on 6 May 1986 in Copenhagen and on 25–27 June 1986 in Bilthoven. Following the former there was limited circulation of a document[7] which contained information for 35 countries on what had been their recommendations concerning the drinking of milk and it was clear that there was a wide variation, not only in activity levels but also on what actions were recommended.

Table 11.3 gives the data on the activities measured. In Sweden the deposition of ^{131}I on grass was 6000–170 000 Bq/m^2, with the highest values in northern Sweden. It was established that a deposition level of 10 000 Bq/m^2 corresponded to an expected milk concentration of 2000 Bq/l.

In the longer term, when ^{131}I was not a problem, care still had to be taken in the USSR in terms of ^{137}Cs contamination. Reduction in the contamination of milk, defined as an activity greater than 370 Bq/l, for Ukraine 1988–94 is given[9] in table 11.4. It was also reported for 1994, the last year in table 11.4, that in Belarus less than 1% of dairy products and only a few tonnes of meat exceeded the intervention levels[10].

11.4 Contamination in wild animals and fish

The zone around the NPP, having minimal disturbance by humans, compared with former times, has now become, in effect, a wildlife reserve[11a,12], with the wildlife populations increasing by 1998 by as much as a factor ten compared to the levels of 1986. Even a family of lynxes have been observed[11a]. The current population in what is a very ancient forest, is estimated to be 3000 foxes, 600 moose, 450 deer, 40 wolves and perhaps upwards of 3000 boars, which in the 1980s had been hunted almost to extinction because they were regarded as such a delicacy[11a], see Plate VII[11b]. However, what is most well known is the reindeer contamination problem in Lappland.

The contamination was caused by the fact that reindeer graze on lichen during the winter period of six months as there is very little fresh vegetation. ^{137}Cs deposition in lichen was studied intensively during the nuclear weapons atmospheric tests in the 1960s. It was found that the half-time elimination of ^{137}Cs from lichen is five to six years. After the accident, the ^{137}Cs distribution was very non-uniform and in 1986–87 could vary in the range 0–20 000 Bq/kg. It was, of course, dependent on rainfall patterns and because of this was most acute in areas of northern Sweden, where thousands of animals were slaughtered, although a large tract of the Finnish Lappland reindeer belt escaped significant radiation levels[13].

In Sweden the limit for meat consumption was initially 300 Bq/kg but this was later increased to 1500 Bq/kg. Even so, up to the spring of 1987 a total of 50 000 reindeer carcasses had been thrown away[13], completely ruining the livelihood of the Lapplanders in Sweden. Sometime later the Swedish government modified their instructions to slaughter and bury the carcasses in pits, so that some of the meat could be fed to foxes and minks in fur farms, since these animals have no place in the food chain.

Table 11.5 gives[9] the contamination of different species of animals and

Table 11.3. Measurements of ^{131}I in milk in May 1986[7]. These data give a good review of which countries in the world received more radioactive fallout than others.

Country	Foodstuff	^{131}I contamination in Bq/l and date
Albania	Fresh milk	<800 [5 May]
Belgium	Farm milk	85–170 [8 May]; 40–80 [10 May]; 28–57 [11–15 May]
Bulgaria	Farm milk	100 [19 May]
Czechoslovakia	Dairy milk	500 [4–5 May]
	Farm milk	1000 [4–5 May]; 1570 [11 May]
France	Dairy products	80–110 [20 May]
Germany, FR	Milk	150–600 [Munich, 1 May]; 500 [14 May]
Greece	Cow's milk	100–400 [10 May]
	Sheep's/Goat's milk	2000–8000 [10 May]
	Dairy milk	150 [10 May]
Hungary	Milk, cows on pasture	100–700 [1–2 May]; up to 1250 [3 May]; up to 2600 [4 May]
	Milk, cows not on pasture	100–200 [3 May]; 200–800 [4 May]
Ireland	Milk	Mean of 21 [May]
Israel	Cow's milk	0.7 [3 May]
	Goat's milk	22 [3 May]
Italy	Dairy milk	55–550 [2–8 May]; later peak levels 3000–6000
Japan	Milk	0.4–3 [4–6 May]
Netherlands	Farm milk	173 [4 May]
Norway	Milk	30 [May]
Poland	Milk	30–2000 [29–30 April], 80–474 [11 May]
Portugal	Milk	0.1 [7 May]
Romania	Milk	450 [May]
	Raw farm milk	>1000 [May]
Spain	Dairy milk	0.3–1.8 [5–7 May]
	Local farm milk	2–65 [5–7 May]
Sweden	Mother's milk	8–25 [Stockholm, 27 April–4 May]
	Milk	2–70 [May] and 700 [Gotland island, May]
Switzerland	Cow's milk	250 [3 May] rising to 1370 [13 May]
	Sheep's milk	5800 [3 May]
	Goat's milk	550 [3 May]

Table 11.3. (continued)

Country	Foodstuff	^{131}I contamination in Bq/l and date
Turkey	Fresh milk	360 [6 May]
	Dairy milk	48 [6 May]
USSR	Milk	No data given in 1986 on milk or other foodstuffs[8]
United Kingdom	Dairy milk	3–240 [9 May]
	Farm milk	370 [205 May] rising to a maximum of 1136
USA	Milk	No ^{131}I activity detected
Yugoslavia	Milk	<400 [2–31 May]

Table 11.4. Quantities in million tonnes of contaminated milk in Ukraine by ^{137}Cs.

Year	Collective farms	Private farms
1988	78.1	246.0
1989	60.8	111.1
1990	62.0	82.6
1991	1.1	27.5
1992	0	7.0
1993	0	0.5
1994	0	0

fish within the 30 km zone around the Chernobyl NPP. For comparison, data are also available, table 11.6, for contamination of lake fish in Swedish Lappland for the period September 1986 to July 1987[13]. The contamination of perch in lake Lugano, Switzerland, were also measured[14] and for 1986–87 was in the range 1–2 kBq/kg which by 1993 had reduced to 0.05 kBq/kg. Algae were also monitored for radionuclide content, as well as fish, and table 11.7 is for data[15] in August 1986 for five different water bodies, including the cooling pond at the NPP. However, table 11.8 shows the percentage contributions to the total dose to algae and fish from three different classes of radionuclide, grouped by half-lives, with measurements made within the 10 km zone in the first two weeks after the accident[15]. This shows that even in this early period after the accident the significance of short-lived radionuclides did not exceed 20% of the total.

The photographs in Plate VII were taken in the Polissya Ecological

Reserve which was established over an area of 220 000 hectares in Belarus in 1986 after the disaster, and is situated between the rivers Dnieper and Pripyat. The sites of some 100 abandoned villages are within this area and the only persons permitted to work there are scientists. The lack of normal activities, such as farming, has enabled the wild animals to experience a population explosion: elks, boars and particularly grey wolves which are an almost extinct species elsewhere in Europe. They are considered to be dangerous, so much so that the authorities have started to attempt to reduce their numbers and recently a total of 100 were shot in a single oblast. The most effective method is to shoot them from helicopters from a height of 20 m from the ground. However, as Leszek Sawicki found[11b] this method has its problems because 'first it took several hours to make the engine start as it was a 25 year old socialist technical disaster'. On his flight, about 100 boars, some 'real giants' and about 50 elk were seen. All the wild animals which are culled are measured for radioactivity and the boar shown had a radioactive content per unit volume 300 times higher than would have been expected before the accident. Boars are the most contaminated because they dig for their food in the forest ground.

Also shown in Plate VII is the entrance to this area, surrounded by barbed wire fences and the notice *Stop, Show Your Pass!* and *Polissya State Radioactive/Ecologic Reserve.*

Table 11.5. Contamination of animals and fish within the 30 km exclusion zone during the period 1986–94.

Species	Mean contamination level (kBq/kg)				
	1986	1988	1990	1992	1994
Boars	470	115	100	115	5
Birds	50	4	4	2	2
Rodents	80	2.5	100	1	2.5
Tritons	7.5	18	28	6	20
Perch	4	1	4	0.2	0.2
Reptiles and lizards	50	50	2.5	100	1
Frogs	27	40	5.5	70	1

11.5 Temporary permissible food contamination levels in USSR

Rural populations may be at greater risk of high exposure after deposition of radioactivity because of a tendency to consume a greater proportion of

Table 11.6. Contamination of lake fish in Swedish Lappland during the period 1986–87.

Fish	Mean contamination (kBq/kg)	Maximum contamination (kBq/kg)	Sample no.
Perch	4.3	28.0	374
Pike	2.0	8.3	254
Whitefish	1.6	9.3	157
Salmon trout	3.3	25.0	235

Table 11.7. Radionuclide content in algae in units of kBq/kg net weight, August 1986.

Radionuclide	Location				
	Cooling pond	Pripyat estuary	Dnieper river (Kiev)	Desna river (Chernigov)	Gulf of Finland (Loviisa)
^{90}Sr	36	3	0.03	0.006	0.005
^{95}Zr	244	38	0.24	—	0.003
^{90}Ni	470	52	0.52	—	0.006
^{137}Cs	160	5	0.13	0.035	0.183
^{144}Ce	296	41	0.30	0.05	0.006

Table 11.8. Contributions to the total dose, from different radionuclides, during the period 26 April to 10 May 1986.

Radionuclide group	Contribution to the total dose (%)	
	Algae	Fish
Short-lived radionuclides with half-life <3 days $^{132,133,135}I$, ^{99}Mo, ^{132}Te, ^{239}Np	6	13
Half-life 8–64 days ^{131}I, ^{140}Ba, ^{141}Ce, ^{103}Ru, ^{95}Zr, ^{95}Ni	60	57
Long-lived radionuclides ^{137}Cs, ^{90}Sr, ^{144}Ce, ^{106}Ru	34	30

locally produced food. Subsistence economies, in which virtually all food is home produced or gathered will be especially susceptible to high levels of transfer to the population in the event of contamination[16].

Subsistence farming was fully practised in Ukraine, Belarus and Russia at the time of the Chernobyl accident. Major agricultural production was collectivized and intensive but many of the villages operated small-scale private production units. Typically this consisted of a vegetable plot, and a few domestic animals such as a cow or goat, pigs and chickens. Generally the land allocated for private farming was of poorer quality than that of the collective farms and, in particular, livestock often grazed on marginal land of little value to the collective. Such land is usually unimproved and gives a higher transfer of radioactivity than does intensively farmed land. Additionally, rural residents often supplement their diet by gathering fungi, nuts and berries from the forests and catching freshwater fish in local lakes and rivers[16].

Data for the USSR were not available for some time, but eventually it was disclosed[17] that the Soviet Ministry of Health approved a provisional regulation for a maximum permissible radioactivity concentration in milk of 3.7 kBq l^{-1}, table 11.9. The levels were set on 5 May 1986 for the standards of temporary permissible ^{131}I concentration, not only in milk but also in other major components of the human diet[18]. Those for ^{134}Cs and ^{137}Cs were established on 30 May 1986. These levels were modified as time progressed and those for 1986–93 are given in table 11.10.

Table 11.9. Temporary permissible levels for ^{131}I content[18].

Foodstuff	Permissible concentration		Assumed average maximum consumption per day
	μCi/l or (μCi/kg)	kBq/l or (kBq/kg)	
Drinking water		0.1 (3.7)	1 l
Milk		0.1 (3.7)	1 l
Cream cheese	1.0 (37.0)		100 g
Sour cream	0.5 (18.5)		200 g
Cheese	2.0 (74.0)		50 g
Butter	2.0 (74.0)		50 g
Fish	1.0 (37.0)		100 g
Green vegetables	1.0 (37.0)		100 g

Table 11.10. Temporary permissible levels for radioactive caesium content[18]. Data for 1986–91 is for the USSR and for 1993 is for Russia only.

Foodstuff	Permissible concentration (kBq/kg)			
	1986	1988	1991	1993
Milk	0.37	0.37	0.37	0.37
Meat, fish, eggs	3.7	1.88–2.95	0.74	—
Fat	7.4	0.37	0.185	0.37
Potatoes, roots, vegetables, fruits	3.7	0.74	0.6	0.6
Cereal products	0.37	0.37	0.37	0.37
Wild berries, mushrooms	18.5	1.74	0.148	0.6
Food for infants	—	0.37	0.185	0.185
Drinking water	0.37	0.0185	0.0185	—

11.6 International recommendations for intervention levels

The WHO and the FAO coordinated meetings from whence derived intervention levels were formulated. Each country was then required to determine what level was considered safe for their own populations, based on the international consensus that had been developed[19].

The earliest European forum for a discussion of the effects of Chernobyl, including the consequences for the food chain, was that on 8–9 January 1987 by the Members of the European Parliament. They could not agree on intervention levels and by the end of 1987 there was still no agreement. Table 11.11 gives the European Commission recommendations of this period[20].

Later, in 1988, WHO produced a set of guidelines for derived intervention levels for radionuclides in food[21]. The value of the derived intervention level (DIL) for food varies inversely with the mass of food consumed and the dose per unit intake factor, and directly with the reference level of the dose used so that

$$DIL = RLD/(md)$$

where RLD is the reference (intervention) level of dose in units of Sv/y, m is the mass of food consumed annually in units of kg/y, and d is the dose per unit intake in units of Sv/Bq.

Figure 11.1 shows how for ^{137}Cs the intake of contaminated food that gives a dose of 5 mSv to an adult varies with the level of contamination. For low intakes, defined as less than 20 kg, high contamination is required before the reference level of dose is exceeded. For low consumption food

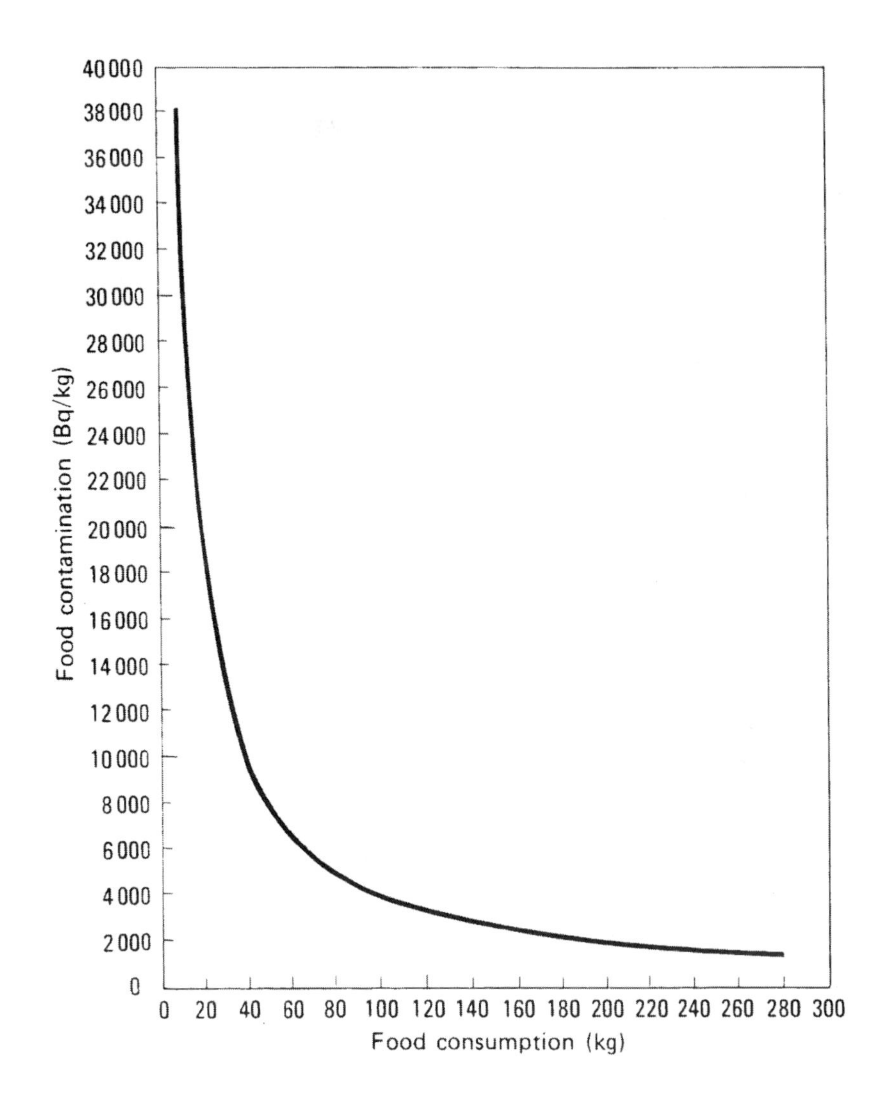

Figure 11.1. Graph of food contamination *versus* food consumption for ^{137}Cs to give a dose of 5 mSv. (Courtesy: WHO.)

Table 11.11. European Commission recommendations which were considered for new limits to replace the emergency EC limits which expired on 31 October 1987[20].

Foodstuff	Activity limits (kBq/kg) for different radionuclides			
	Caesium	Iodine	Strontium	Plutonium
Dairy products	1.0	0.5	0.5	0.02
Other than dairy products	1.25	3.0	3.0	0.08
Drinking water	0.8	0.4	0.4	0.01
Animal foodstocks	2.5	—	—	—

items the contribution to the total radiation dose is therefore small and guideline values for derived intervention levels will be extremely high. This graph[21] of food contamination *versus* food consumption for ^{137}Cs to give a dose of 5 mSv assumes a dose conversion factor of 1.3×10^{-8} Sv/Bq.

Figure 11.2. Warning side by the side of the road within the 30 km zone. Sign reads: DANGER: RADIATION! ON HARD-SHOULDER. (Courtesy: V Zufarov.)

Chapter 12

Decontamination

Introduction

The first emergency measures were the fire fighting and the attempts to stabilize the reactor, and this was followed by distribution of stable iodine to the population to prevent accumulation of ^{131}I in the thyroid, and by the evacuation described in chapter 6. Decontamination of buildings and forests was then attempted and a supply of clean water was also organized. A system of meteorological and radiological monitoring was organized to survey the contamination levels in the area surrounding the NPP and beyond with aerial radiological monitoring carried out by aircraft and helicopters[1]. Measures were also taken to prevent the affected population ingesting contaminated food. This chapter details the decontamination strategies and the measures which had to be adopted to provide clean water supplies.

The Soviet system had manuals for the medical treatment of radiation victims, particularly for investigation and treatment of what was termed 'persons who work in the production of different kinds of sources of radiation: uranium, polonium and plutonium' and in August 1986 publishing approval was given by the State for a limited number of copies[2]. However, there were no such manuals or any organizational infrastructure for countermeasures to solve the problem of environmental contamination over such a large area, and within the exclusion zone at such a high level: as indeed there were no manuals or organization for planning the evacuation of 115 000 persons.

In addition, no remote control robots were available from Soviet sources for removal of the contaminated material on the roof of the turbine hall and within Unit No. 4. West Germany later supplied some manufactured by Kerntechnische Hilfdienst GmbH but the dose rates were so high on the turbine hall roof that the electronics were affected and the robots could not be used at this site[3]. Nor were remote controlled bulldozers for earth moving available locally and, for example, in June 1986, a 19 tonne bulldozer was obtained from the Chelyabinsk Tractor Works. This machine was operated by a driver looking through a narrow slit in an armoured vehicle several dozen metres distant. It was delivered from Chelyabinsk to Kiev on board an Ilyushin-76 jumbo jet and modified with remote control electronics by specialists from Kiev Institutes.

12.1 Decontamination sequence for the NPP area

The decontamination of the plant area was undertaken in the following sequence[4] of five operations.

- Removal of refuse and contaminated equipment from the site.
- Decontamination of roofs and outer surfaces of buildings: sometimes pastes were put onto the walls so that they established a quick drying film which, when peeled off, had radioactive particles sticking to the film.
- Removal of a 5–10 cm layer of soil and its transfer in containers to the waste disposal dump at Unit No. 5 and No. 6 site.
- Laying, if necessary, of concrete slabs on the soil or filling with clean soil.
- Coating of the slabs and of the non-concrete areas with film-forming compounds.

As a result of these measures it was possible to reduce the gamma radiation background in the area of Unit No. 1 to an exposure rate of 20–30 mR/hr. The work of the decontamination teams were such that surfaces were cleaned at a rate of 15 000–35 000 m^2 per 24 hours.

12.2 Turbine hall roof

The Chernobyl NPP site was contaminated over a wide area and radioactive material was scattered on the roof of the turbine hall, the roof of Unit No. 3 and on the many metal pipe supports. The graphite blocks which had landed on the roof often contained embedded fuel and typical blocks weighed 40–50 kg. Each liquidator, because the robots did not function, carried four blocks and then threw them over the side of the roof into a pit.

The average gamma ray dose rate on the roof (the neutron dose rate is not known) was 20 R/min and the dose limit for each liquidator, who were termed for this task *bio-robots*, was 20 R. However, this was almost certainly an underestimate since the graphite blocks must have been carried close to the chest, whereas the dose rate was measured at the point at which the liquidator stepped out onto the roof. Some of the firemen prior to the removal of the graphite block debris worked on the roof for at least one hour and must have received doses considerably in excess of 20 R. Later, when the dose rates were not as high as they were initially, remote control robots were employed. However, they still could not function, this time not because of the dose rate, but because they became trapped in the debris[5].

12.3 Radioactive waste disposal

The initial solution was to bury the waste in the pit excavated for the building of Units No. 5 and 6 but this was not adequate for very long[6]. Eventually a total of some 800 burial sites for contaminated soil, debris and machinery were used within the 30 km zone[6]. Apparently not all these locations are well documented and these form a major long-term threat to contamination of the ground water.

Figure 12.1. Radioactive waste disposal pit for contaminated vehicles. (Courtesy: Chernobylinterinform.)

Some of the sites are still open, as in figure 12.1, and there are also fields full of contaminated machinery, buses, helicopters and lorries lined up in rows looking like vast parking lots, see Plates II and III.

12.4 Forests

Attempts to decontaminate forests were initially made from helicopters, biplanes normally used for crop spraying and from lorries, figure 12.2. This was unsuccessful because the winds and the rain continually re-distributed the radioactivity

Figure 12.2. Unsuccessful attempt to decontaminate part of the red forest, May 1986. (Courtesy: TASS.)

One of the major concerns was the possibility of forest fires and in June 1986 there was a conflagration in the Gomel region of Belarus reported in

Figure 12.3. The *red forest* adjacent to the NPP. (Courtesy: V Zufarov.)

Komsomolskaya Pravda where the fires extended along a one mile front and the situation was such that the firemen needed special breathing apparatus.

The *red forest* near the NPP, figure 12.3, was so-called because of the colour of the pine cones, needles and leaves, following their irradiation by the fallout. Lethal radiation doses were received by coniferous trees within the 10 km zone during the first few weeks after the accident and dose rates measured[7] in pine needles and in mosses are given in table 12.1. Much of the *red forest* was buried, but the amount of timber was so enormous that significant amounts are still left piled up at the side of roads in 1998 and will remain there for many years.

The accident occurred in the early growing season for plant life and within two weeks some 500–600 ha of trees in the vicinity of the NPP received a dose of 80–100 Gy. Over a larger area of approximately 3000 ha where doses exceeded 8–10 Gy some 25–40% of coniferous trees died and 90–95% of pine trees showed significant damage to their reproductive tissues[8].

By the autumn of 1986 dose rates had fallen by a factor of 100 and by 1989 the natural environment had begun to recover, as seen in figure 6.7. Nevertheless the possibility of long-term genetic effects and their significance remains to be studied[8].

Table 12.1. Estimated dose rates in pine needles and in mosses[7].

Date	Location	Dose rate (10^{-2} mGy/day)	
		Pine needles	Mosses
Oct 1987	Janov, 2 km from NPP	34 000[¶]	—
Oct 1987	Chistogalovka, 4 km from NPP	400	—
Aug 1986	Dymer, Kiev region, Ukraine	54	500
Aug 1986	Gomel, Belarus	4.3	220
Aug 1986	Klintsy, Bryansk region, Russia	5.2–48	70
Aug 1986	Leningrad region, Russia	18	—

¶ Dead needle.

12.5 Monitoring and decontamination of transport

Within the 30 km zone it was decided to establish two separate zone, a *special* zone around the NPP and a 10 km zone. Strict dosimetric monitoring of all transport was organized and decontamination points were organized by the militia. At the zone boundaries, arrangements were made to transfer working personnel from one vehicle to another in order to reduce the possibility of transferring contamination across zones.

A *Novosti* report in July 1986 from the city of Zhlobin, some 180 km from Chernobyl, gives some idea of the traffic build-up. The journalist was travelling to Minsk which normally took some two and a half hours.

> It took somewhat longer, and past Bobruisk which was 80 km from Zhlobin, a line of lorries, refrigerator cars and Ladas caught our eye. They had been stopped by traffic militia for a radiation check-up. The militia captain said that they thoroughly hose down vehicles coming from the south when the radiation exposure is over 0.3 mR/hr. In the first days following the accident, he had to hose down nearly every car. Today, in July, it has only been 30 and the overall radiation level is 0.025 mR/hr.

12.6 Work of the Chemical Forces

The following summary of the decontamination achievements of the Chemical Forces of the USSR in the year after the accident to April 1987 was given by their commander, Colonel-General Vladimir Pikalov[9].

> More than 500 residential communities, nearly 60 000 buildings and structures, and several tens of million square metres of exposed surfaces of technological equipment and internal surfaces at

the NPP itself have been decontaminated. Tens of thousands of cubic metres of contaminated soil has been removed and the same amount brought in and several thousand insulating screens have been laid down. Dust has been suppressed on vast territories and several thousand samples have been taken for radioactive isotope analysis. Today, in April 1987, higher than admissible readings of soil contamination with the long-lived radioactive isotopes of caesium, strontium and plutonium are registered mostly on the territory of the NPP and within the 5 km zone surrounding it, as well as in several pockets of territory in Belarus. Soil radioactivity flushing by flood waters has not exceeded 1% and therefore no radical changes in the current level of soil contamination are expected.

12.7 Major problems

Examples[4] of some of the major problems encountered in the major programme of decontamination which was implemented, were as follows.

- Localization of the radioactive contamination, especially because of the very large number of vehicles necessarily involved in the clean-up operations.
- Loose ^{137}Cs contamination.
- Disposal of contaminated clothes and provision of new clean clothes, especially since at one stage near the NPP there were some 1000 persons in protective clothing, together with all sorts of equipment, including many concrete mixers.
- Although thousands of square metres were sprayed daily with inexpensive non-toxic substances, wind erosion of, firstly, roads then soil and crops was substantial.

12.8 Post-accident studies on decontamination strategies

Taking the experience gained at Chernobyl in the early period into account, future strategies of decontamination have been devised in an experimental collaboration between the European Commission, Belarus, Russia and Ukraine[10]. Selected data from this publication are given in tables 12.2, 12.3 and 12.4 with expected results from various techniques stated in terms of expected dose rate (EDR) reduction factors and decontamination factors (Df).

Table 12.2. Expected external dose rate (EDR) reduction factors of techniques applicable for urban surfaces[10].

Technique	Target	EDR	Comments
Turning flagstones manually	Flagstones	6	
Replacement of wall paper	Walls (paper)	>100	
Road planing	Road	>100	Grinding off surface
Fire hosing	Roads	1.1	Water rinsing
Vacuum sweeping	Roads	1.4	Dust close to operators
Roof washer	Roof	2	Rotation brush, air compressor
Manual change of roof cover	Asbestos roof	∞	—
Clay coating	Roof, walls	1.2–3.6	Dry and collect clay films
Hand held electric plane	Wooden walls	5	Upper layer (dust) mechanically removed
High pressure water hosing	Roof	1.3	120 bar pressure
	Walls	2.2	
	Asphalt and concrete	1.7–2.2	
Sandblasting	Walls	4 (dry) 5 (wet)	
Ammonium nitrate spraying	Walls	1.3	Surface rinsed with clean water
Detached polymer paste	Smooth surfaces	4–30	Temperature required of >5 °C
Polymer coatings	Walls: not wooden	4–5	Temperature required of 20–30 °C
Triple digging with shovel	Garden soil	4–15	For virgin soil, bury top layer of 30–40 cm

Table 12.3. Expected external dose (EDR) reduction factors or decontamination factors (Df) of techniques applicable for agricultural soils[10].

Technique	Target, EDR, Df and Comments
Front loader.	Soil, Df = 28, Scraping top soil, 10–30 cm, removes fertile soil layer
Bulldozer.	Soil, Df = 10–100
Grader.	Top layer of ground, Df = 4–10, Scraping of soil surface
Shovel.	Garden soil, EDR = 6, Digging to about 30 cm depth
Turf harvester.	Undisturbed grass, soil, private and forest pasture, urban grassed land, EDR = 3–6, Df = 3–20, Removes the 3–5 cm top soil
Ordinary plough and tractor.	Arable soil, EDR = 9–12, Ploughing virgin land to 25–45 cm depth
Deep ploughing.	Arable soil, Df = 2–4 (crop), Ploughing virgin land soil layer of 25–35 cm
Skim-and-burial plough.	Arable soil, EDR = 10–20, 5 cm of virgin land topsoil buried at 45 cm depth
Addition of potassium.	Arable land, Df = 2–3
Addition of phosphorus.	Arable land, Df = 0.8–1.3, In combination with potassium and nitrogen
Liming using special trucks for spreading.	Acid arable land , EDR = 1–3, Requires a soil pH of 4.5–5.5 and addition of potassium
Radical improvement of pasture by draining, cleaning, disking three times, etc.	Pasture Df = 4–16 (peat), Df = 4–9 (podzol)
Liming and fertilizing.	Forest pasture, Df = 1.5, Manual work. Enrich poor soils with calcium and potassium
Cyanoferrate bolus or Prussian blue.	Cows, Df = 2–3, 3 boli over 3 months
Clean fodder to animals before slaughter.	Cows, Df = 2–3 (on meat), 2 months before slaughter
Prussian blue salt licks.	Cows and bulls, Df = 2–3, Salt lick duration 3 months
Cyanoferrate filters for milk.	Milk, Df = 10, Private farm use if contamination >400 Bq l^{-1}

Table 12.4. Expected external dose reduction factors (EDR) or decontamination factors (Df) of techniques applicable for forests[10].

Technique	Target, EDR, Df, and Comments
Mechanical brush.	Forest litter, EDR = 3.5–4.5, Not wet forest areas of forest <30 years old. Litter layer removal.
Grinding mower.	Under-wood forest, shrubs, Df = 1, Wood stem diameter <8 cm. Not wet forest areas of forest <30 years old. Cleaning of under-wood.
Wood sawing plant.	Timber, Df = 2–4, Not wet areas. Mechanical removal of bark and phloem.
Twin-screw extruder.	Contaminated wood, Df = 50–100, Special wood pulp treatment from wood chips extracts Cs and Sr from the pulp.

Figure 12.4. Sculpture at the side of the road to the Chernobyl NPP which marks the boundary of the territory of the power plant, 1986. (Courtesy: TASS.)

Chapter 13

Water Contamination

Introduction

Some countermeasures have already been described, such as the building of the cooling slab beneath the shattered reactor in order to prevent the so-called China syndrome. This chapter details additional countermeasures as well as the emergency provision of alternative water supplies, and data on contamination levels of ground water and surface water in the river system near the NPP and Kiev, figure 13.1, and on sea water in the Black, Aegean and Mediterranean seas.

13.1 Countermeasures

Construction of a complex of hydraulic engineering structures began with a view to protecting the ground water and surface water from contamination. This included the following.

- A filtration-proof wall in the soil along part of the perimeter of the site of the NPP.
- Wells lowering the water table.
- A drainage barrier in the cooling pond.
- A drainage cut-off barrier on the right-hand bank of the river Pripyat.
- A drainage interception barrier in the south-western sector of the NPP.
- Drainage water purification facilities[1].

To protect the Kiev reservoir an underwater dam was built 450 m in length. This contains a hollow 100 m wide by 16 m deep, the purpose of which is to *catch* radioactive contamination which may get through the Pripyat river tributaries. Other silt traps were also made[1].

Countermeasures were also made to provide clean drinking water in the event, which initially was unknown, that the river system around Chernobyl was to become highly contaminated. These measures involved mooring the

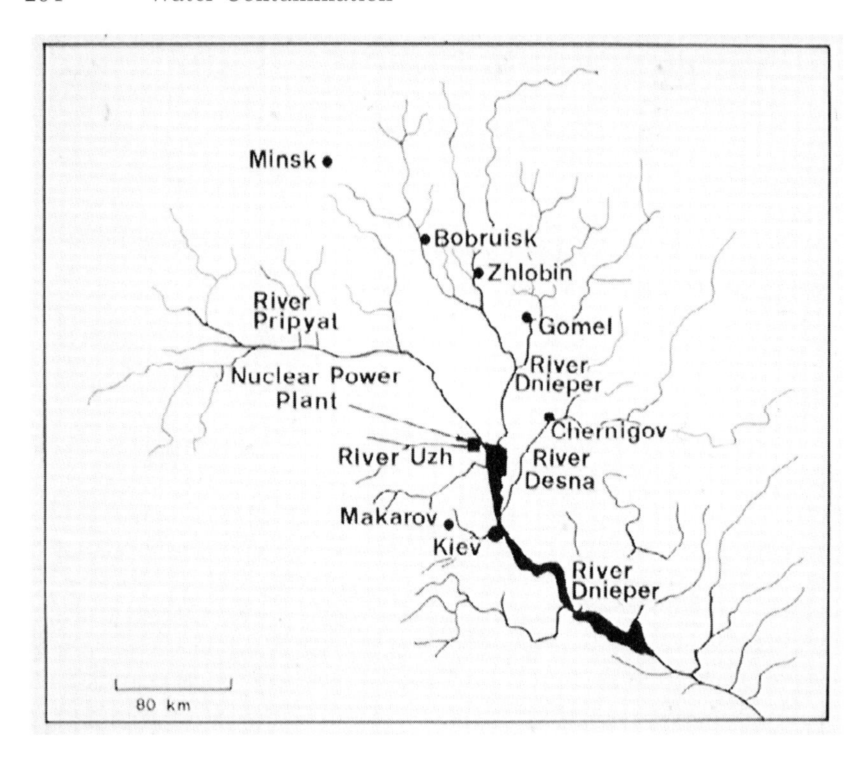

Figure 13.1. The river system of the Dnieper, Pripyat, Desna and Uzh, which are the major rivers in the region of the NPP, the Pripyat marshes and Kiev.

pump ship Rosa-300 in the river Desna, figure 13.1, to provide clean water for the city of Kiev. Two emergency water mains were built within a month, each 6 km in length and, to achieve this engineering feat, pipelines had to be thrown over 18 major obstacles including rivers, bridges, tunnels and roads. In addition, some 400 artesian wells were dug to replace, if necessary, water taken from the river Dnieper.

13.2 Contamination levels

The ^{131}I contamination of surface waters in the Dnieper and Pripyat rivers is given[2] in table 13.1. Since that time there have been changes with water intake from the Kiev reservoir and since 1986 the level of ^{137}Cs contamination has steadily declined[3] and has now reduced to some tens of Bq/litre. Levels of ^{90}Sr contamination have reduced more slowly but these have now stabilized around 1 Bq/litre.

Table 13.1. ^{131}I contamination of the Dnieper and Pripyat rivers 1986–87[2]. (The conversion factor is 1 μCi = 37×10^3 Bq.)

Date	Contamination	
	(μCi/litre)	(Bq/litre)
Before April 1986	1×10^{-6}	0.037
Maximum on 3 May 1986	3×10^{-2}	1 110
Mid-June 1986	1×10^{-4}	3.7
May 1987	1×10^{-5}	0.37

One of the current problems is the 800 radioactive waste sites, originally planned to be temporary but which have remained for more than a decade. They were created during the acute post-accident phase without any proper engineering preparations[4]. These now have a total water volume of some 1 million m^3 with a total activity level approaching 15 PBq and with contamination levels ranging from 1 million to 100 Bq/litre of ^{90}Sr at the storage site to 1000 to 100 000 Bq/litre in the vicinity of certain storage sites[3].

Further afield than the area in figure 13.1, contamination measurements have been made in several seas, the nearest of which to Chernobyl is the Black sea, table 13.2. The contamination of this sea was caused in the short-term by atmospheric deposition but in the long-term by transfer from the Kiev reservoir and the catchment areas of the rivers Dnieper, Dniester and Danube. The main Chernobyl contribution to the Mediterranean arrived by exchange of waters with the Black sea, which essentially has acted as a radioactive source[5].

Table 13.2. Contamination in seas[5].

Sea	Contamination
Baltic	From a few Bq/m^3 to 2 400 Bq/m^3 in 1986
Black	Maximum in 1986 was in its northern area at 500 Bq/m^3
	Mean in 1990 was 52 Bq/m^3
Irish	5.2 Bq/m^3 in 1990
Aegean	5–15 Bq/m^3 surface contamination from ^{90}Sr and ^{137}Cs
Mediterranean	5.7 Bq/m^3 mean ^{137}Cs concentration in 1990

Post-accident experimental and modelling studies have been made within a project of the European Commission, Belarus, Russia and Ukraine, for the transfer of radioactive materials to and in water bodies around

Chernobyl[6]. These have included measurements in the Chernobyl NPP cooling pond, table 13.3, which not surprisingly is one of the most polluted water bodies. The dimensions of the cooling pond, see figure 1.2, are as follows.

- Volume 0.15 km^3
- Surface area 22 km^2
- Length 11 km
- Mean width 2 km
- Mean depth 6.6 m
- Maximum depth 18 m

Table 13.3. Average annual concentrations of ^{137}Cs and ^{90}Sr in the water of the Chernobyl NPP[6].

Year	Concentration (Bq/litre)	
	^{137}Cs	^{90}Sr
1986	1000	20
1987	70	7
1988	30	15
1989	25	15
1990	15	9
1991	7	7
1992	5	5
1993	2	5
1994	4	2

Contamination measurements on fish muscles have also been made in lake Glubokoye which is only 10 km from the Chernobyl NPP. The levels for pike and perch are 120 and 55 kBq/kg of ^{137}Cs which figures can be compared with data for Sweden, see table 11.6, for which the maxima are 8.3 and 28.0 kBq/kg for pike and perch respectively[7].

Plate I

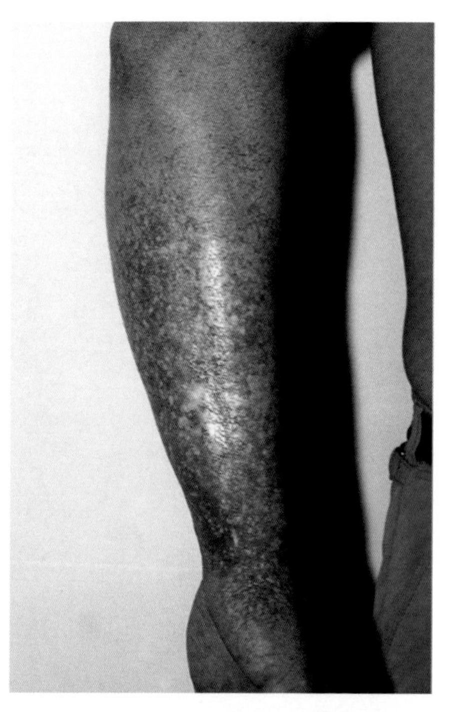

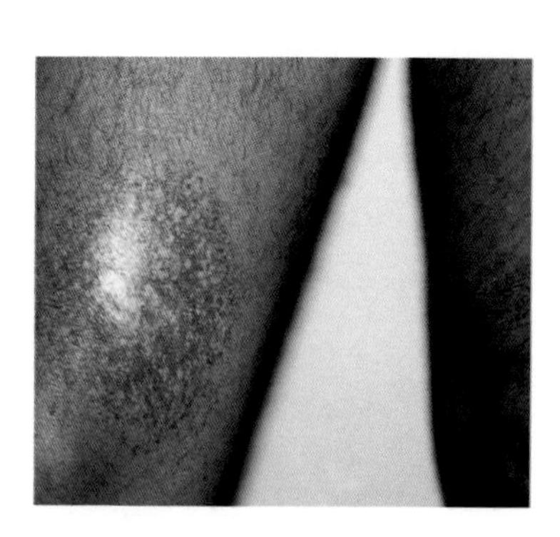

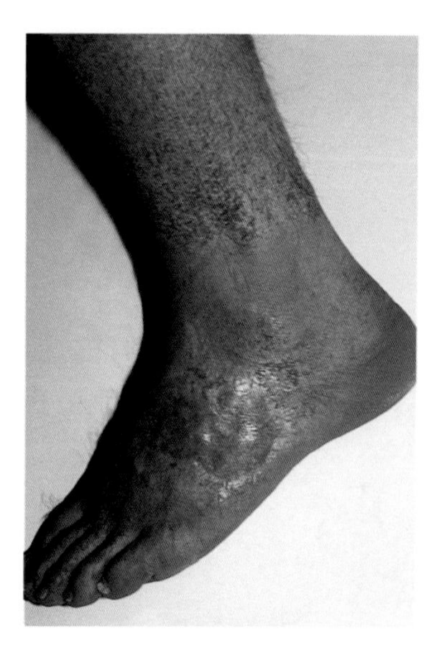

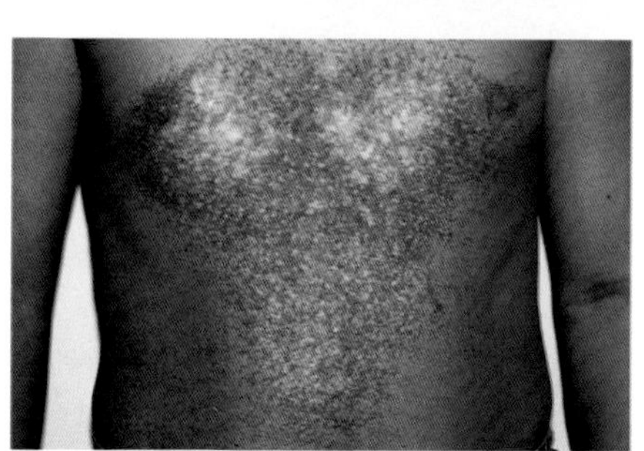

Plate II

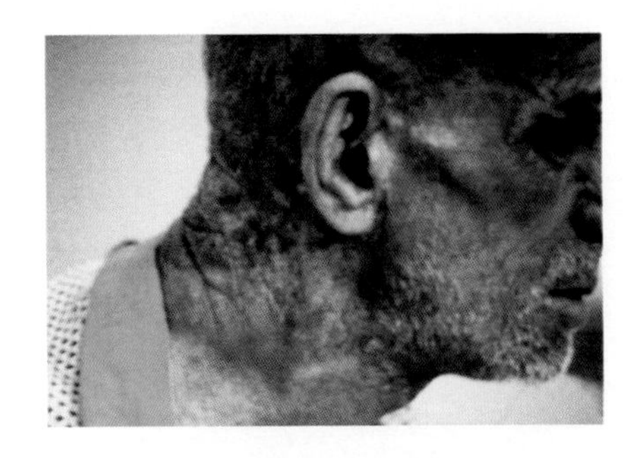

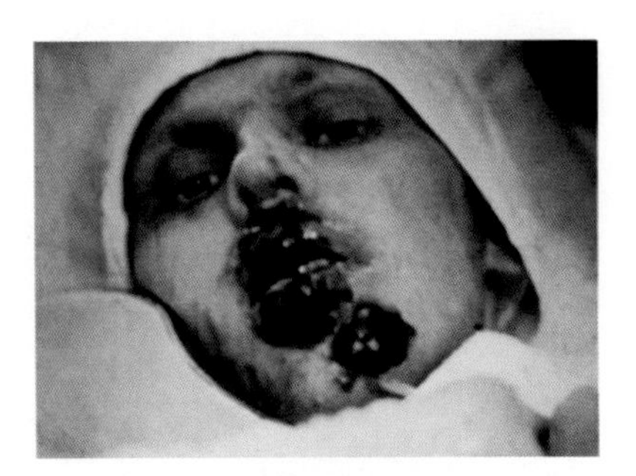

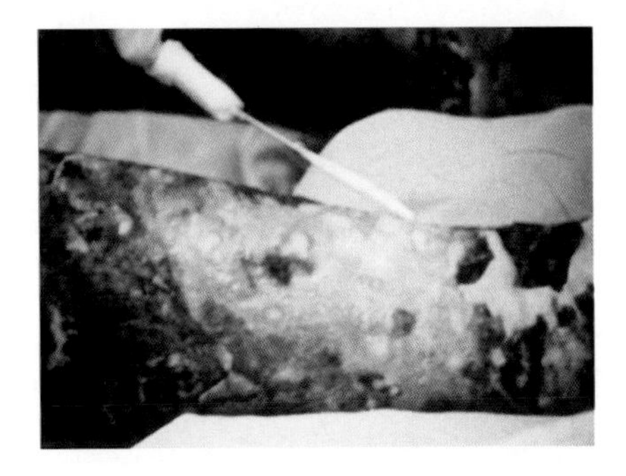

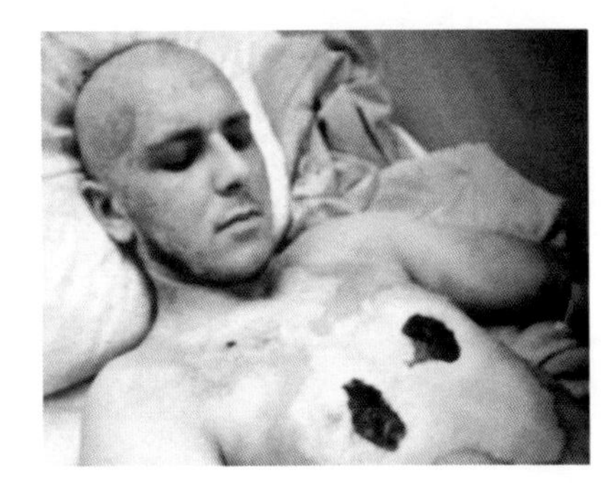

Plate III

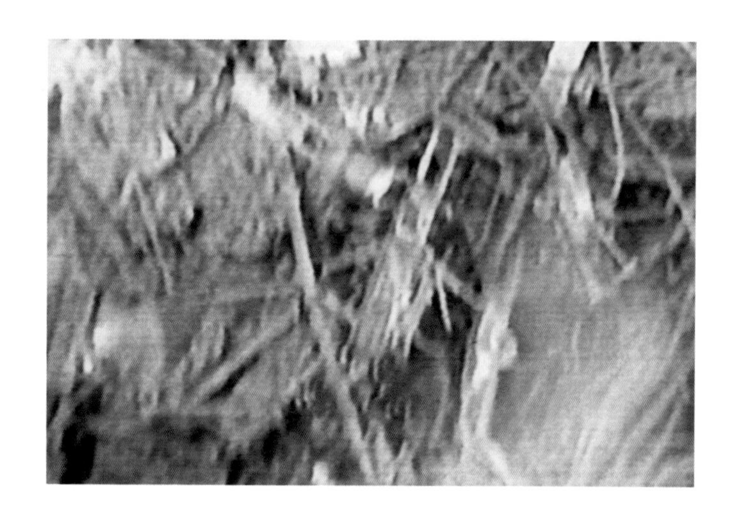

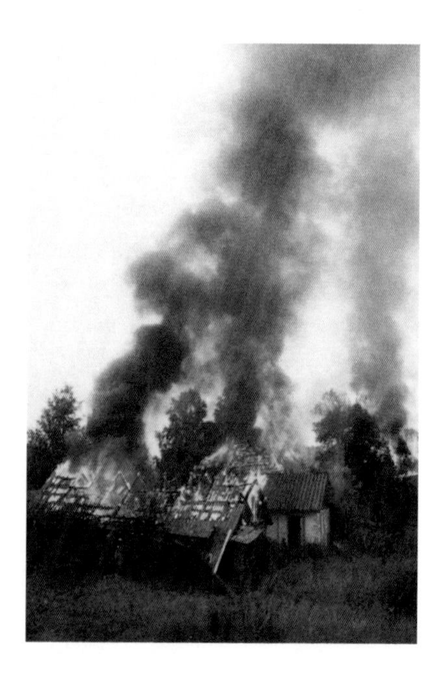

Plate IV

Plate V

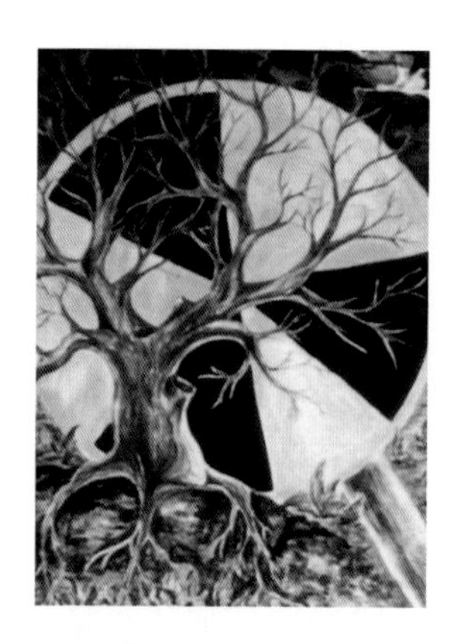

Plate VI

24325

ПОЛЕССКИЙ
ГОСУДАРСТВЕННЫЙ
РАДИАЦИОННО-
ЭКОЛОГИЧЕСКИЙ
ЗАПОВЕДНИК
СТОЙ!
ПРЕДЪЯВИ
ПРОПУСК

Plate VII

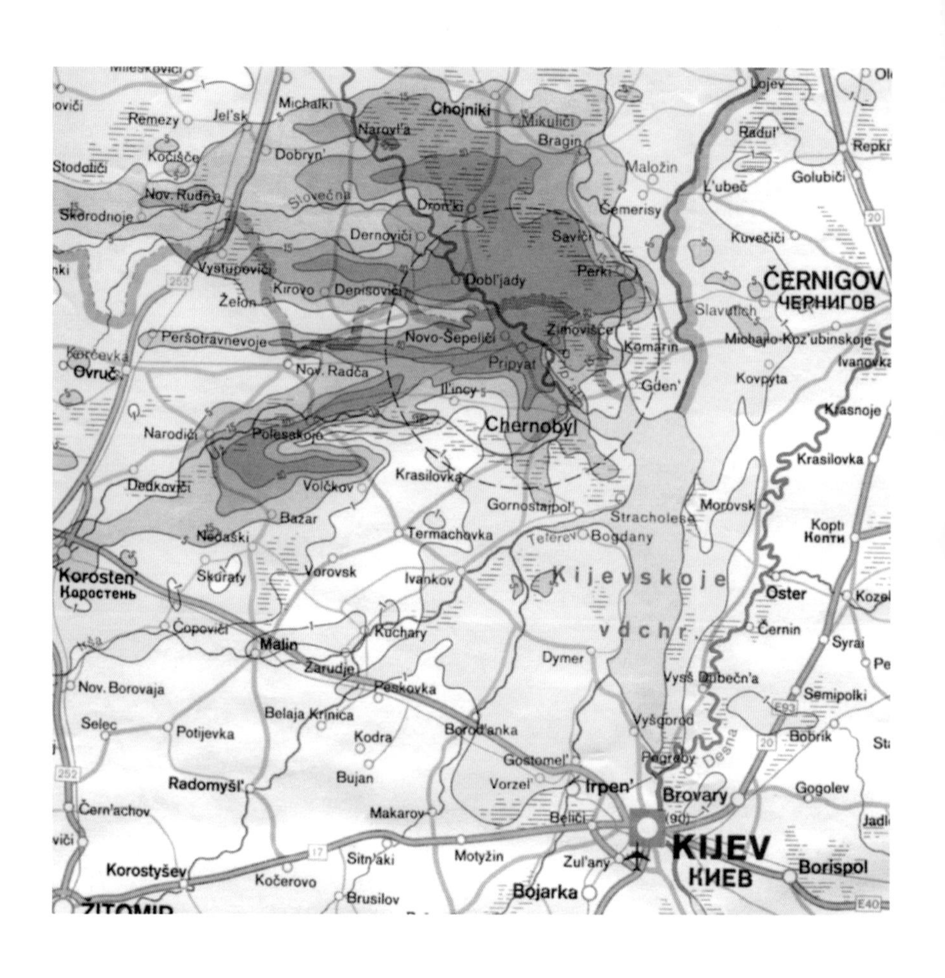

Isopleths of surface ground contamination
Изоплеты поверхностного загрязнения почвы

——1——	1 Ci/ km^2 1 Ки/$км^2$	(37 kBq/ m^2) (37 кБк/$м^2$)
——5——	5 Ci/ km^2 5 Ки/$км^2$	(185 kBq/ m^2) (185 кБк/$м^2$)
——15——	15 Ci/ km^2 15 Ки/$км^2$	(555 kBq/ m^2) (555 кБк/$м^2$)
——40——	40 Ci/ km^2 40 Ки/$км^2$	(1480 kBq/ m^2) (1480 кБк/$м^2$)

Contamination zones
Зоны загрязнения

	1–5 Ci/ km^2 1–5 Ки/$км^2$	(37–185 kBq/ m^2) (37–185 кБк/$м^2$)
	5–15 Ci/ km^2 5–15 Ки/$км^2$	(185–555 kBq/ m^2) (185–555 кБк/$м^2$)
	15–40 Ci/ km^2 15–40 Ки/$км^2$	(555–1480 kBq/ m^2) (555–1480 кБк/$м^2$)
	Above 40 Ci/ km^2 свыше 40 Ки/$км^2$	(above 1480 kBq/ m^2) (свыше 1480 кБк/$м^2$)

– – – – 30 km zone around the Chernobyl nuclear power plant
30-километровая зона вокруг Чернобыльской АЭС

Plate VIII

Chapter 14

Ground Contamination

Introduction

UNSCEAR[1] defines contaminated areas as those where the average ^{137}Cs deposition densities exceed 37 kBq/m^2 (1 Ci/km^2). The choice of ^{137}Cs as the reference radionuclide was made for three reasons.

- Substantial contribution to the lifetime effective dose.
- Long radioactive half-life.
- Ease of measurement.

On this basis, the total areas of ground contamination[2] were 57 900 km^2 in Russia, 46 500 km^2 in Belarus and 41 900 km^2 in Ukraine, making in overall total an area of 146 300 km^2. Since 1 km^2 = 247 acres and 1 mile^2 = 2.59 km^2 this is equivalent to 36.14 million acres or 56 486 miles^2.

Some three million Russian inhabitants live in the contaminated territories of whom 2500 are in SCZs with a level exceeding 1480 kBq/m^2, 200 km^2 of agricultural land has been taken out of production as have 600 km^2 of forests; and agrotechnical measures, some of which have been detailed in section 12.8, have been applied to 30 000 km^2. A total of 17 settlements have been entirely abandoned within the 30 km zone and a population of 50 000 has either been evacuated or left voluntarily[3].

In Belarus, more than 6000 km^2 of land, including 3000 km^2 of fertile arable land has been removed from economic use[4], whereas in Ukraine the agricultural land withdrawn from use is 1,800 km^2 and in addition some 40% of the forested area of Ukraine has been contaminated. A population of some three million, including one million children, are living, or have lived, on contaminated land[5].

The environmental contamination stages are listed[6] in table 14.1 and the behaviour of the radioactive plume and the reported initial arrival times of detectable activity in air[1] are given in figure 4.2.

Table 14.1. Environmental contamination stages[6].

Time scale	Stage
Few days	Atmospheric transfer and deposition of radionuclides occurring on components of terrestrial and water media with their further absorption in biota, soil and sediments.
One month	Dynamics characterized by radionuclide accumulation (bioassimilation) in biota, their retention in buffer abiotic components of ecosystems (soil, sediments), and a reduction of radioactive contamination levels due to the decay of short-lived nuclides.
Years	Secondary redistribution of radionuclides between ecosystem components as a result of transfer of radionuclides from buffer storage and their re-inclusion in exchange cycles.

^{137}Cs concentration does not reduce quickly with time and this is demonstrated by the data in table 14.2 which are for Sosnovyi Bor, a town in a remote zone of the affected area[6]. Background levels for 1985 and the data for 1990 are only for the first six months of that year. Data for ^{137}Cs levels in milk from collective and private farms for the period 1988–94 are given[7] in table 11.4. The conclusions which can be drawn with regard to soil contamination[8] are given in table 14.3.

Table 14.2. Reduction of ^{137}Cs concentration with time in Sosnovyi Bor[6]. (The translation of Sosny is 'pine' and Bor is 'forest'.)

Ecosystem component	Unit of measurement		1986					
		1985	Mean	Max	1987	1988	1989	1990
Surface air	Bq/m^3	5	240	—	67	37	23	18
Atmospheric fallout	Bq/m^3/y	5	9300	—	166	33	32	—
Soil	Bq/kg¶	43	340	—	124	160	110	110
Moss	Bq/kg¶	—	5800	—	2170	1400	1700	—
Pine needles	Bq/kg¶	5	3500	9700	240	190	170	—
Fungi	Bq/kg¶	40	1200	4700	270	600	220	130
Milk	Bq/l	0.15	17	—	3.9	0.7	2.0	—

¶ Wet weight.

Table 14.3. Conclusions which can be drawn with regard to soil contamination[8].

• There is very pronounced heterogeneity of contamination.
• There is radionuclide accumulation in the upper 10 cm.
• The soil clearance half-lives, which fluctuate in the range 10–25 years for Cs and in the range 7–10 years for Sr according to the soil type, indicate that contamination, especially that due to Cs, will remain detectable for a very long time: from several decades to more than 100 years, according to the contamination levels and if there is no intervention.

The contribution of short-lived radionuclides to the total activity released was some 20–30% with a dominant contribution from ^{131}I accounting to 40–50% of the total short-lived contribution[6]. However, because of the short half-life, the variable of interest with ^{131}I is not so much ground contamination but more the reconstruction of thyroid doses from ^{131}I. As an example, figure 14.1 shows, for all Ukraine[9], the distribution by district of the average thyroid dose to persons aged 0–18 years in 1986. The highest average dose is 1.618 Gy for Narodychi and Kaniv. It was the experience of monitoring children in the village school in Narodychi which was described in section 5.2.2. The inhomogeneous distribution of thyroid doses in figure 14.1 reflects the geographical inhomogeneity of the ^{131}I fallout from the radioactive plume.

For longer lived radionuclides such as ^{137}Cs, ^{90}Sr, ^{239}Pu and ^{240}Pu ground contamination is most important and contamination maps are shown in sections 14.1–14.3.

14.1 ^{137}Cs contamination

The boundaries of the contamination zones used as a basis for evacuation and resettlement are given[10] in table 6.1 and are defined by the levels 37–555, 555–1480 and >1480 kBq/m^2, which are sometimes expressed as 1–15, 15–40 Ci/km^2 and >40 Ci/km^2. The lower band is also sometimes divided into two bands with the additional boundary level at 185 kBq/m^2 (5 Ci/km^2).

Areas contaminated by levels greater than 185 kBq/m^2 and the populations living in these contaminated territories in more than 2000 villages and settlements in Belarus, Russia and Ukraine as of the end of 1989 are given in table 14.4. The number of villages and settlements with levels above 555 kBq/m^2 in 1989 were 786.

In table 14.5 data are given for levels above 10 kBq/m^2 for selected European countries, from the *Atlas of ^{137}Cs Contamination of Europe after the Chernobyl Accident.* This atlas was compiled under the Joint

Figure 14.1. Thyroid doses from ^{131}I received by the Ukraine population born between 1968 and 1986[2]. (Courtesy: European Commission.)

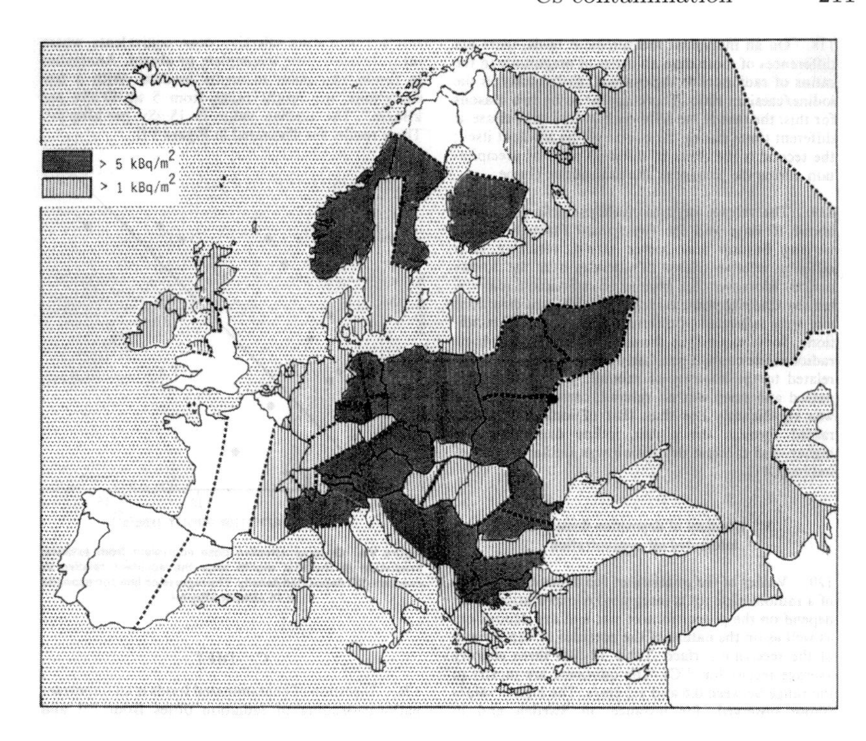

Figure 14.2. ^{137}Cs contamination in countries or larger sub-regions of Europe[1]. Contamination levels shown >5 kBq/m^2 and >1 kBq/m^2. (Courtesy: UNSCEAR.)

Study Project of the CEC/CIS Collaborative Programme on the Consequences of the Chernobyl Accident[2]. For data[1] for Europe for the two levels >5 kBq/m^2 and >1 kBq/m^2, see figure 14.2.

For the 60 km zone surrounding the Chernobyl NPP, figure 14.3 shows[2] a map of lines of equal contamination level for 185, 555, 1480 and 3700 kBq/m^2 which in the map are labelled as 5, 10, 40 and 100 Ci/km^2. Slavutich on the far right is the new permanent town built to replace Pripyat: the temporary town was Zeleny Mys, which is shown in figure 14.9 but not in figure 14.3, is on the boundary of the 30 km zone.

Figure 14.4 shows[12] the main areas of ^{137}Cs contamination, which in this figure are defined as >555 kBq/m^2. These three areas, labelled in the figure as **C**, **B** and **K**, are termed the Central, Bryansk–Belarus and Kaluga–Tula–Orel areas. **C** was formed during the initial, active stage of the radionuclide release, which was predominantly to the west and northwest. **B**, which is centred to the north-north-east of the NPP, was formed on 28–29 April 1986 as a result of rainfall in the Bryansk region of Russia and

Table 14.4. ^{137}Cs contamination measured at the end of 1989[11] and, given in brackets, updated[2] values in 1996.

Contamination level (kBq/m^2)	Area of contamination (thousands of km^2)			
	Belarus	Ukraine	Russia	Total
185–555	10.2	2.0 (3.2)	5.8 (5.7)	18.0 (19.1)
555–1480	4.2	0.8 (0.9)	2.1	7.1 (7.2)
>1480	2.2	0.6	0.3	3.1
>185	16.6	3.4 (4.7)	8.2 (8.1)	28.2 (29.4)
Contamination level	Population (thousands of persons)			
185–555	267	204	113.1	584.1
555–1480	105	29.7	80.9	215.6
>1480	9.4	19.2	4.6	33.2
>185	381.4	252.9	198.6	832.9

the Gomel and Mogilev regions of Belarus. In the most highly contaminated spots in **B** the levels were comparable to **C** and reached 5000 kBq/m^2 in some villages[13]. **K** is some 500 km to the north-east of the NPP and was formed by the same radioactive cloud that produced **B**, as a result of rainfall on 28–29 April. However, the levels in **K** were usually less than 600 kBq/m^2.

In addition to these three main hot spots of contamination there were many areas in the range 40–200 kBq/m^2 and in the territory of the USSR there were[2,14] some 3100 km^2 contaminated above 1500 kBq/m^2, table 14.5.

Plate VIII is a surface contamination map from the *International Chernobyl Project*[15] which is centred on area **C** and figures 14.5 and 14.6 are contamination maps which include areas **B** and **K**.

14.2 ^{90}Sr contamination

Deposition of ^{90}Sr was mainly limited to the zone near the NPP, figure 14.7, and virtually no areas outside the 30 km zone were contaminated at a level exceeding 100 kBq/m^2. Contaminations exceeding 37 kBq/m^2 were almost all within a distance of less than 100 km from the NPP. Relatively few areas of 37–100 kBq/m^2 were located in the region of Gomel, Mogilev and Bryansk[15].

Table 14.5. Areas in European countries[2] contaminated by ^{137}Cs in the range 10–185 kBq/m^2, listed in terms of ranking of percentage contamination deposited in Europe.

Country	Area in 1000 km^2 contaminated above a specified kBq/m^2 level			Percentage of contamination deposited in Europe
	10–20	20–37	37–185	
Belarus	60	30	29.9	33.5
Russia	300	100	48.8	23.9
Ukraine	150	65	37.2	20.0
Sweden	37.4	42.6	12.0	4.4
Finland	48.8	37.4	11.5	4.3
Bulgaria	27.5	40.4	4.8	2.8
Austria	27.6	24.7	8.6	2.7
Norway	51.8	13.0	5.2	2.3
Romania	14.2	43.0	—	2.0
Germany	28.2	12.0	—	1.1
Greece	16.6	6.4	1.2	0.8
Slovenia	8.6	8.0	0.3	0.5
Italy	10.9	5.6	0.3	0.5
Moldovia	20.0	0.1	0.06	0.45
Switzerland	5.9	1.9	1.3	0.35
Poland	8.6	1.0	—	0.23
Czech Republic	3.4	0.36	—	0.09
Estonia	4.3	—	—	0.08
Slovak Republic	2.1	—	—	0.05
Lithuania	1.2	—	—	0.02

14.3 ^{239}Pu and ^{240}Pu contamination

It is difficult to detect plutonium radionuclides[15] and the only area located with levels exceeding 4 kBq/m^2 was within the 30 km zone, figure 14.8. Near Gomel, Mogilev and Bryansk the levels were 0.07–0.7 kBq/m^2 and in the Kaluga, Tula and Orel area were 0.07–0.3 kBq/m^2.

14.4 Exclusion zone area of the Ukraine

The *International Chernobyl Project* of the IAEA published their results[15] in 1991, including contamination maps for ^{137}Cs, ^{90}Sr and $^{239}Pu/^{240}Pu$ which have become the standard reference maps, see Plate VIII and figures 14.5–14.8. However, what has not been well publicised, because of the difficulty of obtaining the book outside Kiev, is that in 1996 the Na-

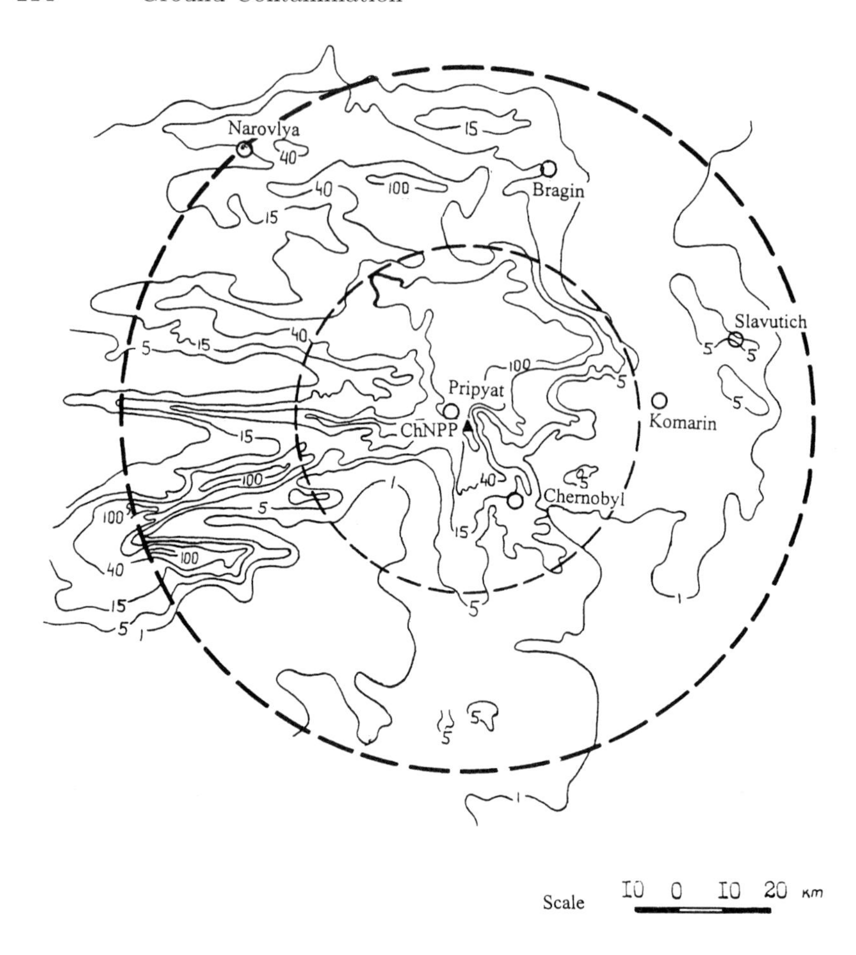

Figure 14.3. ^{137}Cs contamination in 1989 in the 60 km zone around the Chernobyl NPP[2]. Contamination levels shown: 185, 555, 1 480 and 3 700 kBq/m^2 (5, 15, 40 and 100 Ci/km^2). (Courtesy: European Commission.)

tional Academy of Ukraine published[16] an *Atlas of Chernobyl Exclusion Zone*, which includes maps for both beta radiation contamination and for gamma radiation contamination, together with a commentary in English. Other maps within this *Atlas* include those showing radionuclide transportation pathways and effective ground water flow velocities and of the influence of the meteorological conditions on radioactive pollution over the territory of the Ukraine. What is not included is data relating to Belarus and therefore the town of Gomel and part of the 30 km zone are not included. Nevertheless, it is a valuable publication and a summary of major

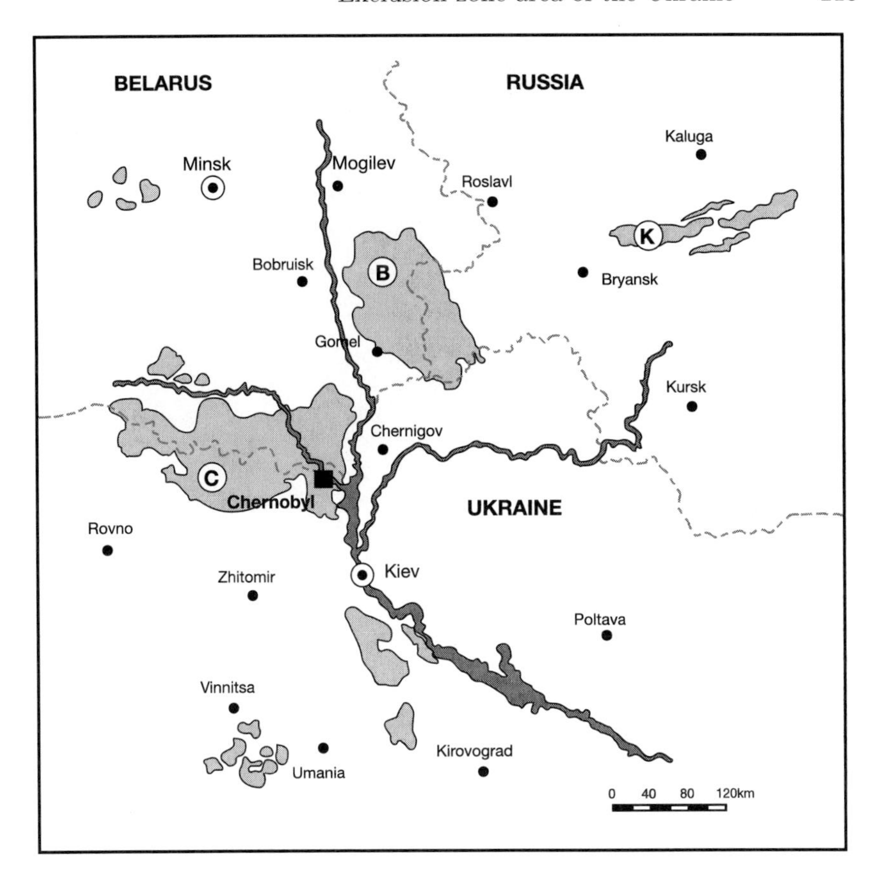

Figure 14.4. ^{137}Cs contamination *hot spots* defined as contamination >555 kBq/m2. **C**: central around the NPP, **B**: Bryansk–Belarus, **K**: Kaluga–Tula–Orel areas. After the map[12] published by Nuclear Energy Agency Organisation for Economic Co-operation and Development.

aspects of the statistical data are given below. Some of this information has not been previously reported, whereas in some instances the information is an update from what was previously available.

The exclusion zone area within the borders of the Ukraine was initially 2044.4 km^2 but following the evacuation of the population groups most at risk, a further 1800 km^2 was added to this zone. Forty-eight Ukranian villages were completely evacuated and 37 partially evacuated and as a whole this area exceeds 1.5 times that of Luxembourg.

The surface contamination of the exclusion zone territory was predominantly in the upper 5 cm of soil and consisted of 110 000 Ci of ^{137}Cs,

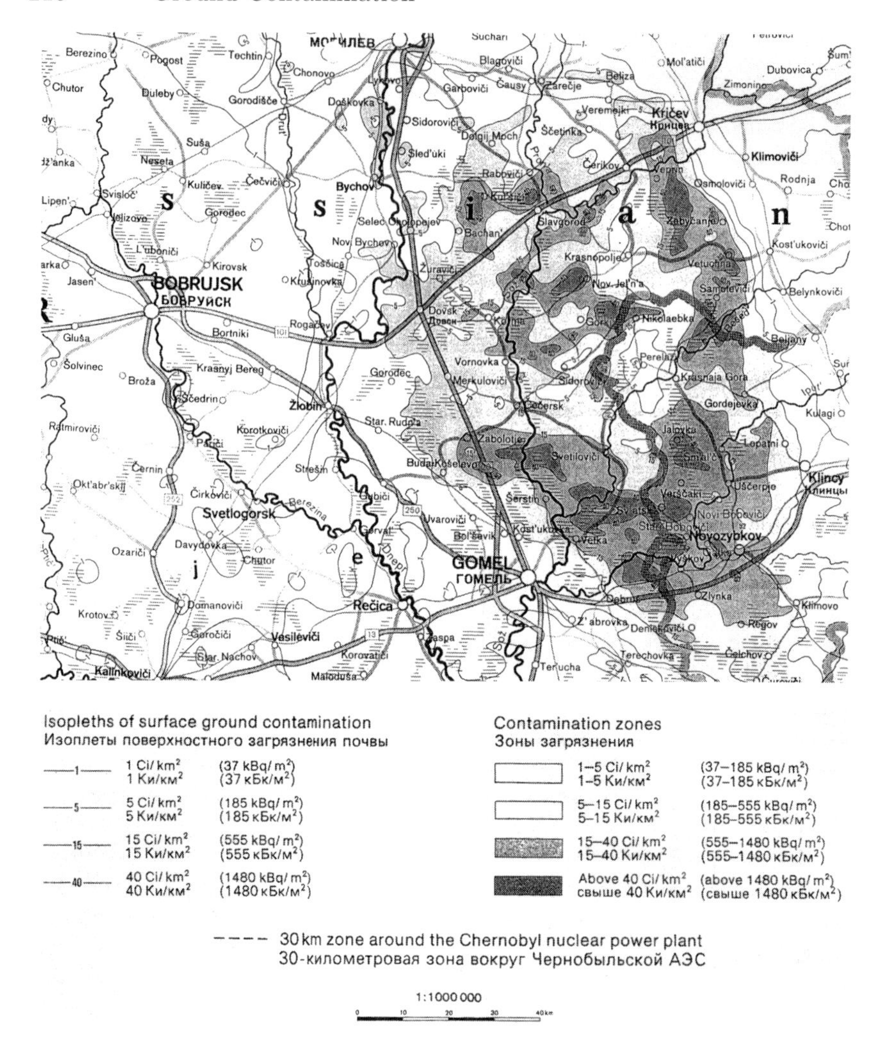

Figure 14.5. ^{137}Cs contamination map including the Bryansk–Belarus *hot spot*: an International Chernobyl Project map[15]. (Courtesy: IAEA.)

127 000 Ci of ^{90}Sr and 800 Ci of plutonium radioactive isotopes. In addition 20 million Ci of radioactive material is still within the Sarcophagus. The number of temporarily planned radioactive waste sites exceeds 800. Originally, there were to be three permanent waste burial sites. The activity registered in the sites is some 380 000 Ci and this is in a volume of 1 million m^3. In addition, up to 3500 Ci of ^{137}Cs, up to 800 Ci of ^{90}Sr and a few Ci of plutonium are concentrated in the bottom sediment of the

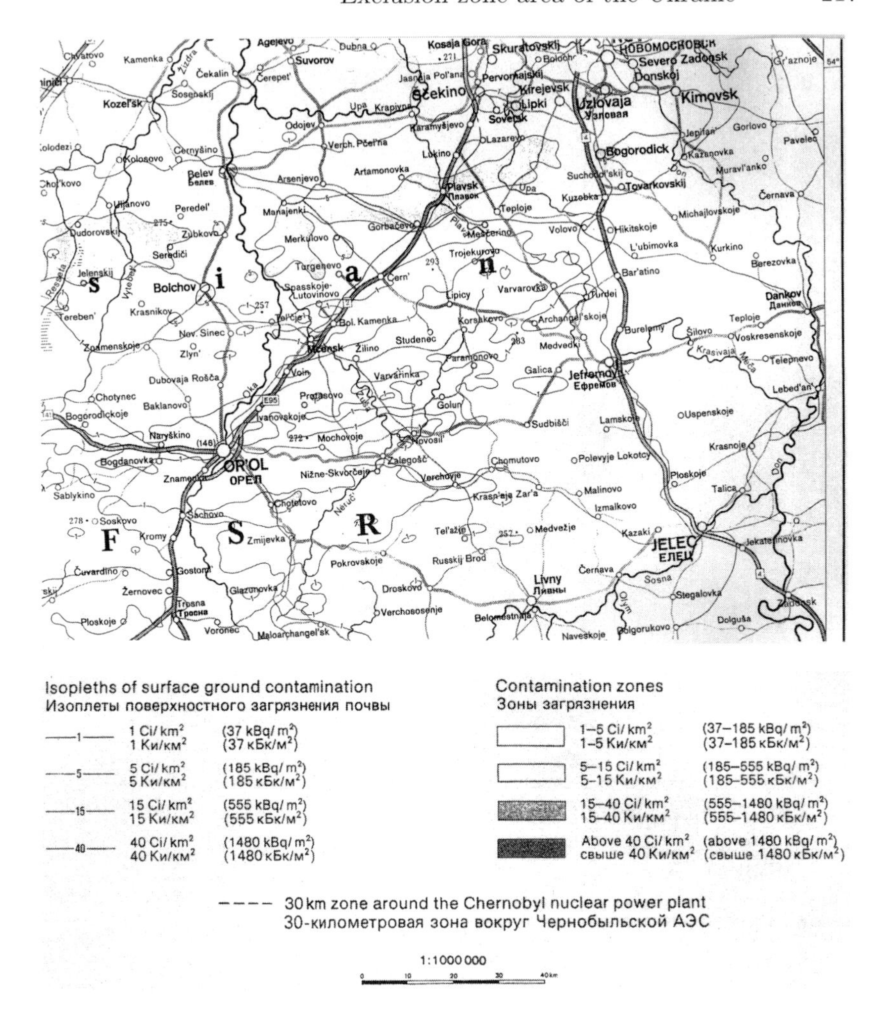

Figure 14.6. ^{137}Cs contamination map including the Kaluga–Tula–Oral *hot spot*: an International Chernobyl Project map[15]. (Courtesy: IAEA.)

cooling pond of the NPP which has a water volume of 160 million m^3.

In 1989–93 the mean annual radioactive removal *via* Pripyat river to the Kiev reservoir was 112–426 Ci/year for ^{90}Sr which included removal from the zone of about 60% of the initial activity. The data for ^{137}Cs was 52–125 Ci/year with a removal from the zone of about 20% of the initial activity. Migration of radionuclides outside the zone is considered to be insignificant: less than 1 Ci/year. Removal of radionuclides by wind transport is also not significant: dense vegetation makes a wind removal

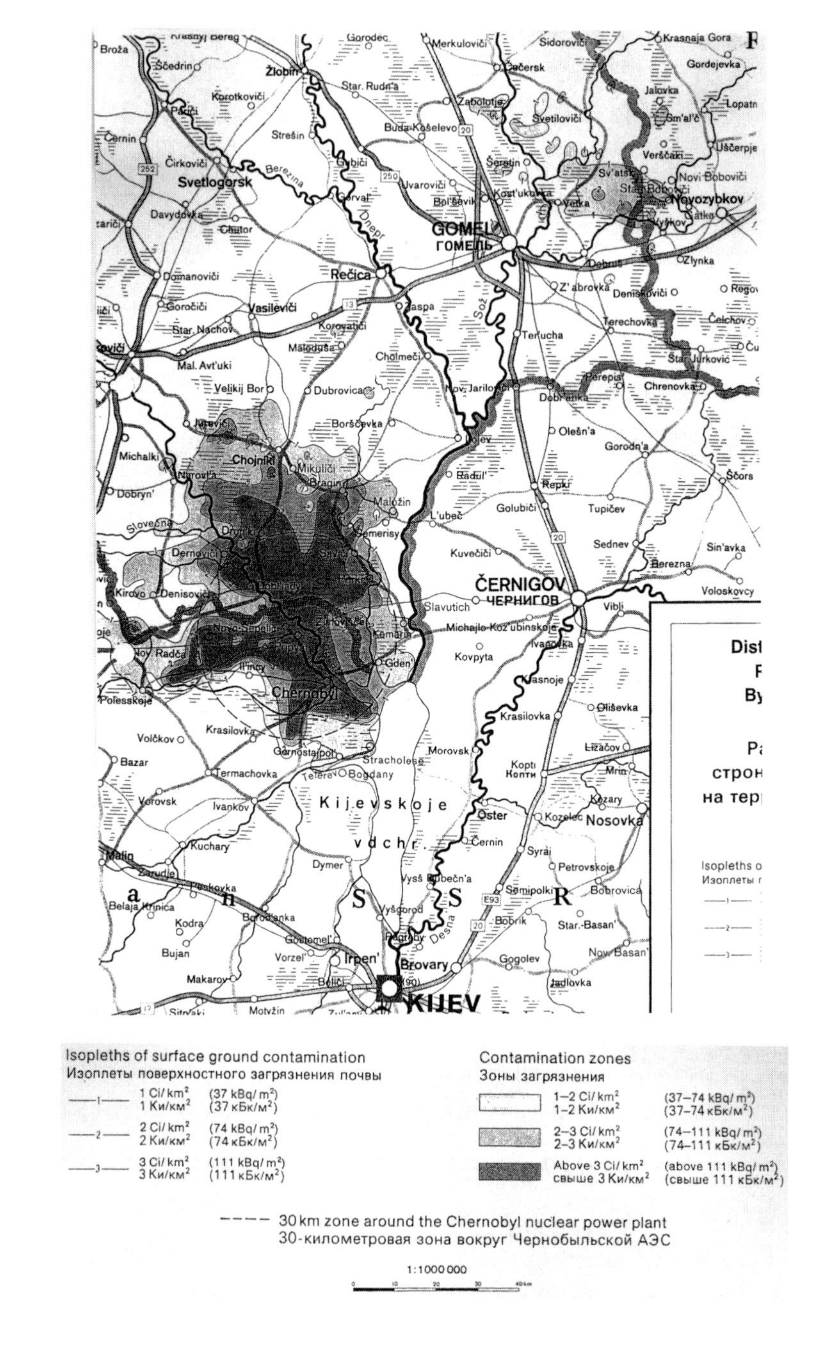

Figure 14.7. ^{90}Sr contamination map: an International Chernobyl Project map[15]. (Courtesy: IAEA.)

Figure 14.8. ^{239}Pu and ^{240}Pu contamination map: an International Chernobyl Project map[15]. (Courtesy: IAEA.)

pathway impossible. The most intensive contamination pathway into the ground waters is from the temporary waste disposal pits.

Currently, the most biologically hazardous radionuclide is ^{137}Cs which is responsible for more than 90% of the external intake dose received by workers at the Chernobyl NPP, scientists and medical staff working in the zone, and the returnees.

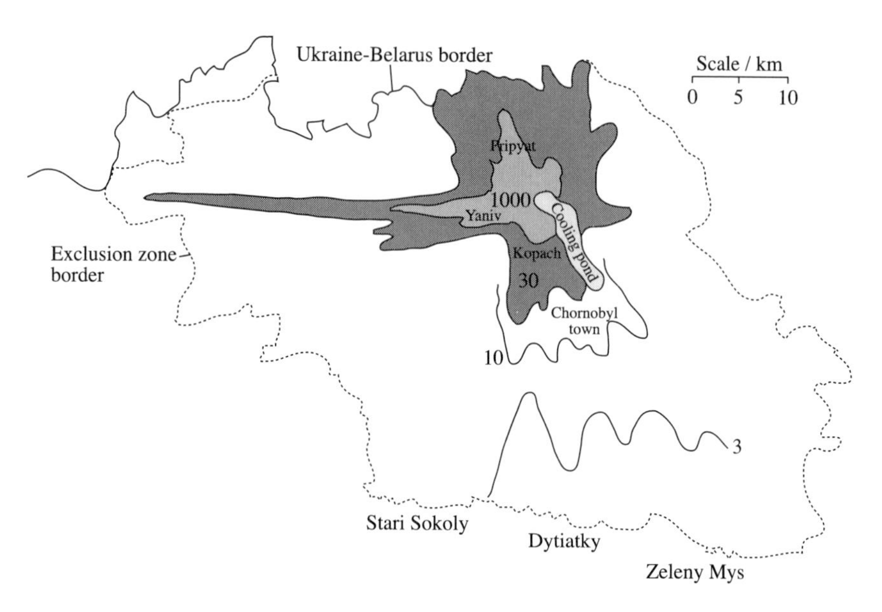

Figure 14.9. Beta contamination map of the exclusion zone in the Ukraine: after the map published by the National Academy of Science of Ukraine[16].

As seen in figure 14.4 the ^{137}Cs deposition is very non-uniform. In particular there is an area called[16] the *Western finger*, see figure 14.9, which extends from the NPP towards the west (in figure 14.9 this is towards the left) is very narrow, only some 1.5–2.0 km to 5 km in parts, and along an axial distance of 70 km the ^{137}Cs deposition decreases from more than 10 000 Ci/km^2 to 10 Ci/km^2. The maximum deposition was noted in the *red forest* which before it was cut down was approximately 2.5 km west of Unit No. 4.

There is a wide *Southern finger* which gradually separates into five separate *jets*[16], of which the most easterly travels along the Pripyat river bed before sharply turning west. On the whole, this Southern finger has the highest deposition of ^{90}Sr and of the transuranic radioactive isotopes. The *Northern finger* is much wider than the southern and was formed mainly because of extensive rainfall. Its area is 25 × 60 km^2 and is immediately to the north of the Ukraine–Belarus border. At its central part the ^{137}Cs deposition is 500–1 500 Ci/km^2. Caesium deposition is a factor of 2–3 times higher than strontium deposition. The *South-west finger* is a complex vortex-like pattern and contains high levels of ^{137}Cs of 100–500 Ci/km^2 and above. For example, in the village of Dibrova in 1994, levels of 600–900 Ci/km^2 were measured.

In addition to the areas of high values of ^{137}Cs activity there are also low values in several villages in the 450 km^2 area south-west of the NPP at some 16–18 km distance from it. Deposition measured in three of the cleanest villages were only 0.28, 0.30 and 0.35 Ci/km^2.

A global range in the zone for the decrease in ^{137}Cs since 1986 is 25–50% of which the main causes of the decrease are as follows.

- Natural decay of ^{137}Cs which accounts for some 20–21%.
- Partial washout of the surface layer of soil from elevated areas and slopes.
- Deactivation of various villages and settlements.
- Penetration of some radionuclides into the ground at depths > 10–20 cm from the soil surface.

Forests occupy half of the exclusion zone territory and of the remainder some 30% is arable land, 8.5% water objects and only 6% is taken up by settlements and roads. Table 14.6 summarizes the effect of the radiation on trees for the period from 1986–88. For the later period of 1989–95 tree growth was classed as normal for degrees of absorbed dose 1, 2 and 3. For the two higher dose groups, 4 and 5, for 1991–95 the effect was classed as *formation of a new plant community*.

The term *fingers* used by the National Academy of Sciences of Ukraine[16] is by analogy with the hand-shaped beta contamination which is seen in figure 14.9. This map indicates the boundary of the exclusion zone, a term often used synonomously with 30 km zone, although in fact the 30 km zone is a geometrically defined area whereas the actual zone boundaries are irregular as can be seen here. The lines of equal beta radiation contamination, marked 1000, 30, 10 and 3 are given in units of 100 decays per cm^2 per minute. Since 1 Bq equals 1 decay/second, the 1000 beta radiation line represents a containation boundary of 6000 kBq/cm^2. The check points at the border of the zone are also given in figure 14.9, Dytiatky, Zeleny Mys and Stari Sokoly as well as the site of the village of Kopach, see figures 14.10 and 14.11, and the village of Yaniv where the railway station nearest to the NPP was located, see table 6.3 and figure 14.12.

Table 14.6. Radiation doses received by forest areas and the effects on tree growth[16].

Degree of absorbed dose in 1986	Effects on tree growth during 1986–1988		
	1986	1987	1988
1. Noticeable (0.1–0.5 Gy)	Growth change	Normal	Normal
2. Low (0.6–5.0 Gy)	Suppression of growth	Occasional morphoses	Normal
3. Intermediate (6–10 Gy)	Strongly suppressed growth, morphoses, occasional perishing of separate trees	Partial forest restoration, morphoses, absence of flowering	Restoration of timber growth, morphoses
4. High (11–60 Gy)	Absence of timber growth, browning of needles, perishing of a part of trees	Survival of separate tree groups	Restoration of timber growth, numerous foliage development morphoses
5. Acute: completely destructive (>60 Gy)	Total forest destruction	Needles fall off and splintering of bark	Bark splintering, appearance of shoots and herbaceous cover

Figure 14.10. Only the village sign remains of Kopach. The small grass hillocks in the background of this 1998 photograph are the buried remains of the burnt houses. (Photograph: R F Mould.)

Figure 14.11. Typical wooden house to be found in villages within the 30 km zone. Many have been burnt down and buried such as in the village of Kopach but a few still remain. In 1998, when looking through the windows, one could still see the signs of an urgent evacuation: such as pans and plates on the kitchen table and an unmade bed. (Photograph: R F Mould.)

Figure 14.12. View of the power station from the bridge over the railway line at Yaniv, 1998. This line was so contaminated by the accident, see Yaniv in figure 14.9 within the beta contamination boundary of 6000 kBq/cm^2. The railway was so badly contaminated in April 1986 that it could not be used for evacuating the population. (Photograph: R F Mould.)

Figure 14.13. The Ditjatki Control and Passport Point at the entry to the exclusion zone, see map in figure 14.9. (Photograph: R F Mould.)

Chapter 15

Psychological Illness

Introduction

Technological disasters in the period 1984–86 have included not only that at Chernobyl but also the explosion at the chemical plant in Bhopal, India and the crash of the American space shuttle Challenger. In the wake of these, people have grown more sceptical of new technologies and more fearful of familiar technologies around them, particularly when they perceive that there is an impact on health.

The Chernobyl accident is often described in terms such as *the greatest man-made catastrophe* and this is correct in terms of public concern and anxiety in a period of peacetime. However, Chernobyl was not the greatest catastrophe in some other respects, such as the amount of toxic emissions (for example, much less ^{131}I was released from Chernobyl than from nuclear weapons testing) and the number of fatalities[1–3], table 15.1.

There is, though, no doubt that Chernobyl is the greatest psychological disaster of the 20th century, having had a worldwide impact. In 1986, just after the accident, the major cause of concern was the future expected increase in the incidence of thyroid cancer and leukaemia, particularly in children and adolescents. However, a decade later[4,5] it became apparent that the magnitude of psychological and social problems, of which radiation phobia (see section 6.6) is only one aspect, far outweighed that of radiation-induced cancers, and that the social and associated economic problems of psychosocial illness in Ukraine, Belarus and Russia, due to the Chernobyl accident, would be enormous, both now and in the 21st century.

15.1 Liquidators

Most of the 106/134 power station workers and firemen who recovered from acute radiation syndrome (ARS) continued to have emotional and sleep disturbances[5,6] and all survivors experienced, as a post-ARS symptom,

Table 15.1. Major industrial disasters in the 20th century before the Chernobyl accident[1–3].

Year	Accident	Site	No. of deaths
1921	Explosion in chemical plant	Oppau, Germany	561
1942	Coal dust explosion	Honkeiko colliery, China	1572
1947	Fertilizer ship explosion	Texas City, USA	562
1956	Dynamite truck explosion	Cali, Colombia	1100
1957	Nuclear reactor fire	Windscale, UK	0
1959	River dam collapse	Fréjus, France	421
1963	Surge of 10^8 m^3 water from a reservoir	Vaiont, Italy	2600
1975	Mine explosion	Chasnala, India	431
1976	Chemical leak	Soveso, Italy	0
1979	Biological/chemical warfare plant accident	Novosibirsk, USSR	300
1979	Nuclear reactor accident	Three Mile Island, USA	0
1984	Natural gas explosion	Mexico City, Mexico	452
1984	Poison gas leak	Bhopal, India	6954

continual lethargy and chronic tiredness. In a study of the quality of life of liquidators from Kiev[7], it was found that 40% had psychological difficulties, table 15.2. Also, of the 360 000 Ukranian liquidators, who ten years after the accident still required medical treatment or constant supervision, a total of 35 000 are invalids[8].

Table 15.2. Quality of life of liquidators who live in Kiev[7].

Quality of life feature	Percentage with a given feature
Unfavourable housing conditions	70
Unfavourable working conditions	60
Require more long-term rest	40
Require hospital treatment	40
Psychological and emotional difficulties	40
Required an improved diet	38
Required physiotherapy	27
Dietary alterations	20
Required psychotherapy	10
Alcohol abuse	2

It has also been reported that the medical and social rehabilitation problems of the Russian liquidators are getting worse[9] and that the prevalence of mental disorders in this population group is a factor of ten higher than the prevalence of mental disorders in the Russian population as a whole[10].

Data on suicide rates are available for a cohort of 2093 liquidators from Kyrgystan. For the period 1989–94 there were 69 deaths of which 16 were suicides and ten were from alcohol poisoning[11]. The number of liquidators who have died due to accidents is also greater than expected. For example, a cohort of 5446 from Lithuania followed up for the period 1987–95 included 251 deaths, the major causes of which were injuries and accidents: although the overall death rate was not higher than that of the total population[12a]. This pattern is similar for a cohort of 4833 Estonian liquidators* of whom 144 died during the period 1986–93. There were a high number of deaths due to accidents, violence and poisoning and almost 20% of the 144 deaths were due to suicide[13,14].

However, these death rates and the major causes cannot be totally attributed to the Chernobyl accident and ignoring the underlying socioeconomic framework. For example, Kyrgystan includes the Techa river basin in Chelyabinsk province, see section 4.9, where it is known that the quality of life of the population has been poor for many years[11].

15.2 Residents in contaminated territories

The populations considered in this section are those living in Belarus, the Ukraine and Russia, where it has been reported[15] that up to 90% of persons living in contaminated territories thought they had, or might have, an illness due to radiation exposure, compared with up to 75% of the populations living in *clean* territories[15]. The complaints reported in this particular study were widespread. About 80% complained of fatigue and over 40% complained of loss of appetite, regardless of whether their area

* By 6 May 1986 military reservists, predominantly from the Baltic countries, had been brought into the zone. Approximately 4000 Estonians were *conscripted* often at night and with apparently little regard for their domestic situation: some were ill and others had wives about to give birth. Their initial period of work in the zone was set at 30 days with two days vacation during the month. This was extended first to two and then to six months. It was reported in the *Estonian Komsomol* newspaper[12b] that the Estonians had reacted violently and downed tools and gone on strike for an undetermined period. Twelve of these Estonian liquidators were apparently shot for mutiny. A second instance of shooting, which I was advised not to try and corroborate because *official sources* now deny this occurred, happened at a check point to the exclusion zone. A heavily laden man, woman and child described as gypsies (the goods that were being carried had been stolen from evacuated homes) were approaching the point and refused to stop when ordered to do so by the border guard. The man was shot and immediately fell dead on the road, and the guard then told the woman and child that they 'would die a much longer and more horrible death because they had entered the radiation zone'.

was contaminated or clean. Also, 80% in the contaminated areas wanted to relocate elsewhere whereas the corresponding figure for the clean areas was 20%[15]. These figures, and the results of similar studies, are indicators of serious psychological problems.

Psychological illness has been documented in several studies and UNESCO, for example, has summarized[16] the major psychological and social problems, table 15.3. In addition, when asked in a recent survey, teenagers affected by Chernobyl estimated their life expectancies to be 12 years shorter, on average, than teenagers who were unaffected.

Table 15.3. Psychological and social problems reported in various surveys[16].

- Low self-esteem: 50% of the population in one survey.
- A tendency to link all illnesses to Chernobyl.
- High personal anxiety.
- Feelings of being a victim.
- Feelings that there is no future.
- Feeling unable to influence the present or the future.
- Reduced intellectual achievement.
- Social tensions over eligibility for Chernobyl benefits.
- Conflict between healthy living and the need to save or earn money by accepting contamination.
- Mistrust of government experts and information.
- In some cases, individuals or whole communities, hide their fear by cultivating a false optimism, denying that Chernobyl has any negative consequences.
- Escapism through alcohol and drugs¶

¶ Alcoholism was rife for many years in the USSR and was correlated with the incidence of stomach cancer, which until recent years was higher than that for lung cancer, in spite of many of the population being heavy smokers in earlier years. Jokes about vodka, red wine and Chernobyl have been very popular but it is often not realized that there is some basis of truth in these stories. For instance one medical liquidator (see section 5.2.2) was shown by a power plant worker the Radiation Protection Guide for the Chernobyl NPP and it contained the advice 'In case of an accident take red wine'. She also noted that just after the accident all grocery stores around Kiev had stocked boxes and boxes of red wine, but that eventually the Government ordered this practice to be stopped, and also that the advice on red wine was eliminated from a later edition of the Radiation Protection Guide.

However, the results of the studies on psychological problems are not always reported in the same way because the study designs differ, as do the populations chosen for study, but the following results are a good cross-section, and essentially deliver the same *message*: that many, but not all, of the illnesses reported by people living in contaminated territories were psychosomatic.

One segment of the International Chernobyl Project[17,18] included physical examinations of 501 adults who complained of various symptoms. One of the results showed that there were no significant differences ($P > 0.05$) for several conditions, between the illness perception of the residents in contaminated settlements and those in clean settlements who acted as a control group for the study, table 15.4. It is noticeable, though, that where there is a significant difference the illnesses are stress related. The results on illnesses that were confirmed[17] are given in table 15.5 which shows a significant difference for disturbances of the abdomen, which is the organ that is probably most vulnerable to stress effects[18].

Table 15.4. Percentage of respondents who replied *Yes* to possible symptoms in the International Chernobyl Project study[17].

Illness	Contaminated settlements	Control settlements
Significant difference ($P < 0.05$)		
Fatigue	89	81
Appetite loss	53	42
Chest pains	53	43
Thyroid/goitre	25	11
Anaemia	8	5
No significant difference ($P > 0.05$)		
Headache	81	77
Depression	42	42
Sore throat	40	35
Hair loss	26	25
Diarrhoea or constipation	27	25
Weight gain	19	14
Weight loss	15	15
Menstrual irregularity	9	6
Nosebleeds	16	11

A Finnish study[19a] compared data over a period of 14 years for all the residents of a contaminated village and all the residents of a clean village in the Bryansk region of Russia. Mental stress was measured using a four-point scale[20,21].

- Somatic symptoms.
- Anxiety.

Table 15.5. Confirmed illnesses for the two settlement groups[17].

Illnesses	Survey result
Disorders of skin, joints, ears, eyes, nose, pulses, heart, lungs, kidneys, neurological disorders.	*No significant difference* ($P > 0.05$) *between the two groups*
Disturbances of the abdomen	*Significant difference* ($P < 0.05$) *between the two groups*

- Social dysfunction.
- Severe depression.

Perceived symptoms were recorded on a 17-item check list. The results were similar to those of the International Chernobyl Project[17] in that the scores differ between the two villages and reflect anxiety, depression and other indicators of stress, whereas the prevalence of specific symptoms of physical illness was the same.

Table 15.6 presents information about risk perception concerning the consequences of the Chernobyl accident and compares the perceived risk with other causes of premature death[19b]. Population groups from the city of Kiev and from the village of Bogdany, which is on contaminated territory in Kiev region, were asked to respond to a survey in which they had to rank 15 risk situations from the most dangerous to the least dangerous in terms of premature death. The reponses were pooled and the results presented with the most dangerous perceived risk being scored 100 relative to the other perceived risks. As seen from table 15.6, those in Kiev considered car accidents to be the most dangerous with Chernobyl in second place tied with motorcycle accidents and those caused by electrical faults. The rural population of Bogdany placed Chernobyl with the highest risk and indeed, Chernobyl was the only risk score for which the Bogdany perception was higher than that of Kiev. To put the Kiev data in perspective[19b], the main causes of premature deaths in 1993 in the Ukraine were fatal traumatic injury at home (4.2 per 10 000 population), traffic deaths (1.9 per 10 000), fatal traumatic injury at work (9.1 per 100 000) and premeditated murder (7.7 per 100 000).

In conclusion it is noted that the World Health Organization in their summary of their 1995 conference made three comments with regard to psychological illness[5].

- Psychosocial effects, believed to be unrelated to direct radiation exposure, resulted from lack of information immediately after the accident,

Table 15.6. Comparison of perceived risks of premature deaths for 15 different factors for population groups in the city Kiev and in the rural village of Bogdany[19b]

Factor	Perceived risk on a scale 0–100	
	Kiev	Bogdany
Chernobyl accident	40	100
Car	100	20
Motor cycle	40	11
Electricity	40	14
Fire	33	25
Firearms	33	16
Alcoholism	29	20
Smoking	29	10
Railroad	25	8
Other factors	22	13
Swimming	22	13
Surgery	20	11
X-ray diagnosis	15	10
Lightning	14	11
Natural radioactivity	14	7

the stress and trauma of compulsory relocation to less contaminated areas, the break in social ties among community members and the fear that radiation exposure could cause health damage in the future.

- The immediate psychological impact was similar to that of a natural disaster such as an earthquake, fire or flood. A survey after the accident showed that headaches, a feeling of pressure in the chest, indigestion, sleep disturbance, loss of concentration and alcohol abuse were common.
- The national health registries of Belarus, the Ukraine and Russia recorded significant increases in many diseases that are not related to radiation. These have included endocrine diseases, mental disorders and diseases of the nervous system, sensory organs, and digestive and gastrointestinal systems. Congenital abnormalities have also been observed. While present evidence does not suggest that these diseases are radiation induced, it is possible that such problems resulted from the considerable stress caused by the accident.

In addition, similar conclusions were also drawn from the IAEA 1996 conference[22], which commented that it was understandable that people, who were not told the truth for several years after the accident because of the secrecy of the USSR system, continue to be sceptical of official statements and to believe that illnesses of all kinds that now seem to be more prevalent must be due to radiation.

15.3 Atomic bomb survivors

In studies of the delayed effects of atomic bomb exposure, psychological effects are often suggested in the descriptions that have been made of the physical or mental conditions of the survivors. The terms *A-bomb disease* and *A-bomb neurosis* have frequently been used by physicians and by survivors. Whereas A-bomb disease is used to represent such non-specific complaints as easily fatigued, loss of weight during summer, cold-like symptoms, gastrointestinal symptoms, A-bomb neurosis includes fear of leukaemia and of cancer[23]. Nevertheless, although health management programmes have been conducted for the somatic symptoms of the survivors, very little is still known about the mental effects[24] of A-bomb exposure.

Table 15.7. Distribution of observed to expected ratio for suicides as a function of dose[23].

Years	No. of suicides	Dose range (cGy)			
		0–9	10–39	40–179	180+
1950–54	83	1.03	0.48	1.35	0
1954–58	101	1.16	0.67	0.96	0
1958–62	58	1.13	0.79	0.38	0
1962–66	57	1.25	0.24	0.77	0.44

Table 15.8. The incidence of schizophrenia in children irradiated *in utero* at Hiroshima and Nagasaki: variation with time of pre-natal exposure[25].

Pre-natal exposure	Males		Females	
	No. exposed	Schizophrenics No. (%)	No. exposed	Schizophrenics No. (%)
0–13 weeks	336	6 (1.8)	326	2 (0.6)
14–27 weeks	326	5 (1.5)	336	6 (1.8)
28–40 weeks	284	1 (0.3)	318	1 (0.3)
0–40 weeks	946	12 (1.3)	980	9 (0.9)

Suicides among ATB survivors have often been reported in newspapers but epidemiological studies[23] have shown that any increase in suicide rates is not correlated with radiation exposure or subsequent anxiety, table 15.7. Survivors of the atomic bombs also suffered poor living conditions and lack of social support for a long time. These could have been contributing factors.

It has, however, been shown[25] that there is a trend towards a greater risk of schizophrenia when the pre-natal exposure was 0–27 weeks compared to exposure at 28–40 weeks, table 15.8.

Table 15.9. Clinical diagnosis by psychiatrist's interview using the ICD-10 system[24].

Major category of ICD-10	No. of diagnoses
Mood disorders (F3)	29
Neurotic, stress-related and somatoform disorders (F4)	24
Behavioural syndromes associated with physiological disturbances and physical factors (F5)	9
Mental and behavioural disorders due to psychoactive substance use (F1)	4
Other disorders	5
No diagnosis of psychological disorder	88

In a detailed extensive study of Nagasaki ATB survivors[24], the proportion of high scores was greatest in the age group <50 years (12.5%) and lowest among in the age group >60 years (7.8%). The mean score was also greater among those proximally exposed, defined as 2 km or less from the epicentre, than among the distally exposed. Results of clinical diagnosis by psychiatrist's interview on 153 survivors are given in table 15.9. Six survivors were given complex diagnoses which is why the total in table 15.8 is 159 and not 153.

Table 15.10. Estimated prevalence of ATB survivors with mental disorders[24].

Score	No. of survivors	Prevalence rate (%)	Estimated No.
Low	3776	8.2	310
Middle	480	42.9	206
High	409	62.7	256
All scores	4665	16.7	779

It was also estimated in this study, table 15.10, that about 17% of ATB survivors had some mental disorder diagnosed by a psychiatrist. The scores refer to those used in a WHO collaborative study protocol of mental illness in general health care[26].

Figure 15.1. In February 2000 the CEO of British Nuclear Fuels resigned following the discovery that quality control records at the BNFL Sellafield (formerly called Windscale) reprocessing plant had been falsified. They related to the safety of a shipment of uranium and plutonium mixed oxide fuel (Mox) bound for Japan. Then a month later, reported in the *Sunday Telegraph* of 26 March, a saboteur, who was one of the workforce of 10 000 at Sellafield, hacked through wires on robot arms, forcing the shutdown of a vitrification unit where highly active nuclear waste is treated and stored. In this unit, using the remote control robots, the liquid waste is put in glass canisters which are embedded in concrete for safety and storage.

Chapter 16

Other Non-Malignant Diseases and Conditions

Introduction

For the Chernobyl accident health effects in terms of non-malignant diseases and conditions, the most detailed studies have been on psychological illness, brain damage *in utero*, reproductive health, haematological diseases and thyroid diseases. These are the four classes of disease which are considered in sections 16.2–16.5.

Other diseases and conditions have, of course, been reported, such as those of the cardiovascular system and of the immune system, but no correlations have been established with radiation exposure from the Chernobyl accident. There are so many confounding factors including stress due to the accident, socioeconomic conditions and an inadequate diet that it is extremely unlikely even with long-term detailed follow-up that it will be possible to prove any significant correlations. An additional major problem is that a unified infrastructure for such a study, such as those undertaken by the RERF in Japan, does not exist, with the three republics of the Ukraine, Belarus and Russia all working independently on their own national programmes, with relatively little funding available.

16.1 Atomic bomb survivors

There have been studies of non-cancer diseases in ATB survivors[1] and these have been more extensive than for Chernobyl because of the long-term follow-up available and the centralized Life Span Study (LSS) by the RERF which commenced in 1958 with 120 000 subjects and 1950 as the year of the base population[1,2] and the Adult Health Study (AHS) with 20 000 subjects drawn from the LSS. A total of 19 diseases were studied including coronary heart disease, stroke, uterine myoma and chronic liver disease (chronic hepatitis and chronic cirrhosis). Figure 16.1 shows[1,3] the estimated relative risk at 1 Gy of the incidence of non-cancer diseases in ATB survivors.

16.1.1 Cardiovascular diseases

The major study was of myocardial infarction, table 16.1, although data were also analysed using the various endpoints of atherosclerosis: incidence of cerebral infarction, prevalence of aortic arch calcification, prevalence of isolated systolic hypertension and pulse wave velocity. All these endpoints showed a positive dose response, which supports the possibility of a real, though weak, association between radiation exposure and atherosclerosis[1].

Table 16.1. Study of myocardial infarction in ATB survivors[1,3].

- 163 males and 125 females identified as new cases of myocardial infarction 1958–90.
- Incidence rates compared by dose groups after adjusting for risk factors of blood pressure and total serum cholesterol levels, as well as age, sex and city.
- Significant increase in myocardial infarction incidence in the heavily exposed survivors.
- Estimated relative risk was 1.17, with a 95% confidence interval (CI) of 1.01–1.36 and $P = 0.02$.
- Excess most significant among those who were less than 40 years old when exposed.

16.1.2 Uterine myoma

Uterine myoma is a benign tumour of the uterus and follow-up studies for 1958–86 showed[1,3,4] a remarkable dose response between incidence and ATB radiation exposure. The estimated relative risk at 1 Gy, see figure 16.1, was 1.46 with a 95% CI of 1.27–1.70 and $P < 0.001$.

16.1.3 Chronic liver diseases

A significant dose response was found between chronic liver diseases and atomic bomb radiation[1,3,4]. The estimated relative risk at 1 Gy, see figure 16.1, was 1.14 with a 95% CI of 1.04–1.27 and $P < 0.006$.

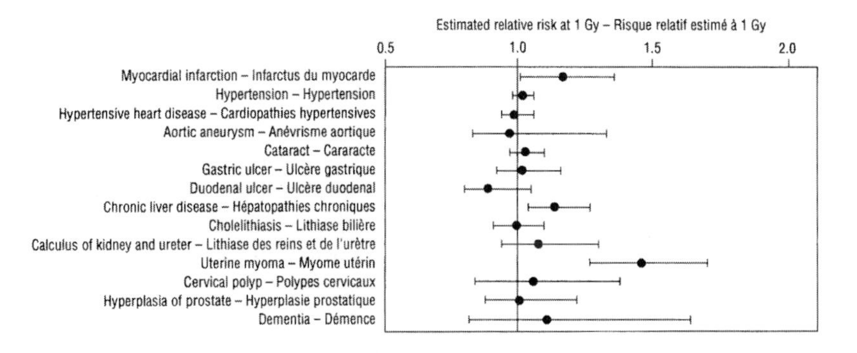

Figure 16.1. Relative risk for the incidence of non-cancer diseases[1]. (Courtesy: WHO.)

16.2 Brain damage *in utero* and mental retardation

It is known from studies of children treated by radiation therapy for intracranial tumours[5] or for acute lymphoblastic leukaemia[6,7] that some are adversely affected in that their IQ is at an educationally subnormal level. This has, for example for leukaemic children, been demonstrated at three years after treatment[2] and also after an even longer period of nine years[7]. These are not *in utero* irradiations, but both for radiotherapy treatments and *in utero* exposures any subsequent mental retardation is assessed using an IQ scale.

The IQ distribution for the general population is Gaussian with a mean of 100 and a standard deviation of about 15 IQ points and the region to the left of two standard deviations from the average, that is, values less than 70 IQ points, corresponds to the clinical designation of severe mental retardation. The mechanism of mental retardation reduction is thought to be the production of a dose-dependent lack of functional connections of neurons in the cortex of the brain[8].

The value of this downwards shift in IQ is estimated to be about 30 IQ points per Sv, and the radiation-induced shift in IQ for a dose of 1 Sv would result in severe mental retardation in about 40% of the exposed individuals[8]. These figures are based on an *in utero* study of a cohort of 2800 prenatally exposed ATB survivors[9].

This showed that the incidence of survivors with both small head circumference and severe mental retardation increased with increasing uterine absorbed dose: with a strong influence when the foetal brain was irradiated at 8–25 weeks of gestation. Approximately 80% of the mental retardation was caused by radiation exposure at 8–15 weeks of gestation: the most radiosensitive period of foetal brain development. The threshold appears to be in the range[9] 0.12–0.23 Gy. For the leukaemic children[7] the radiation therapy dose to whole brain was either 18 Gy or 24 Gy, and for the intracranial tumours[5] was 45 Gy for children under the age of three years and 50–55 Gy for those older.

One of the WHO IPHECA pilot projects[10] was devoted to a study of brain damage *in utero* and set out to examine all children born within a year of the accident to women evacuated from the 30 km zone and to women living in SCZs where the territory was contaminated. Children born in uncontaminated areas were matched for age, socioeconomic background, residential environment and educational level, and served as controls for the project. Details of this study of 4210 children and its findings are given in table 16.2.

In a much smaller study[11] of 100 Ukranian children exposed to what was termed acute prenatal irradiation only nine were clinically assessed as free of neuropsychiatric problems and of the 93 with diseases of the nervous system, two were assessed as mentally retarded. In the control group 45% suffered from diseases of the nervous system. However, the study size is small and the authors comment that it would be unwise to attribute all the changes to the Chernobyl accident and that there are confounding factors which can contribute to the findings.

The mental development of children exposed to ionizing radiation *in utero* or in infancy has also been investigated in a separate study in Belarus[12]. Thyroid doses due to ^{131}I were estimated for two cohorts: 130 who were exposed and a control group of 176 age-matched children in clean areas. As with other studies, IQs were measured using an adapted Wechsler Intelligence Scale[13]. It was concluded that in the exposed children a considerable reduction in intelligence was found. Nevertheless, it was pointed out that thyroid hormone deficiency could, in turn, lead to a delay in CNS development and to a reduction of intelligence.

16.3 Reproductive health patterns

TASS reported[14] that in Gomel there were increases in the incidence in newborn children of meningocele, Down's syndrome and hydrocephalus; and it distributed photographs of children born with thalidomide-like limb abnormalities, figure 16.2. Nevertheless, TASS accompanied this information with the statement that there is no direct proof that the increases in these diseases amongst newly born Belarussian children are caused by radiation.

This is supported by the results of a study[15] in four contaminated regions of Belarus—Gomel, Mogilev, Brest and Vitebsk—where pregnancy outcome data were analysed for the period 1982–90, table 16.3. The evidence is neither strong nor consistent for a Chernobyl effect on maternal

Table 16.2. The WHO IPHECA brain damage *in utero* project[10].

Population studied

- Belarus, 906 in contaminated areas and a control group of 962 in clean areas.
- Russia, 725 in contaminated areas and 300 in clean areas.
- Ukraine, 558 in contaminated areas, including 115 who with their mothers were evacuated from Pripyat and Chernobyl, and 759 in clean areas.

Preliminary results

- The incidence of mental retardation in the exposed children was higher than in the control group.
- There was an upward trend in behavioural disorders and emotional problems in exposed children.
- The incidence of borderline and nervous and psychological disorders in the parents of the exposed group was higher than in the controls.

Conclusions

- On the basis of the investigations conducted so far, it is impossible to reach any definitive conclusions about the relationship between a rise in the number of mentally retarded children and the ionizing radiation due to the Chernobyl accident.
- The results obtained are difficult to interpret and require verification. The stress and concern of parents, for instance, could have influenced the results.
- While the infrastructure for research has been established, it is necessary to continue well planned epidemiological investigations and dosimetric follow-up.

and child health but certain trends that are plausibly related to radiation stand out. For example, the rise in maternal anaemia in Gomel and Mogilev, and the congenital malformations in Gomel. These findings require validation through other studies. The post-natal death rate fell significantly for the four regions by 44%, 37%, 31% and 34% and this is probably, in part, an effect of improved health care in these affected regions. It must, though, be noted that environmental risks to reproductive health are not only radiation, but also industrial waste and agrochemical pollution and these have to be taken into account.

The International Chernobyl Project also reported[16] on infant and perinatal mortality levels and found that these were relatively high in the three republics as a whole as well as in contaminated territories. These levels were similar before the accident, but now appear to be decreasing. No statistically significant evidence was found of an increase in the incidence of foetal anomalies as a result of radiation exposure.

Table 16.3. Reproductive health patterns 1982–90 in contaminated regions in Belarus[15].

Statistic	Gomel	Mogilev	Brest	Vitebsk
Foetal death rate (i.e. stillbirths per 1000 live births plus stillbirths)	Slightly increased (6% change)	Declined by 11%	Declined by 14%	Stable
Low birth weight frequency	Stable	Rising noticeably	Stable	Stable
Infant mortality	Declined in all regions from earlier to later period			
Maternal morbidity	Anaemia rates stable pre-1986		Anaemia rates rose slowly pro-1986	
	Anaemia rates rose markedly in all regions in the late 1980s, but this rise occurred earlier and more dramatically for Gomel and Mogilev. Kidney pathology and eclampsia present similar patterns but in Gomel and Mogilev are less pronounced than with anaemia			
Neo-natal morbidity	Intra-uterine hypoxia became somewhat more frequent in the early 1990s			
	Rate of perinatal infections stable.			Steep rise in the early 1990s (16% increase)
Congenital anomalies	Markedly more frequent after 1986 (nearly doubled)	Moderately more frequent after 1986		Stable 1982–90

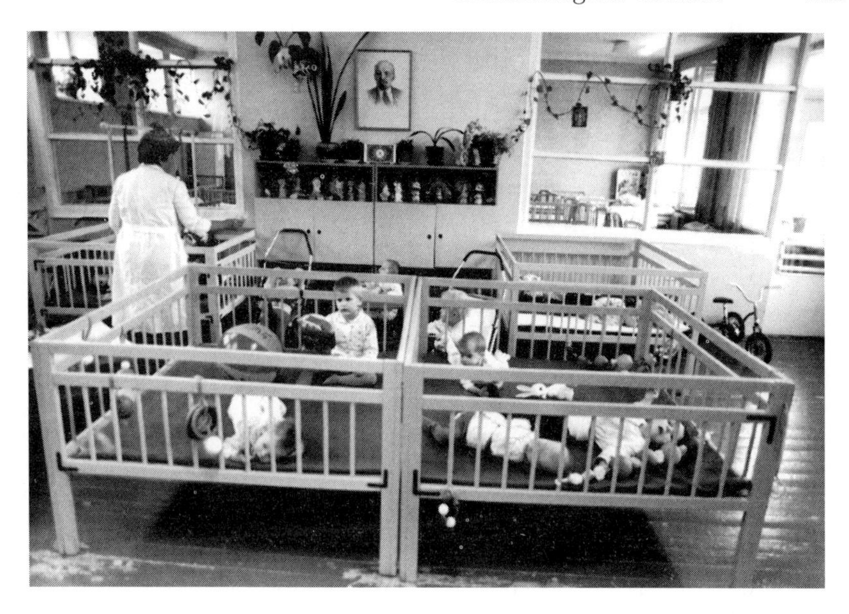

Figure 16.2. A group of Gomel infants in a hospital ward allocated to 'invalid children'. (Courtesy: TASS.)

In conclusion, one unexpected statistic reported[17] by the Director of the Department of Obstetrics of the Ukranian Institute of Mother and Child in Kiev, Anatoly Zakrevsky, was that for the more than 2000 babies born in the first year after the accident to mothers evacuated from contaminated territories, 6% were twins compared with the expected 0.5%. However, as with other reviews and studies, it was reported that these babies born in the year to April 1987 had no pathologies that might be the result of the accident.

16.4 Haematological diseases

Non-malignant diseases of the blood and blood-forming organs has been the group terminology most often used in Chernobyl studies, seldom with a subdivision into specific diseases and, for example[18], table 16.4 gives prevalence rates for Russian liquidators and for the Russian population as a whole with a morbidity ratio as an index top show the difference. For haematological diseases, excluding leukaemia because it is a malignant disease, it is seen that the morbidity in liquidators is 3.6 times as likely as in the general Russian population.

Table 16.4. Non-malignant disease prevalence rates per 100 000 persons for 152 325 Russian liquidators and for the Russian population as a whole[18].

Disease class	Prevalence rate		Morbidity ratio
	All Russia	Liquidators	
Blood and blood-forming organs	94	339	3.6
Circulatory system	1472	6306	4.3
Digestive system	2635	9739	3.7
Endocrine system	327	6036	18.4
Mental disorders	599	5743	9.6

The WHO IPHECA project on haematology[10] was established to detect and treat leukaemia and related blood disorders in a population of some 270 000 living on the contaminated territories of the SCZs. In the course of this project a number of blood disorders were identified in areas of ^{137}Cs contamination of 555 kBq/m^2. These included relatively few cases of histiocytosis X, agranulocytosis, aplastic anaemia and myelodysplastic syndrome, table 16.5. No region of Belarus has shown an increase in the incidence of anaemia between the two time periods, before and after the accident.

In the example of myelodysplastic syndrome, identification of cases was not possible before 1987–88 because, prior to that time, there were no laboratories capable of diagnosing this syndrome in Belarus. This is reflected in table 16.5. Data for adolescents and adults, defined as aged >15 years, are given in table 16.6. Observations in Ukraine and in Russia are similar in that there is no correlation with radiation exposure from the Chernobyl accident.

An important outcome from the IPHECA Project is the significant improvement in the diagnostic facilities in the health services of Belarus, the Ukraine and Russia, and in the qualifications and training of medical personnel. This may have led to more effective early detection not only of the non-malignant diseases in tables 16.5 and 16.6 but also of leukaemia. Possible reasons for an improvement in diagnosis are given in table 16.7 which subdivides possible causes of disease into methodological and actual[19].

Screening results are listed in table 16.7 and in the period 1991–96 the Sasakawa Memorial Health Foundation of Japan funded the largest screening programme of children in five medical centres in Belarus, the Ukraine and Russia[20]. Haematological examinations were carried out for 118 773 children and the major findings are given in table 16.8.

Table 16.5. Incidence of aplastic anaemia and myelodysplastic syndrome reported in the IPHECA haematology study[10]. Data are for children aged 0–14 years in all Belarus, Minsk City and selected oblasts.

Population	1979–85		1986–92	
	No.	Incidence	No.	Incidence
Aplastic anaemia				
All Belarus	65	0.41	72	0.44
Minsk	11	0.49	18	0.68
Brest	12	0.51	8	0.32
Vitebsk	6	0.29	4	0.19
Gomel	10	0.37	15	0.55
Mogilev	8	0.41	8	0.40
Myelodysplastic syndrome				
All Belarus	1	0.01	29	0.18
Minsk	1	0.00	7	0.27
Brest	0	0.00	4	0.16
Vitebsk	0	0.00	4	0.19
Gomel	0	0.00	5	0.18
Mogilev	0	0.00	4	0.20
Histiocytosis X				
All Belarus	9	0.06	26	0.16
Agranulocytosis				
All Belarus	4	0.03	6	0.04

16.5 Thyroid diseases

It has already been mentioned in table 16.7 that improved screening can result in an apparent increase in disease which, in reality, is due only to methodology. This effect is probably an explanation for part of the increase in non-malignant thyroid disorders which were reported after the accident, some of which are seen in table 16.4 where it is seen that the morbidity rate of endocrine system diseases is almost 20 times greater for the cohort of Russian liquidators than for the Russian population as a whole[18].

It should also be noted that even before 1986, endemic goitre and dietary iodine deficiency were to be found in most of the areas which were later contaminated by the accident: although there is no evidence that iodine deficiency alone could have caused the increase in childhood thyroid cancer[10].

Table 16.6. Incidence of aplastic anaemia and myelodysplastic syndrome reported in the IPHECA haematology study[10]. Data are for adolescents and adults in all Belarus.

Population	1979–85		1986–92	
	No.	Incidence	No.	Incidence
Aplastic anaemia				
Males	63	0.38	84	0.49
Females	123	0.61	136	0.67
Myelodysplastic syndrome				
Males	2	0.01	19	0.11
Females	2	0.01	18	0.09
Histiocytosis X				
Males + females	2	0.01	2	0.01
Agranulocytosis				
Males + females	70	0.19	73	0.19

Table 16.7. Possible causes of increases in the incidence of diseases[19].

Methodological

- Improved screening of the population, enabling earlier diagnoses of diseases.
- Ignoring possible demographic changes.

Actual

- Change in the tenor of life and habitual diet.
- Psychological stresses and anxiety resulting in physical symptoms and affecting health.
- Radiation exposure effects.

The International Chernobyl Project found no abnormalities in either the thyroid stimulating hormone (TSH) or thyroid hormone (free T_4) in the children studied. The mean thyroid sizes and the size distributions were the same for contaminated and clean settlements. Thyroid nodules were extremely rare in children although they occurred in up to 15% of adults in both the contaminated and clean areas[16].

The Sasakawa study[20] examined some 160 000 children during the five years 1991–96, and all of this population were born between 26 April 1976 and 26 April 1986. The results are summarized in table 16.9 where it is

Table 16.8. Sasakawa haematological screening programme results[20].

- Prevalence of anaemia was higher in girls than boys, range 0.2–0.5%.
- It is suggested that one-third of the anaemia cases were due to iron deficiency.
- Prevalence of leukopenia was lower in girls than boys, range 0.2–1.1%.
- No sex difference for leukocytosis, range 2.8–4.9%.
- No sex difference for thrombocytopenia, range 1.0–1.3%.
- No sex difference for eosinoiphilia, range 12.2–18.9%.
- Incidence of haematological disorders was independent of ^{137}Cs contamination of residency.

seen that 45 905 thyroid abnormalities were diagnosed in 119 178 children. The five regions studied were Mogilev and Gomel in Belarus, Bryansk in Russia, Kiev and Zhitomir in the Ukraine.

Table 16.9. Sasakawa thyroid screening programme results[20].

No. of children screened	No. of children with a given diagnosis					
	Goitre	Abnormal echogenity on ultrasound study	Cystic lesion	Nodular lesion	Cancer	Anomaly
119 178	41 930	2597	502	577	62	237

Hypothyroidism can occur in extreme cases of iodine deficiency and can also be caused by high levels of radiation exposure. Its prevalence is defined by a low T_4 and a high TSH, and prevalence was particularly high in Gomel at 34.6 per 10 000 population whereas in other areas the prevalence was in the range 4.7–16.8 per 10 000 population[21]. The prevalence of hyperthyroidism was much more consistent, in the range 4.5–9.4 per 10 000 population[21].

16.6 Ocular disease

Radiation retinopathy as a complication of radiation therapy was described as early as 1930 and can be produced by external beam radiotherapy at doses as low as 15 Gy but is more common at fractionated doses of 30–35 Gy. Above this range, at 36–72 Gy, radiation optic neuropathy can also occur[22]. However, cataracts are the most frequent delayed complication in the eye following radiotherapy and in a series of 138 patients treated by radiotherapy for cancer of the orbit, using 220 kV x-ray deep therapy with direct shielding of the affected eye[23], the incidence of bilateral cataracts was 6/138 and of unilateral cataracts 10/138.

Shielding of the eye in the treatment of head and neck cancer such as tumours of the antrum and particularly of those of the nasopharynx was very difficult in the deep x-ray therapy era when wide fields were used[24]. However, with the advent of the megavoltage radiation era, which commenced with ^{60}Co teletherapy machines, far more effective shielding techniques were developed[25a] and radiotherapy-induced cataracts virtually disappeared, except after total body irradiation[25b].

ATB cataracts form the second largest body of data on radiation-induced cataracts after that of radiotherapy but both types are clinically similar. ATB cataracts have been divided into four degrees of severity[26], from *minute* to *severe*, but visual disturbances only occur in the severe degree. It is also noted that the RERF makes the point that care must be exercised with diagnoses since similar diagnostic observations are also evident with age-related cataracts.

The frequency of ATB cataracts and the severity of the opacity are radiation dose-dependent. Symptoms developed from several months to several years after exposure, with severe cases occurring soon but mild cases not appearing until after a latency period[26]. The frequency increases with proximity to the hypocentre, with the maximum exposure distance at which ATB cataracts are formed believed to be 1.6–1.8 km. The RERF estimated threshold radiation dose value is 0.6–1.5 Gy.

The threshold dose estimate is stated slightly differently by UNSCEAR[8]: the figures for acute exposures to the lens opacities sufficient to result, after some delay, in vision impairment are: 2–10 Gy for sparsely ionizing radiation and 1–2 Gy for densely ionizing radiation. UNSCEAR also state that the threshold dose rate is not well known for long-term chronic exposure, but is likely to exceed 0.15 Gy per year for sparsely ionizing radiation. The ICRP recommended[27] annual dose limits to the lens for occupational workers and for the public are, respectively, 150 and 15 mSv, table 1.1.

Data on ocular disease in Chernobyl liquidators are rarely to be found in the literature. It has been reported[28] that opthalmological examinations are given once or twice per year for the cohort of liquidators whose medical surveillance is undertaken in Moscow: for the eight cases of radiation cataract that have been found, all had had a dose greater than 5 Gy. The total cohort number is unknown and therefore no frequency can be estimated.

The only wide-ranging study on eye pathologies in liquidators[29] has been made by the Institute of Occupational and Environment Health of the

Medical Academy of Latvia, Riga*. A total of 6475 Latvians participated as liquidators during 1986–91. Of those employed between 1986 and 1987, 1320 worked close to the reactor and turbine hall, 1130 at installations next to the reactor site and 2213 in the general environment[30]; 87% were aged 20–39 years and 56% worked during 1986. Their documented doses are not high, estimated to be in the range 0.01–0.05 Gy, and they were exposed to radiation for a relatively short time, 1–4 months. Subsequently they have been living in areas not contaminated by radioactive fallout. This Latvian cohort therefore differs from the liquidator cohorts of the Ukraine, Belarus and Russia who still live in contaminated territories[30].

By 1996 most of the liquidators had registered with at least one disease, with the most frequent new diagnosis being a digestive system disease which for the years 1992, 1994 and 1996 represented annual totals of 17.2%, 18.6% and 21.8%. Data for nervous system and sensory organ diseases (these include ocular diseases) were, respectively, 15.3%, 16.0% and 18.4%. Mental disorders were also high for this period, 1992–96: 13.2%, 13.8% and 11.7%. The annual number of disease registrations for the three years were 6003, 4342 and 6460[30].

A total of 668 eye pathologies were classified using ICD-9 numbers 361.3 to 379.20, and represented data for 571 liquidators out of a population of 4841 who had been exposed to ionizing radiation. The most common pathology was retinal angiopathy (41.5%) followed by myopia (21.4%) and cataract (8.4% of 668). Of the total of 56 cataracts diagnosed, 38 occurred in liquidators working in 1986, 10 in 1987 and eight in 1988. The overall

* This Institute has also made an interesting study of the biological monitoring of metals as indicators of pollution[31], including the study of blood, hair and urine samples of a cohort of Chernobyl liquidators. The levels of lead in the blood (μg/dl) of liquidators was of the same order as in the controls, who were a random selection of inhabitants of industrial towns, 6.75 ± 3.36 *versus* 5.15 ± 1.12, but significantly less than workers occupationally exposed to lead, 28.03 ± 3.14 μg/dl. Specialists are still arguing as to what is a dangerous level in blood, Centres for Disease Control define lead levels above 25 μg/dl as lead poisoning although some experts would like to see this level reduced to 10–15 μg/dl[32]. The concentration of lead in urine (mg/l) of the Latvian liquidators although not significant, gave a trend showing higher values for the liquidators, 0.07 ± 0.01 than for the occupational workers, 0.05, and the control group, 0.03 mg/l: WHO guidelines for exposed employees are 0.05 mg/l. However, it is not possible to associate elevated lead levels in Latvian liquidators with the Chernobyl accident because of the confounding fact that in recent years, metal poisoning ranks third among occupational diseases in Latvia[31]. Plutonium, ^{239}Pu and ^{240}Pu, has been studied in Lithuania as an environmental pollutant, but not for biological monitoring purposes. Environmental concentrations in south-western and western regions of Lithuania were found to be 500–8400 mBq/kg in comparison to 100–460 mBq/kg in other areas. This was caused by the Chernobyl radioactive plume crossing Lithuania[33]. In conclusion it is noted that currently there is much interest in the measurement in urine of a heavy metal other than lead, namely uranium, since this is relevant to studying the causes of Gulf War Syndrome and the possible role, particularly for the American soldiers who were injured by so-called *friendly fire*, which was played by depleted uranium which was a component of artillery shells[34–36].

incidence of cataract in the cohort was 1.2% (56/4841)[29].

All diagnoses were confirmed by opthalmologists and age-related factors could largely be ignored since only 9.5% of liquidators were aged 40 years or more when working at Chernobyl. However, in a comparison of 200 liquidators with a control group of 200, although there was a significant difference in the incidence of retinal angiopathy (28 *versus* 9) there was no significant difference in the incidence of myopia (13 *versus* 16) or cataract (9 *versus* 4)[29].

Figure 16.3. Stained glass window in the Central V I Lenin Museum in Revolution Square, Moscow, 1980. Lenin's image and name was everywhere during the Soviet Union era, for example within the Lenin Rooms in schools, given as the name of NPPs such as at Chernobyl, and even placement of his portrait in hospital rooms, see figure 16.2.

Chapter 17

Cancer Risk Specification

Introduction

This chapter defines the terminology which is used for the specification of risks of cancer of the thyroid, other solid cancers and leukaemia for populations irradiated as a result of the Chernobyl accident or atomic bomb explosions.

There are many risk factors for the various site-specific cancers, and these are usually subdivided into host factors and environmental factors. Examples of host factors are sex, age, genetic predisposition and pre-cancerous lesions. Environmental factors include tobacco, diet, industrial chemicals in the work place and two factors relevant to the Chernobyl accident: radiation and socioeconomic conditions.

The cancers for which ionizing radiation is a high risk factor are as follows[1]: lung, bone, ovary, thyroid and leukaemia. X-rays and radium gamma-rays were observed to be a hazard soon after their discovery at the end of the 19th century[2] but a realization of the possible deleterious effects were not fully accepted until the 1920s. This was too late for the patients treated in the early years of the 20th century for non-cancerous conditions such as haemangioma, ring worm, goitre and for the removal of facial hair as a beauty treatment. A significant number of these cases eventually, often some 20 years later, developed skin cancers. The specification of risk in these early days was not quantitative and was limited only to statements to the effect that a treatment for a non-malignant condition was not advisable.

However, qualitative statements of risk have now passed into history and the terminology in this chapter is now the standard for specifying risks of radiation-induced cancers from atomic bomb radiation or from the Chernobyl accident.

With irradiation received from the bombs at Hiroshima and Nagasaki and as a result of the Chernobyl explosion there are of course no benefits, only risks. However, in the medical diagnostic or therapeutic situation the physician must balance risk against benefit. The guiding principle is to keep the radiation exposure as low as reasonably achievable: the ALARA principle. With cancer treatment, for example, the benefits can be increased survival and improved quality of life and these far outweigh any risk, even though the radiation doses in radiation oncology treatments are several magnitudes greater than in x-ray diagnostic investigations.

17.1 Absolute risk

Absolute risk is the excess risk attributed to irradiation and is usually expressed as the numerical difference between irradiated and non-irradiated populations: for example, one excess case of cancer per one million people irradiated annually for each Gy. Absolute risk can be given using an annual basis or a lifetime, 70 year, basis.

It is the magnitude of risk in a group of people with a certain exposure, but it does not take into account the risk of disease in unexposed individuals. It cannot therefore help to discover whether the exposure is associated with an increased risk of the disease.

The NCRP report[3] *Induction of Thyroid Cancer by Ionizing Radiation* defines an absolute risk coefficient (R) in the formula

$$R = (C/n) \times (10^6/[D \times y])$$

as the number of cases attributable to radiation exposure per million person-rad-years at risk. C is the number of cases attributable to radiation exposure, n is the number of subjects at risk in the irradiated population, D is the average radiation dose (in rad) to the thyroid gland, and y is the average number of observed years at risk per subject.

17.2 Relative risk and excess relative risk

Relative risk (RR) is the ratio between the number of cancer cases in the irradiated population to the number of cases expected in the unexposed population. A relative risk of 1.1 indicates a 10% increase in cancer due to radiation, compared with the *normal* incidence in the baseline/reference group. Excess relative risk (ERR) is relative risk minus 1.0. Relative risk is more appropriate to use than absolute risk when considering selected population groups.

Relative risk, unlike absolute risk, gives an estimate of how strong an association exists between exposure to a factor and the development of a disease. Examples of relative risks are seen[4] in figure 16.1 for the incidence of non-cancer diseases in ATB survivors.

Table 17.1 shows data[5] for leukaemia among ATB survivors for the period 1950–85 to illustrate the use of absolute and relative risks.

Table 17.1. Estimated relative risks at 1 Gy and absolute risks stated as an excess risk per 10^4 person-years per Gy (PY/Gy) for leukaemia in ATB survivors[5] for the period 1950–85. The numbers in brackets are 90% confidence intervals.

RR at 1 Gy	Excess risk per 10^4 PY/Gy
4.92 (3.89–6.40)	2.29 (1.89–2.73)

17.3 Attributable fraction

The term attributable fraction (AF) is sometimes used in the presentation of risk estimates and when expressed as a percentage is defined as

$$AF = \frac{\text{Excess deaths}}{\text{Total deaths from the same cause}} \times 100$$

and an example is shown in table 17.2 for leukaemia and solid cancers for a cohort of 200 000 liquidators[6].

Table 17.2. Predicted background and excess deaths for a lifetime period, from solid cancers and leukaemia for 200 000 liquidators[6].

Cancer type	Mean dose (mSv)	Background no. of cancer deaths	Predicted excess no. of cancer deaths	AF (%)
Solid cancers	100	41 500	2000	5
Leukaemia	100	800	200	20

17.4 Mortality ratio

Relative and absolute risks are used in predictive modelling beyond the period of observation but it is only from the 1980s that much work has been undertaken with BEIR[7] and other models. Prior to such modelling for ATB survivors virtually all that was stated was essentially mortality ratios (MRs) for various subgroups of survivors where MR was defined as follows.

$$MR = \frac{\text{Observed deaths}}{\text{Expected deaths}}$$

Table 17.3 shows data for leukaemia from one of the earlier RERF reports[8]. The MR peaks for the period 1950–54 which is reflected in the graph[5] in figure 17.1 for the period 1945–80.

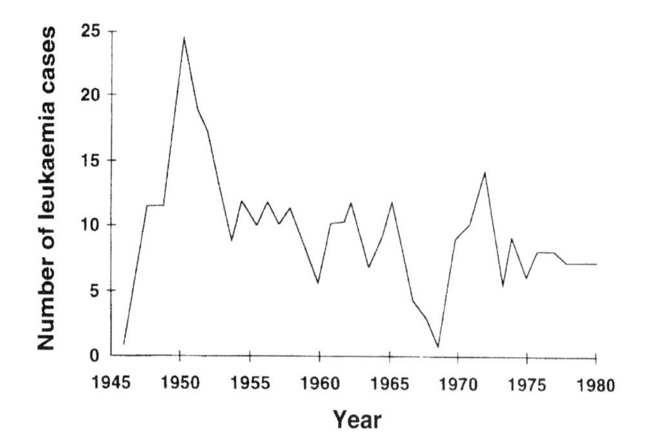

Figure 17.1. Leukaemia incidence[5] among atomic bomb survivors who were proximally exposed within 2000 m from ground zero and who received a dose in excess of 1 rad. (Courtesy: National Institute of Radiological Sciences, Tokyo.)

Table 17.3. Leukaemia mortality ratios for atomic bomb survivors[8].

Time period	MR
1950–54	5.9
1955–59	3.5
1960–64	1.7
1965–69	1.8

17.5 Confidence interval and confidence limits

The mean or the proportion observed in a sample is the best estimate of the *true* value in the population and the distribution of the values obtained in several samples would be approximately Gaussian for large samples. The Gaussian distribution is an alternative name for the *normal distribution*, and is symmetrical and bell shaped. The mathematical formula for the *general* normal curve is

$$y = \left\{1\Big/\left(\sigma\sqrt{2\pi}\right)\right\}\exp\{[-(x-\mu)^2]/2\sigma^2\}$$

where μ is the mean and σ is the standard deviation. The *standard* normal curve has $\mu = 0$ and $\sigma = 1$ and a total area beneath the curve equal to 1 between the limits $x = -\infty$ and $x = +\infty$. The probability of an observation which is normally distributed showing at least as large a deviation from the population mean as 1.645σ is 0.10 (0.05 in each *tail* of the distribution), and for 1.96σ is 0.05 (0.025 in each *tail*). It should also be noted that when the sampling experiment for the measurement of a mean is repeated n times, and the mean of the n means is m, the standard deviation of m is given a special name, standard error (SE) where SE = sample standard deviation/$\sqrt{n}$

A confidence interval (CI), which is between two confidence limits, is a range of values which we can be confident includes the true value. A CI extends either side of the mean by a multiple of the standard error (SE). For example, the CI between (mean − 1.645.SE) and (mean + 1.645.SE), where SE is the notation for standard error, is the 90% CI, whereas the CI between (mean − 1.96.SE) and (mean + 1.96.SE) is the 95% CI. The 90% CI will not include the true population value 10% of the time, and the 95% CI will not include the true population value 5% of the time.

An example of the use of CI is in section 16.1.3 where it is stated that the relative risk for ATB survivors at 1 Gy, of chronic liver diseases, is 1.14 with a 95% CI of 1.04–1.27.

17.6 Prediction modelling

There have been several studies using prediction models for various cancers[9], both for Hiroshima/Nagasaki and for Chernobyl, but it must be remembered that they do not *guarantee* accurate predictions, and confidence intervals on the results can be large. Nevertheless they are a useful method for estimating, within specified limits, what might be likely to occur if the time-dependent data on which they are based, continue in the short term along similar trends. It is essential to make this caveat because prediction models are based on extrapolation of data during only a limited period in the defined group of persons under study.

Risk estimates are derived from observations of cancer death rates in exposed and unexposed populations, and the excess cancer risk may relate to the spontaneous risk, that is, the *natural* incidence of cancer, in one of two basic ways, using either an absolute risk or relative risk. There are, however, variants with more complex methods of estimating the constants of the models, than those now described.

17.6.1 Additive model

In the absolute risk model the additional risk associated with radiation exposure is assumed to be independent of the spontaneous risk. This model is sometimes termed a *constant additive* prediction model[10], according to which there is a constant number of excess cancers in any given year per unit number of persons exposed per unit dose. That is, the number of excess cancers is fixed, regardless of the baseline risks:

$$R(\text{exposed}, t) = A + R(\text{unexposed}, t)$$

where A is the *absolute excess risk* for all times t which are greater than the latency time. The *baseline risk* $R(\text{unexposed}, t)$ is the number of cancers expected in the cohort had they not been exposed.

The value of A and the baseline risk may be estimated using a regression model fitted to the onset time of every cancer, with the risk estimates then given as a function of age, sex, time since exposure and possibly other risk factors. If this additive prediction model is correct then at any post-latency time the difference between observed and expected cancers, divided by the total PY/Gy observed, will be a constant. Indeed, an alternative to regression modelling for the estimation of the value of A is to make the following computation for the cohort:

$$\frac{(\text{No. of cancers observed}) - \left(\begin{array}{c}\text{Total no. of cancers expected}\\ \text{had the cohort not been exposed}\end{array}\right)}{\text{Total no. of PY/Gy of observation}}$$

17.6.2 Multiplicative model

In the relative risk model the additional risk is assumed to be proportional to the spontaneous risk; that is, the ratio of incidence or mortality rate in the exposed to that in the unexposed population is constant once the latency time has elapsed. Thus

$$R(\text{exposed}, t) = RR \times R(\text{unexposed}, t)$$

where the relative risk RR is constant for all times t greater than the latency period. The value of RR can be estimated[10] by dividing the number of observed cancers at some time t after the latency time by the number of expected cases. If the multiplicative model is descriptively correct, there should be an approximately constant relative risk at any post-latency time in an exposed cohort. A more complex model is a *variable multiplicative risk* model in which the value of RR is estimated from the data by regression methods, with respect to age at exposure, sex, time since exposure and possibly other risk factors.

17.6.3 BEIR V model

Instead of a linear dose–response pattern, the BEIR V committee used a model[7] with a linear-quadratic dose-response function to estimate the risk of radiation-induced cancer. In this model the relative risk function depends on sex, age at exposure to radiation and elapsed time since exposure.

$$R = F \times T \times K \times W$$

where F is a function of dose, T gives the dependence of R on time since exposure, K is the dependence of R on age at exposure and W is the effect of an additive interaction between radiation and other known risk factors[10].

The results of this model for leukaemia and bone cancers give the probability of causation rising rapidly with time after the minimum latency time, reaching a peak the height of which depends on dose and which is maintained for 10–20 years, and then falling to zero at the end of the risk period. For other cancers, the probability of causation is approximately constant at all ages after the minimum latency time has passed but is a function of age at exposure. It is typically higher for young ages at exposure, declines to a minimum for ages 40–50 at exposure and then may rise slightly or stay approximately constant[10].

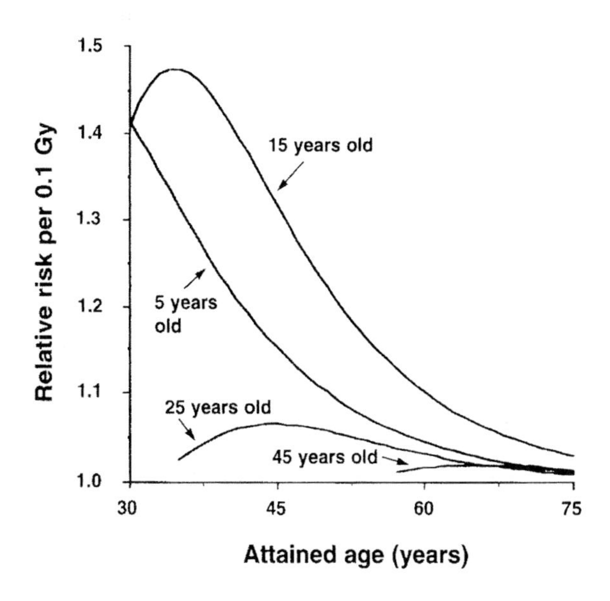

Figure 17.2. Replotted BEIR V data[7,9] of the relative risk of breast cancer per 0.1 Gy as a function of attained age for four different ages (5, 15, 25 and 45 years) at time of exposure. The time since exposure is the attained age minus the age at the time of exposure. (Courtesy: Institute of Physics Publishing.)

Figures 17.2 and 17.3 are examples of replotted BEIR V data[8,9] for breast cancer for relative risk, figure 17.2, and for absolute risk, figure 17.3. Most of the data which have been used to fit the model parameters are from single dose/high dose rate exposure such as for ATB survivors. Hence the BEIR V parameters may not be very accurate for highly fractionated and/or low dose rate exposure.

Figure 17.4, relating to Russian liquidators, is the result of RNMDR

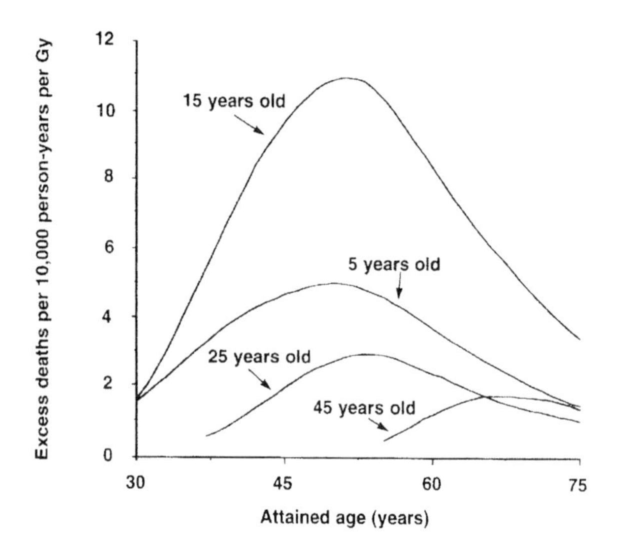

Figure 17.3. Replotted BEIR V data[7,9] of excess deaths from breast cancer as a function of attained age for four different ages at the time of exposure. This is an absolute risk per 10 000 women per year per Gy. (Courtesy: Institute of Physics Publishing.)

modelling[11], and shows the cumulative distribution of deaths from radiation-induced cancers which for all types of cancer reaches a maximum just below 500. No confidence limits are given with these predicted estimates but they are likely to be large.

17.7 How does radiation cause cancer?

Human tumours have a wide range of growth rates and a wide range of histological types. The link between cancers in humans and occupational exposure first came to public and professional attention with the discovery of scrotal cancer in chimney sweeps* and urinary bladder cancer in

* The first direct link between a specific occupational exposure and a specific cancer was in 1775 by Percival Pott in his book *Chirurgical Observations* when he pointed out the association between chimney sweeping and cancer of the scrotum, a disease very rare in the general population but very high among chimney sweeps. Scrotal cancer was again linked with an occupational environment in the early part of the 20th century, and given the name *mule spinner's cancer*. In this instance the carcinogenic material was shale oil used to lubricate cotton spinning machines (mules) some 120 feet long. During this work the mule spinners had cause to lean over the machine and consequently the groin was regularly soaked by the shale oil. Over a period of many years this caused cancer of the scrotum.

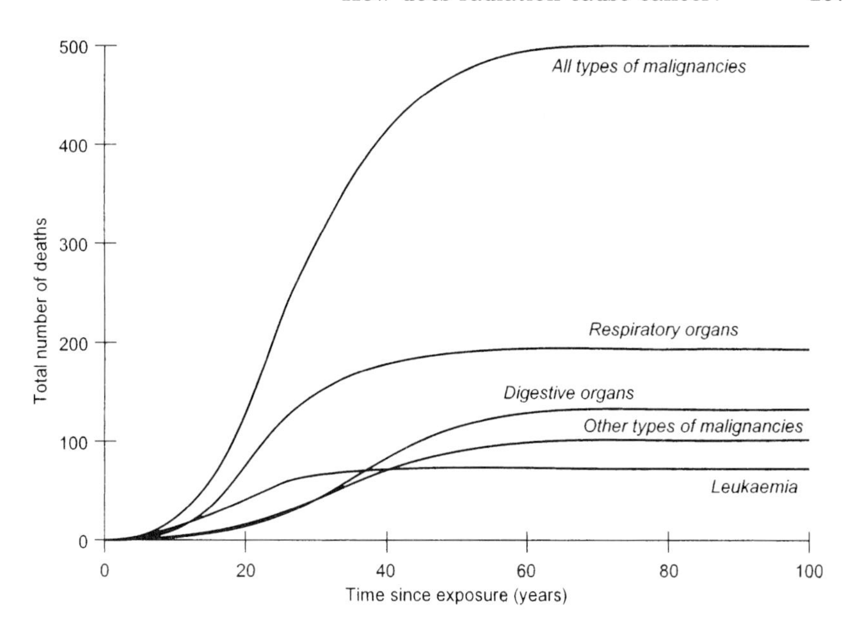

Figure 17.4. Cumulative distribution of the expected number of radiation induced cancer deaths in Russian liquidators as a function of time since exposure within the 30 km zone[11]. (Courtesy: WHO.)

chemical industry workers. Subsequently a number of industrial chemicals were found to cause cancer. In the 1940s the link between asbestos fibres and mesothelioma of the lung was proven, and later the association between smoking and lung cancer. However, according to current estimates no more than 5% of all cancers, but more likely a lesser percentage, can be traced directly to occupational or environmental exposure[12]. Observations show that anywhere in the world, lifestyle and lifestyle-related behaviour are more likely to cause or promote cancer development.

That ionizing radiation was a causative factor for cancer was not immediately realized following the discovery of x-rays in 1895 and of radium in 1898, and their widespread applications in medicine for x-ray diagnosis and therapy and for radium therapy. Adverse radiation reactions, such as radiation burns, were soon reported but it took a longer time to associate x-rays and radium with the production of skin cancer on the hands and forearms of physicians, engineers and technologists, following radiation dermatitis and chronic ulceration.

The problem in the field of ionizing radiation and cancer at the end of the 20th century is no longer whether radiation is a causative factor but to provide an answer to the question 'How does radiation cause cancer?'

The following summarizes the current state of knowledge as reviewed by Brenner[13] during the 41st Annual Scientific Meeting of the American Society for Therapeutic Radiology and Oncology (ASTRO) in San Antonio in November 1999.

While it is known that radiation does cause cancer, and parts of the mechanism are understood, the full picture is still unclear. Very broadly, we know the following.

- Radiation is quite efficient at inducing chromosomal aberrations such as deletions and translocations.
- Deletions can cause inactivation of tumour suppressor genes which, in turn, is often associated with the induction of solid tumours. Examples of tumour suppressor genes are p32 and Rb, and it is clear that radiation is capable of inactivating these genes.
- Chromosomal translocations can result in the activation of an oncogene which, in turn, is associated with haematopoetic cancers. To this point, however, the link has not yet been demonstrated between radiation oncogenesis and activation of a particular oncogene.

Radiation oncogenesis is a multi-stage process. Despite their frequently clonal orgin, as the cells of cancer grow and divide, progressive stages can be identified, which extend from pre-neoplasia to malignancy. These steps are usually operationally described as initiation, promotion and progression. The progressive nature of cancer has been known for many years. It was first described in phenomenological terms for skin cancer in animals, and more recent evidence for the multi-stage nature of cancer has come from studies of clinical progression of colorectal cancer. These studies have demonstrated an association between (1) the clinical progression of the cancer from a benign state, through non-malignant adenomas, to full-blown cancer; and (2) the activation of oncogenes, the loss of anti-oncogenes and other chromosomal changes.

In summary, radiation carcinogenesis, in common with any other form of tumour induction, is likely to be a complex multi-stage process. Whilst this statement is almost certainly true, it is as yet only a qualitative observation. Our current limited state of knowledge as yet precludes systematic quantitative modelling of all the various stages from early subcellular lesions to observed malignancy.

All that we currently know quantitatively about the risks of radiation-induced cancer in man are a result of the following studies.

- Extrapolation from animal data.
- Relevant human cohorts: ATB survivors (which is by far the major source of information); tuberculosis patients receiving multiple fluoroscopic examinations for pneumothorax; children epilated for treatment of tineas capitis; and radiotherapy patients treated for cancer using x-rays, radium or radionuclides such as ^{60}Co, ^{137}Cs and ^{192}Ir.

However, there are also three emerging population cohorts which might provide useful quantitative data in the future:

- Chernobyl
- Chelyabinsk
- Airline pilots

There are, though, many problems in estimating the risks of radiation-induced cancer in man, table 17.4.

Table 17.4. Problems in estimating radiation induced cancer risks in man[13].

- Dose reconstruction.
- Statistics: including obtaining statistical significance.
- Controls.
- Latency.
- Dose extrapolation.
- Dose-rate extrapolation.
- Age and time dependencies.
- Transfer models, e.g. from Japan ATB survivors to Chernobyl or USA.
- Neutrons at Hiroshima: at Chernobyl it was only α, β and γ radiation.

At low doses the Japanese ATB dose–response relationships are consistent with a *linear* relationship between dose and cancer risk. Although a matter of some controversy, such a linear model, implying that there is no threshold for risk, is most likely valid[13], table 17.5.

Table 17.5. Arguments for a linear model[13].

- Most tumours are probably of monoclonal origin.
- High doses of ionizing radiation can produce sufficient damage in a given cell to start the process of oncogenesis.
- As the dose is decreased, because of the random nature of energy deposition by ionizing radiation, the number of single cells in which this *sufficient damage* occurs will simply decrease linearly with decreasing dose, even at extremely low doses.
- Therefore a linear extrapolation for the risk of radiation carcinogenesis down to arbitrarily low doses is justified.

The third point in table 17.5 would not apply, for example, to chemical carcinogens, and so the likely existence of thresholds for some chemical carcinogens does not imply the same for ionizing radiation which has a unique stochastic mode of energy deposition.

Chapter 18

Cancer

Introduction

Thyroid cancer and leukaemia are not the only cancers that have been studied for the ATB survivors, Chernobyl liquidators and for populations living on contaminated territories in Belarus, Ukraine and Russia. However, the incidence of thyroid cancer is the most notable health effect of the Chernobyl accident, especially for children and adolescents in the Gomel region of Belarus. This is why the majority of this chapter is devoted to thyroid cancer. In terms of other cancers in Chernobyl populations, no significant increases in their incidence have been observed.

18.1 Incidence and mortality

A crude annual cancer incidence rate is a measure of the new cases of a cancer in a particular year. It is usually quoted as a proportion per 10^5 of a defined population at risk, but can also be quoted per 1000 or per million population at risk. The adjective *crude* refers to the fact that the rate is not modified to take into account such factors as age or reference year. For example, crude incidence and mortality rates per 10^5 for lung cancer in males in 1990, are respectively:

$$\frac{\text{Number of new cases of lung cancer in males in 1990}}{\text{Average number of males at risk in 1990}} \times 10^5$$

and

$$\frac{\text{Number of deaths from lung cancer in males in 1990}}{\text{Average number of males at risk in 1990}} \times 10^5$$

and must be stated for a defined population, for example the male population of England and Wales.

Age-specific rates refer to a population in a specified age range, usually a five-year or ten-year range. The definition of the age groups will depend on the nature of the cancer and its distribution in the population. For example, age-specific incidence and mortality rates per 10^5 for lung cancer in males aged 40–45 years in 1990 are respectively:

$$\frac{\text{Number of new cases of lung cancer in males aged 40–45 years in 1990}}{\text{Average number of males aged 40-45 years at risk in 1990}} \times 10^5$$

and

$$\frac{\text{Number of deaths from lung cancer in males aged 40–45 years in 1990}}{\text{Average number of males aged 40–45 years at risk in 1990}} \times 10^5$$

Table 18.1 gives crude annual incidence rates for selected cancers[1]. The number of excess cancers required to be able to demonstrate statistical significance will be, for example, much larger for lung cancer which has a high incidence than for thyroid cancer and leukaemia which have a low incidence compared to that for lung cancer.

Table 18.2 gives mean annual age-specific incidence rates for thyroid cancer for Belarus 1988–92 and for comparison, for selected populations in the former USSR (Estonia), Scandinavia (Denmark), Europe (Germany) and the USA[1]. These figures emphasize that thyroid cancer is a very rare cancer and also that the incidence is higher in females than in males.

Some of the most recent data on cancer mortality for countries in the European Economic Community[2a] reported that mortality rates per 10^5 males ranged from 0.9 in Bavaria to 0.3 in south-east England and that the rates per 10^5 females ranged from 1.3 in Baden-Wurtemberg to 0.5, again in south-east England. These can be compared with the crude incidence rates in table 16.2 of 0.8–3.0 for males and 3.1–7.7 for females: although the country populations are not the same.

The differences between mortality and incidence rates reflect the fact that there is a good prognosis for thyroid cancer following treatment. In the USA and Japan, thyroid cancer constitutes 0.2% and 0.4% of all fatal cancers in males, and 0.8% and 0.1% of all fatal cancers in females. The corresponding data for fatal cancers of the lung in the USA and Japan are 33.2% and 14.2% in males and 7.4% and 29.1% in females[2a]. For the year 1999 the American Cancer Society estimates an incidence of 16 000 new patients with thyroid cancer and approximately 1200 are expected to die of this disease, mainly of anaplastic cancer (500–700) or medullary cancer (300–400) with a true death from differentiated cancer being quite rare. There has, though, been a steady rise in incidence in the United States since 1974 when the annual number of new cases was 8000[2b].

Table 18.1. Mean annual cancer incidence[1] crude rates for selected tumour sites for the population of Belarus 1988–92: from *Cancer Incidence in Five Continents Volume VII.* Comparison with the populations of Hiroshima city 1986–90 and Nagasaki prefecture 1988–92. CR is the mean annual crude rate per 10^5 population. No. is the total number of cases registered for the five-year period.

Tumour site and [ICD-9]	Belarus				Hiroshima			
	Males		Females		Males		Females	
	No.	CR	No.	CR	No.	CR	No.	CR
Solid tumours								
Lung [162]	17 510	73.3	2709	10.0	1193	45.5	469	17.3
Stomach [151]	12.140	50.9	8879	32.8	2508	95.7	1384	51.1
Breast [174]	—	—	10 424	38.5	—	—	1172	43.3
Prostate [185]	3049	12.8	—	—	329	12.6	—	—
Bladder [188]	3207	13.4	655	2.4	388	14.8	114	4.2
Thyroid [193]	357	1.5	1166	4.3	90	3.4	321	11.8
Leukaemia								
Lymphoid [204]	1480	6.2	1119	4.1	46	1.8	25	0.9
Myeloid [205]	615	2.6	627	2.3	105	4.0	71	2.6
Monocytic [206]	25	0.1	25	0.1	3	0.1	2	0.1
Other [207]	56	0.2	72	0.3	0	0.0	1	0.0
Unspecified [208]	289	1.2	298	1.1	12	0.5	8	0.3
Average annual population	4774 763		5412 498		524 101		541 845	

18.2 Thyroid cancer

18.2.1 Survival

Rates per 10^5 population are demographic statistics which are reported for countries or large regions. Survival results, typically five-year survivals, are reported for series of cancer patients treated at a particular hospital, and are often subdivided by prognostic factor such as disease stage or histology. Because of the relatively small number of thyroid cancer patients, survival rates are not reported in major reviews such as the EUROCARE study of survival of cancer patients in Europe[3], because the case numbers just do not exist in sufficient quantity. Data for survivals must therefore be obtained from individual papers in the medical journals and from reviews of such papers in oncology textbooks, table 18.3.

Table 18.1. (Continued)

Tumour site and[ICD-9]	Nagasaki			
	Males		Females	
	No.	CR	No.	CR
Solid tumours				
Lung [162]	2584	70.0	1070	25.9
Stomach [151]	4164	112.8	2503	60.6
Breast [174]	—	—	1612	39.0
Prostate [185]	619	16.8	—	—
Bladder [188]	681	18.5	215	5.2
Thyroid [193]	76	2.1	389	9.4
Leukaemia				
Lymphoid [204]	64	1.7	42	1.0
Myeloid [205]	133	3.6	130	3.1
Monocytic [206]	6	0.2	3	0.1
Other [207]	5	0.1	5	0.1
Unspecified [208]	48	1.3	25	0.6
Average annual population	738 020		826 530	

18.2.2 Pathology

Pathology of well differentiated papillary and follicular carcinomas are given in table 18.3 and these represent the most common pathologies for thyroid carcinoma, table 18.4. Well differentiated tumours grow slowly whereas anaplastic tumours are fast-growing and all the latter are classified as stage IV disease, which has the worst prognosis in the range of stages I–IV, regardless of the extent of the disease. Most studies show a near zero two-year survival and it is known that there is a greater incidence of anaplastic thyroid carcinoma in countries with endemic goitre[4].

Thyroid tumours demonstrate unique features which include the following[2b].

- Age has a very important prognostic bearing with an excellent outcome in young individuals while in the elderly it has a poor outcome. This is the only tumour in the entire human body where age is included in the AJCC Staging Classification.
- The presence of nodal metastases has very little prognostic bearing: although the incidence of nodal metastases is very high. Again, this is the only tumour in the body where nodal metastases has little bearing.

Table 18.2. Mean annual age-specific incidence rates per 10^5 population for thyroid cancer in selected populations[1].

Age Group	Belarus 1988–92		Estonia 1988–92		Denmark 1988–92	
	Males	Females	Males	Females	Males	Females
0–	—	0.3	—	—	—	—
5–	1.7	2.1	—	—	—	—
10–	1.7	1.9	—	—	0.2	0.1
15–	0.9	1.0	—	0.4	0.3	0.3
20–	0.5	2.0	—	—	0.4	1.6
25–	0.5	2.4	0.7	1.1	0.8	2.1
30–	0.9	4.5	0.3	3.0	0.3	4.6
35–	1.3	5.4	—	4.1	0.4	3.6
40–	2.1	8.0	1.6	4.2	1.1	3.6
45–	1.4	6.5	0.9	5.5	1.6	3.0
50–	2.2	6.2	1.3	6.3	1.0	4.3
55–	2.3	6.9	2.5	2.0	1.4	2.8
60–	2.4	7.9	3.4	8.6	3.6	3.7
65–	4.7	6.7	0.9	9.0	2.7	4.2
70–	5.6	7.4	3.0	10.2	2.4	5.4
75–	5.2	4.9	1.8	10.3	3.9	7.7
80–	1.1	5.9	9.5	12.9	4.1	7.8
85+	3.8	4.0	6.6	8.5	6.9	9.4
Crude rate	1.5	4.3	0.8	3.8	1.1	3.1
Population						
Male	4774 763		730 516		2535 740	
Female	5412 498		831 952		2609 420	
Total cases						
Male	357		31		142	
Female	1166		157		400	

- Distant metastases are fairly common in young individuals but even in these cases the overall outcome and long-term survival is good.
- Multicentricity is fairly common.

18.2.3 Liquidators

In a follow-up review of 167 862 Russian liquidators to the end of 1994, a total of 47 thyroid cancers were diagnosed of which 42.8% were follicular,

Table 18.2. (Continued)

Age Group	Germany[¶] 1988–89		USA–SEER Programme 1988–92[W]		USA–SEER Programme 1988-92[B]	
	Males	Females	Males	Females	Males	Females
0–	—	0.1	—	—	—	—
5–	0.1	0.1	—	—	—	0.2
10–	—	0.2	0.3	0.8	0.4	—
15–	0.5	1.4	0.5	2.7	—	0.7
20–	0.2	1.3	1.2	7.0	0.7	2.6
25–	0.6	1.7	2.0	9.1	0.5	1.5
30–	0.5	2.9	2.9	10.2	1.1	4.3
35–	1.6	3.8	4.0	11.8	2.4	4.7
40–	1.2	4.5	3.5	11.8	0.5	5.9
45–	2.0	5.4	4.6	12.3	1.7	6.6
50–	2.5	4.3	5.2	11.1	2.7	7.4
55–	3.5	4.5	4.9	10.7	4.3	4.9
60–	4.2	7.8	6.8	10.2	7.7	9.3
65–	4.5	7.8	7.8	9.9	4.0	9.7
70–	6.9	7.2	7.2	10.0	4.8	10.6
75–	5.9	9.4	8.1	9.1	5.9	7.9
80–	6.2	9.4	5.8	8.7	5.7	9.1
85+	3.1	6.7	5.6	7.1	8.3	14.5
Crude rate	1.5	3.8	3.0	7.7	1.3	3.6
Population						
Male	7956 150		9316 445		1222 851	
Female	8691 850		9635 329		1353 081	
Total cases						
Male	243		1384		80	
Female	656		3719		243	

[¶] Only the Federal States Berlin, Brandenburg, Mecklenburg-Vorpommern, Sachsen-Anhalt and Free States Sachsen and Thuringen.

[W] White population.

[B] Black population.

Table 18.3. Survival results for cancer of the thyroid[4]. Data for well differentiated thyroid carcinoma after surgical resection ± ablation by ^{131}I therapy, or suppression by thyroxine, or both (nk = not known).

Year of publication	No. of cases	Survival rate (%)		
		5y	10y	20y
Papillary carcinoma				
1986	104	97	nk	nk
1986	859	nk	nk	93
1977	576	nk	92.4	nk
1986	121	92	74	nk
Radiation induced papilllary carcinoma				
1985[5]	296	nk	99	nk
Follicular carcinoma				
1986	23	96	nk	nk
1986	170	nk	94	nk
1984	37	nk	84	nk
1985	84	73	43	nk
1986	46	87	66	nk

Table 18.4. Incidence of different thyroid carcinoma pathologies: data for USA, Scandinavia and Europe[4,6].

Histology	Percentage with a given pathology			
	Boston[4]	Mayo Clinic[6]	Stockholm[4]	Basel[4]
Papillary	58	62.3	58	25
Follicular	24	17.6	19	39
Medullary	3	6.5	7	2
Anaplastic	15	13.6	16	26

33.3% were papillary and 14.3% were other types of carcinoma[7]. These cancers are radiation dose related and it was found that 28/47 worked in the high risk period of April–July 1986 and 15/47 were liquidators in 1987 whereas 4/47 worked in 1988–90. A total of 33/47 were diagnosed when the liquidators were in the age range 35–49 years and the time interval between work within the 30 km zone and diagnosis of thyroid cancer was in the range 1–8 years. The mean radiation dose was estimated to be

140 mGy, but doses could only be estimated[7] for 30 of this cohort of 47, table 18.5.

Table 18.5. Dose distribution for 30 Russian liquidators who were diagnosed with thyroid cancer[7].

Dose (mGy)	No. with thyroid cancer	Total no. of liquidators who received a given dose
0–49	11	41 199
50–99	8	30 929
100–199	3	23 734
200–249	3	20 828
>250	5	2626
Total	30	119 346

Table 18.6 gives the results of prediction modelling for risk of developing thyroid cancer in the liquidator population. There is good agreement between the modelling results for the cohort of Russian liquidators for whom the observation period was 1986–94 and the results of the BEIR V recommended model[8], but it should be noted that the confidence intervals are very large: due to the small number of observed thyroid cancers which were used for the RNMDR modelling.

Table 18.6. Thyroid cancer predictions for the Russian liquidator population.

Estimation method	Excess relative risk per 10^4 PY/Gy	Absolute risk per 10^4 PY/Gy
RNMDR modelling[7]	5.31 (95% CI = 0.04, 10.58)	1.15 (95% CI = 0.08, 2.22)
BEIR V modelling[8]	5.8	1.25

There have been other reports of incidences of thyroid cancer amongst the liquidator population but the numbers are smaller than those detailed here[7] and information is not always available about the time periods they worked within the 30 km zone and the doses they received. In addition, not all cases had their thyroid cancer verified histologically. In terms of small numbers, 28 thyroid cancers have been reported from Belarus, Ukraine and Russia[9], and three have been reported from a cohort of 3208 Lithuanian liquidators[10]. However, for thyroid cancer in adults the depth of thyroid screening in a population may greatly influence the observed incidence, and this is particularly relevant to the liquidator population where follow-up of health effects is much more active than, for example, for the residents of contaminated territories which, in turn, is more active than the general populations in Ukraine, Belarus and Russia.

18.2.4 Children and adolescents

In an analysis of 131 thyroid cancers in children in Belarus, 128 of 131 diagnosed in the period 1986 to July 1992 were papillary carcinomas[11]. However, in this study and a more extensive pathological study[12] of a cohort of 84 cases diagnosed 1991–92, the classification papillary included follicular. This latter study[12] found 33% follicular, 14% papillary, 34% solid papillary, 10% mixed papillary and 9% diffuse sclerosing papillary.

These thyroid cancers in children were, however, atypical in that more than 95% were highly invasive with some tumour spread both within the thyroid gland and into surrounding soft tissues (55/131 showed direct extension to the parathyroid tissues), blood vessels, lymph nodes and in a few (6/131) also into the lungs. This unexpected aggressiveness of the thyroid tumours leads to the conclusion that the increase does not entirely result from the screening because if it were, the increased incidence would be of the more usual non-aggressive type.

Most thyroids with a normal aetiology of cancer can be successfully removed with good prospects of complete recovery[10–12]. First choice treatment is always surgery, if at all possible, and the procedure should be a total or near total thyroidectomy. Dissection of the paratracheal groove should also be performed, regional lymph nodes explored and lymph node dissection performed in cases with lymph node metastatic disease[13].

There have been few reports of childhood deaths from thyroid cancer in the irradiated populations. By the year 1992 in Belarus, only one seven-year old child had died although ten were seriously ill[11] and by the end of 1995 in Belarus and the Ukraine combined, a total of three children had died[14].

Protocols have been developed[13] for diagnosis, treatment and follow-up of thyroid cancer in children exposed to radiation, the treatment being an integrated procedure including surgical, radiometabolic and hormonal manoeuvres. Many cases have also been treated in European centres collaborating with Belarus, one example being the work of Reiners in Essen and later in Würzburg 1994–95 where a total of 309 post-operative courses of ^{131}I therapy were delivered to 95 cases of advanced thyroid carcinoma: 55/95 originated from the Gomel region and their ages at surgery were in the range 7–18 years, with a mean of 9.4 years. Follow-up is available in 80/95 children who received more than a single course of ^{131}I therapy and complete remission was achieved in 44/80. Partial remission was obtained in 33/80 whereas there was no change in 3/80. Progression of the disease was not observed in any of these 80 children[13].

Figure 18.1 shows the enormous increase in childhood thyroid cancers in Belarus after the accident, only eight registered during 1974–85 and 574 registered during 1986–97. The map for the post-accident period shows that the majority of the cancers, 305/574, are in the Gomel region which correlates with the rainfall pattern of deposition of ^{131}I from the radioactive plume. Figure 18.2 is the thyroid cancer incidence map of Gomel region subdivided into smaller areas with the maximum incidence, 100/574, in the area containing Gomel town[15a]. Figure 18.3 shows the surface contamination on 10 May 1986 for Belarus[15b] which can be correlated with the thyroid incidence map in figure 18.1 for 1986–97

For this period 1986–97 there were a total of 1798 childhood cancers of which thyroid represented 31.9% with the next highest incidence being brain cancer at 30.8% followed by kidney 11.9%, bone 10.8%, soft tissues 8.6%, eye 5.1% and skin 0.9%. Table 18.7 gives a breakdown of the 574 cases in figure 18.1 by year and regions, which again emphasizes the remarkable increase in the Gomel region and the less marked increase in the Brest region[15a]. The total childhood population in Belarus is 2.3 million[16].

Table 18.7. Incidence of thyroid cancer in children and adolescents[15a], age up to 14 years, in Belarus during the period 1986–97.

Region	1986	1987	1988	1989	1990	1991	1992	1993	1994	1995	1996	1997	1986–97
Gomel	1	2	1	3	14	43	34	36	44	48	42	37	305
Brest	0	0	1	1	7	5	17	24	21	21	25	13	135
Others	1	2	3	3	8	11	15	19	17	22	17	16	134
All Belarus	2	4	5	7	29	59	66	79	82	91	84	66	574

The ages of the children at the time of irradiation due to the accident and at the time of diagnosis are given in table 18.8 and it is seen that three-quarters of this cohort of 574 thyroid cancer cases were irradiated when they were aged less than four years old[15a]. In a small study of 11 children who were treated for their cancer by thyroidectomy, the youngest was born two days after the accident, 8/11 were exposed *in utero* but were older than three months foetal age at irradiation. This is relevant because the foetal thyroid begins concentrating iodine at 12–14 weeks gestation[17].

Similar data to those for Belarus in table 18.8 are also available for children in the Bryansk region of Russia[13], table 18.9, but this is a much smaller population. Subdivision is also given in terms of ^{137}Cs surface soil contamination in Bryansk. In this small cohort of children it is seen that those living in areas of contamination >185 kBq/m^2 were irradiated at

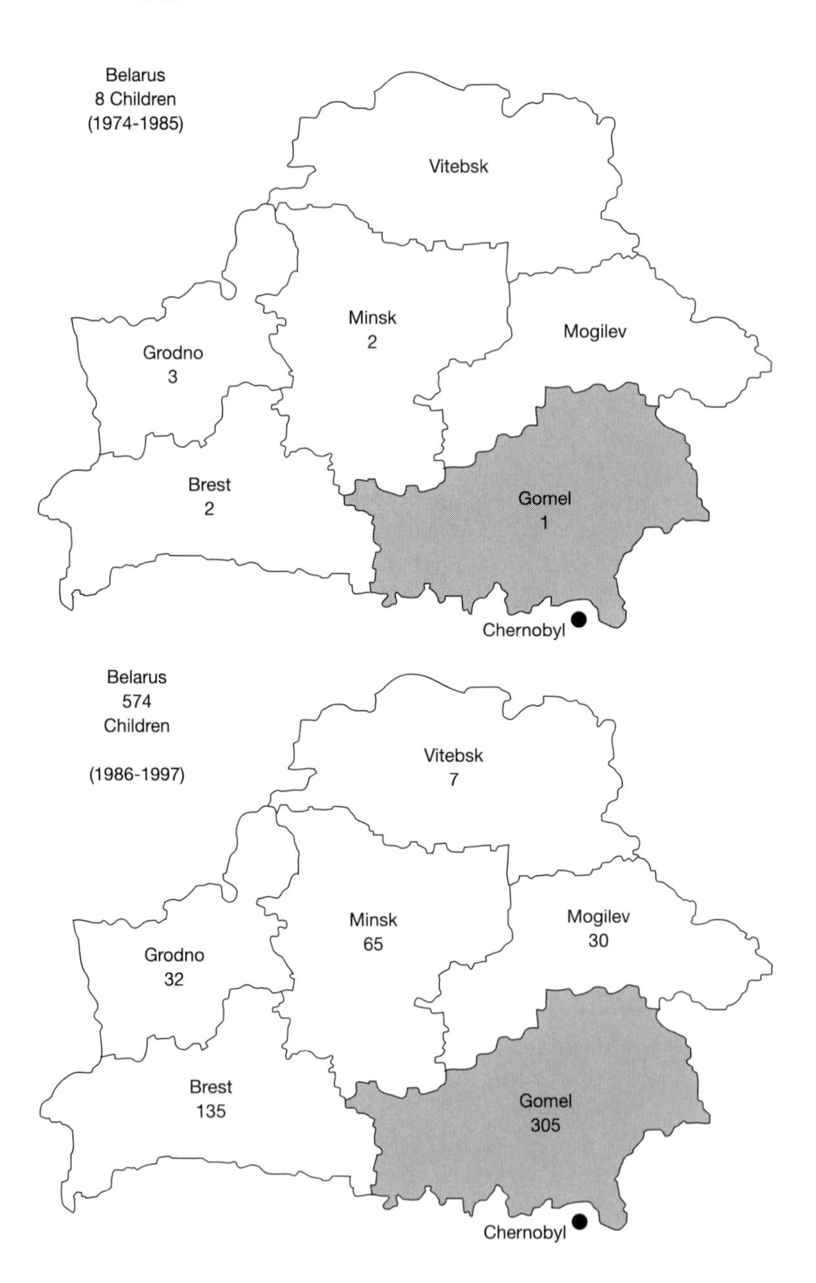

Figure 18.1. Number of thyroid cancers in children before and after the Chernobyl accident in the six oblasts which comprise Belarus[15a]. (Courtesy: A G Mrochek.)

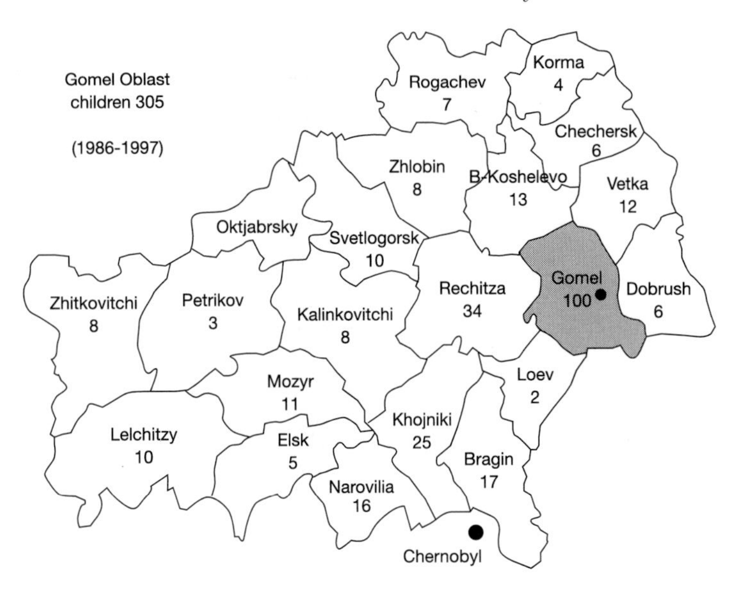

Figure 18.2. Number of thyroid cancers in children 1986–97 in the 20 rayons which comprise Gomel oblast[15a]. (Courtesy: A G Mrochek.)

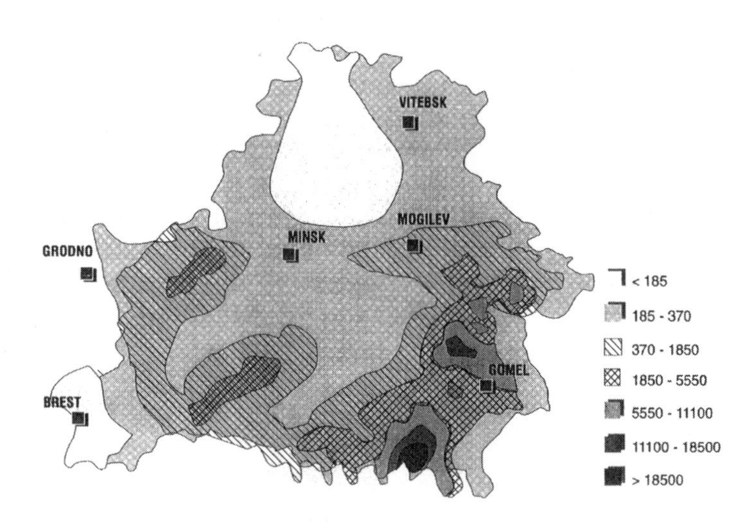

Figure 18.3. Map of ^{131}I surface contamination (kBq/m^2) in Belarus as restored for 10 May 1986[15b]. (Courtesy: IAEA.)

Table 18.8. Age at time of irradiation and age at time of diagnosis of thyroid cancer of 574 Belarus children[15a].

Age group (years)	Age at irradiation (%)	Age at diagnosis (%)
0–4	76.6	1.0
5–9	21.8	29.0
10–14	1.6	70.0

earlier ages than those in areas <185 kBq/m^2, but for both groups the time lapse between irradiation and diagnosis of thyroid cancer was the same, 7.5 years.

Table 18.10 gives the number of cases of childhood cancer for the years 1986–94 for the entire Ukraine[16,18] to end-June 1994, with a childhood population of 12 million; and for Russia the figures are only for the Bryansk and Kaluga oblasts, to end-December 1994. These two Russian oblasts together have a childhood population of 500 000 whereas that of the Ukraine is 12 million. The childhood population of Belarus is 2.3 million. The map in figure 18.4 shows the two Russian oblasts in relation to the most highly contaminated areas of Belarus and Ukraine. Areas which have a contamination higher than 185 kBq/m^2 are marked on this map.

The incidence data in table 18.10 only extend to 1994 and it is likely that the 1994 levels will continue for at least a further two years before starting to decline. Such cancer registry data are also subject to modification in future publications as it is always possible, for example, that some registrations are not recorded as soon as others and also that sometimes a reassessment of pathology leads to a case no longer considered to be cancer. However, these changes are likely to be small and the general trends in tables such as 18.7 and 18.10 can be considered to be correct.

The mean annual age-specific incidence rates for 1988–92 for Belarus[1] are given in table 18.2 and for the age groups 5–9 years and 10–14 years are, respectively, 17 and 21 per 10^6 for males and females, and 17 and 19 per 10^6 for males and females. However, for the sub-population of Gomel for 1991–92 the incidence was 80 per 10^6 children[17]. Expressed slightly differently, for 1990–94, for the most and least contaminated parts of both Belarus and Ukraine, table 18.11 shows the thyroid cancer incidence rates per 10^6 children, the actual number of cancers and the childhood populations[19].

The thyroid collective dose to the Ukranian childhood population, aged 0–15 years at the time of exposure, is estimated[20] to be 16.4×10^4 person-Gy and figure 18.5 shows the thyroid cancer incidence in the Ukraine from 1986–94 in people exposed in childhood by the accident[21]. The increase in

Table 18.9. Age at time of irradiation and age at time of diagnosis of thyroid cancer of 58 Russian children[13].

^{137}Cs contamination > 185 kBq/m^2				^{137}Cs contamination < 185 kBq/m^2			
Age at irradiation (y)	No. of cases	Age at diagnosis (y)	No. of cases	Age at irradiation (y)	No. of cases	Age at diagnosis (y)	No. of cases
0	1	5	1	0	0	0–6	0
1	9	6	2	1	0	8–10	0
2	2	7	1	2	1	10–12	3
3	2	8	3	3	2	12–14	2
4	2	9	2	4	0	14–16	4
5	0	10	1	5	0	16–18	6
6	0	11	3	6	1	18–20	7
7	2	12	1	7	5	20–22	5
8	0	13	2	8	2	22–24	5
9	0	14	0	9	5	24–26	2
10	0	15	1	10	2	26–28	1
11	0	16	1	11	1		
12	1	17	0	12	5		
13	1	18	0	13	1		
14	0	19	2	14	3		
15	3	20	0	15	3		
		21	1	16	0		
		22	0	17	3		
		23	1	18	1		
		24	1				
Totals	23		23		35		35
Mean age = 4.8 Range 0–15 y		Mean age = 12.3 Range 5–24 y		Mean age = 10.6 Range 2–18 y		Mean age = 18.1 Range 10–27 y	

Table 18.10. Number of childhood thyroid cancer in Ukraine and Russia, 1986—94.

Country	1986	1987	1988	1989	1990	1991	1992	1993	1994	1986–94
Russia	0	1	0	0	2	0	4	6	11	24
Ukraine	8	7	8	11	26	22	47	42	37	208

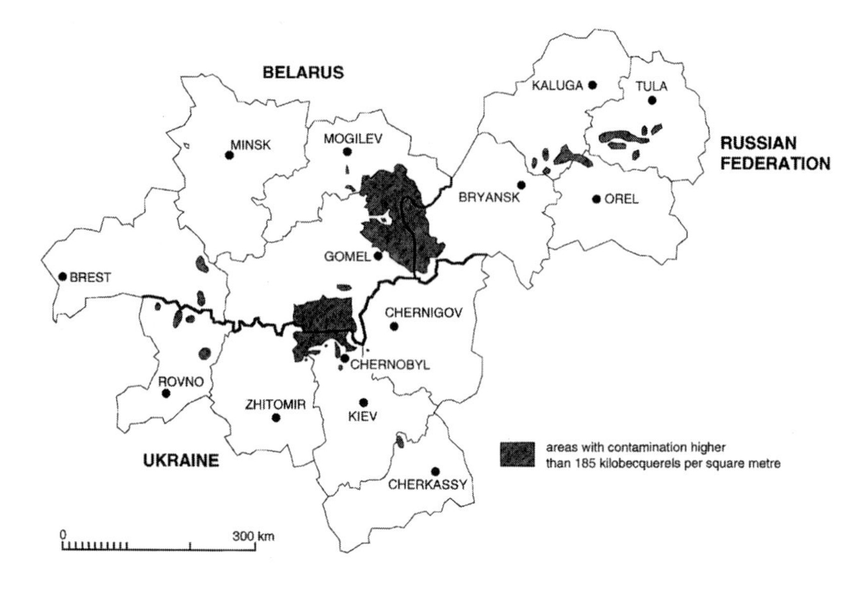

Figure 18.4. Map[16] of the region covered by the WHO IPHECA projects, which are the most contaminated territories in Belarus, Ukraine and Russia including the areas with ^{137}Cs contamination > 185 kBq/m^2. (Courtesy: WHO.)

Table 18.11. Thyroid cancer incidence in children in Belarus and Ukraine, 1990–94.

Region	Population in millions	No. of thyroid cancers	Incidence rate per 10^6
Gomel	0.37	172	92.0
Remainder of Belarus	1.96	143	14.6
Northern Ukraine¶	2.0	112	10.6
Remainder of Ukraine	8.8	65	1.5

¶ The six contaminated oblasts of the Ukraine.

thyroid cancer in the exposed children is seen to have exceeded the expected increase of the baseline rate, even when the variation of baseline rate as a function of age is taken into account.

By the end of 1984 a total of 542 thyroid cancers had been reported in the population who were aged 0–18 years at the time of the accident. In 1986 the age-specific incidence for 0–14 years old was some 0.7 per 10^6 compared to 0.4–0.5 per 10 in Belarus. In 1994 in the Ukraine it had risen to seven per 10^6 children. This is a factor of some 2.5–3.0 times lower than

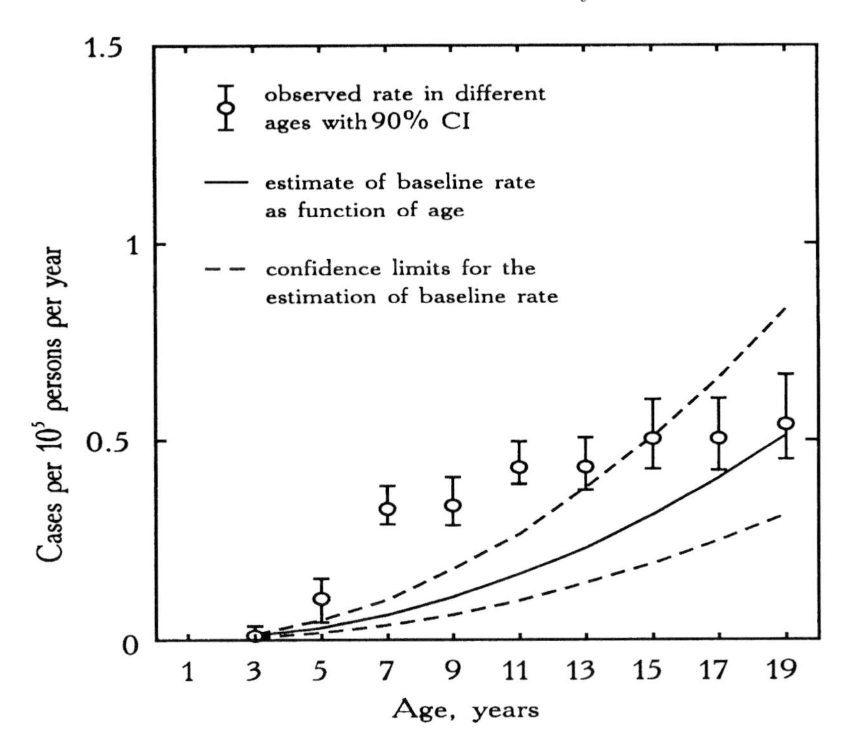

Figure 18.5. Thyroid cancer incidence in the Ukraine from 1986–94 in persons exposed in childhood to the Chernobyl accident radioactive fallout[21]. (Courtesy: BfS.)

for Belarussian children for the same time period. The absolute numbers are of the same order of magnitude but the Ukranian childhood population is some five times larger than that of Belarus.

Projections have been made[9] for the six oblasts which constitute Belarus, and for the most heavily contaminated oblast in Russia, Bryansk, for the numbers of thyroid cancer cases for population groups that were exposed as children when aged in the range 0–14 years, and these are given in table 18.12 for a lifetime period of 95 years and compared with projections for US white populations[22]. As would be expected, these are highest in Gomel oblast. The USA white population from the SEER programme was used because it was considered likely that the Belarus incidence rates underestimate the true incidence, especially before the accident. A five-year latent period was used for the projection model which assumed a constant relative risk.

The attributable fraction, (excess cases/total cases) $\times$ 100, for the various oblasts are similar regardless of whether the Belarus or USA background incidence data are used in the model. It ranges from 75–77 for Gomel to 4–5 for Vitebsk, with the predicted excess thyroid cancer deaths ranging from 1495–4300 for Gomel for, respectively, the USA and Belarus background incidence data.

Table 18.12. Predictions[9] of background and excess cases of thyroid cancer among persons exposed to radioactive isotopes of iodine, including ^{131}I, during childhood. Background incidence rates are for 1983–87 (AF = attributable fraction).

Oblast	Population size	Average dose (mSv)	Background no. of cancer deaths	
			Belarus rates No. (%)	USA rates No. (%)
Gomel	403 000	290	438 (0.11)	1400 (0.35)
Mogilev	294 000	90	352 (0.12)	1000 (0.34)
Brest	377 000	30	452 (0.12)	1300 (0.34)
Minsk	399 000	20	478 (0.12)	1400 (0.35)
Grodno	302 000	15	362 (0.12)	1050 (0.35)
Vitebsk	361 000	5	5 (0.00)	18 (0.00)
All Belarus	2140 000	80	2558 (0.12)	7400 (0.35)
Bryansk	92 000	35	110 (0.12)	300 (0.33)

Table 18.12. (Continued)

Oblast	Predicted excess cancer deaths			
	Belarus rates No. (%)	AF	USA rates No. (%)	AF
Gomel	1495 (0.37)	77	4300 (1.07)	75
Mogilev	350 (0.12)	50	1000 (0.34)	50
Brest	132 (0.04)	23	380 (0.10)	23
Minsk	104 (0.03)	18	300 (0.08)	18
Grodno	53 (0.02)	13	150 (0.05)	13
Vitebsk	<1 (0.00)	5	1 (0.00)	4
All Belarus	2157 (0.10)	46	6200 (0.29)	46
Bryansk	42 (0.05)	28	120 (0.13)	29

18.2.5 Adults living in contaminated territories

From the Belarus national cancer registry which provided data for children and adolescents, figures 18.1 and 18.2, tables 18.7 and 18.8, the records for thyroid cancers in those who were adults at the time of the accident are also available. Comparing the incidence in the two 12-year periods before and after the accident, it is found that there was an increase[15] by a factor of 3.4, table 18.13. This might, in part, be due to the effect of increased medical awareness and better medical surveillance after 1986, and also greater emphasis on cancer registration: although the Belarus cancer registry, founded in 1953, was already well established by 1986. However, for the ATB survivors the highest risk of thyroid cancer was found among those exposed before the age of ten years, and the highest risk was seen 15–19 years after exposure[23].

Table 18.13. Thyroid cancer in Belarus before and after the Chernobyl accident[15] (— = not stated).

Period	Children (see figures 18.1 and 18.2)	Adults						
	All Belarus	Vitebsk	Grodno	Minsk	Brest	Mogilev	Gomel	All Belarus
1974–85	8	—	—	—	—	—	—	1383
1986–97	574	640	395	1586	542	552	932	4647

18.2.6 Screening for thyroid cancer

It has been reported[24] that the effect of heightened medical surveillance of the ATB survivors included in the Adult Health Care Study was a 2.5 times increased risk of thyroid cancer compared with those survivors not given biennial clinical examination. However, in the Sasakawa screening study[25–27] of children in Belarus, Ukraine and Russia it was found that formal screening did not make a significant contribution to diagnosis, table 18.14. Routine examinations in schools were also carried out, including palpation of the neck, and over half the thyroid cancer cases in Ukranian children were found during such school examinations. A further 30–35% were diagnosed because the parents took their child to a doctor.

The WHO IPHECA projects[16] included one devoted to the thyroid which encompassed screening and this has now developed into the International Thyroid Project[28] which is a programme that aims at resolving a

Table 18.14. Screening results of the Sasakawa study.

Oblast	No. of children examined	No. of thyroid cancer cases detected
Belarus		
Gomel	14 000	9
Mogilev	18 000	2
Ukraine		
Zhitomar	19 000	5
Kiev	19 000	6
Russia		
Bryansk	17 500	4

number of public health related questions. The project, coordinated from a designated WHO Collaborating Centre in Minsk, includes international collaborating centres working on epidemiology, pathology and molecular biology, dosimetry, diagnosis, radiotherapy and radiobiology, treatment and psychosocial aspects.

This international project is another major growth point in the understanding, diagnosis and treatment of radiation induced thyroid cancer, a field which began in 1950 with the first report[29] on the development of human thyroid cancer caused by exposure to radiation, a study of 28 children and adolescents, and which was followed by thyroid cancer being the first solid tumour reported among the ATB survivors[30].

18.3 Leukaemia and other cancers

18.3.1 ATB survivors

Radiation-induced cancers are characterized by a latent period which is the time lapse between exposure and diagnosis. For the ATB survivors the minimum latent periods are 2–3 years for leukaemia, shown[31] schematically in figure 18.6, whereas for bone cancer they are 3–4 years and for solid cancers, including thyroid cancer, they are approximately 10 years. The mean latent periods are usually much longer and for solid tumours can be 20–30 years[32].

The latent period depends on the age at exposure and this has been illustrated in figures 17.2 and 17.3 for breast cancer[8]. Also, the radiation induced excess cancers, over and above *naturally* occurring cancers, are often observed with a high frequency at the age when the naturally occurring tumours have the highest incidence. Thus women exposed at age 20 may

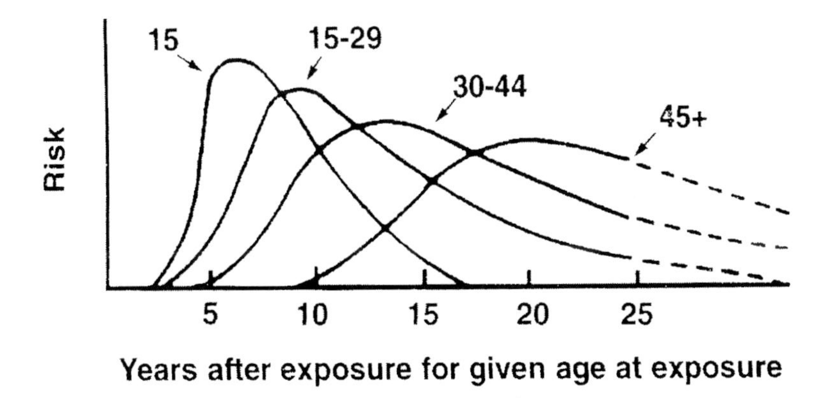

Figure 18.6. Schematic representation of the induction period and risk of leukaemia as a function of age at exposure[31].

develop breast cancer within 10–20 years whereas the female ATB survivors aged 0–9 years at exposure did not demonstrate an excess of breast cancers until they reached 30–40 years of age[32–34].

Mean annual crude incidence rates have been given[1] in table 18.1 for three types of leukaemia—lymphoid, myeloid and monocytic—in Belarus, Hiroshima and Nagasaki. Table 18.15 summarizes data[35] from the USA for the different types of leukaemia and includes a subdivision into acute and chronic. Of relevance to Chernobyl irradiated populations and to ATB survivors, chronic lymphocytic leukaemia which, as seen in table 18.15, has the best prognosis, does not appear to be caused by radiation.

Studies on the ATB survivors, except for those *in utero* survivors, that there is a strong link with the incidence of leukaemia. This is stated[36,37] in table 18.16 which reviews and details the strength of the association between ATB irradiation and the incidence of various diseases.

For the children of ATB survivors, no effects with statistical significance, including borderline significance, have yet been found in relation to exposure. The lack of statistically significant relationships have been confirmed for the following effects: solid tumours, leukaemia, stillbirth, major congenital anomalies, early mortality, chromosomal abnormalities and protein variants[36,37].

Figure 18.7 also summarizes the associations, or lack of them, between selected cancers and ATB radiation by presenting excess relative risk[23] at 1 Sv for solid tumours. If the confidence limits include zero for excess relative risk or include one for relative risk then this is a guide to their being no statistically significant association.

Table 18.17 gives the latest published[38] estimated averaged risk es-

Table 18.15. Incidence, survival and mortality data estimates for leukaemia in the USA[35].

- 28 700 new cases in 1998 evenly divided between acute and chronic leukaemias.
- 1998 incidence of 26 500 in adults and 2200 in children.
- 59% of childhood cases are acute lymphocytic leukaemia.
- In adults the most common types are acute myelocytic (35%) and chronic lymphocytic (25%).
- 21 600 deaths in 1998.
- 63% one-year and 42% five-year relative survivals for all leukaemias combined.
- Improvements from mid-1970s to late 1980s in survival of acute lymphocytic: from 38% to 57% and in children for the same period the improvement has been from 53% to 80%.
- In the USA and Japan leukaemia constitutes 4.1% and 3.8% of fatal cancers in males and 3.0% and 3.9% of fatal cancer in females[2a].

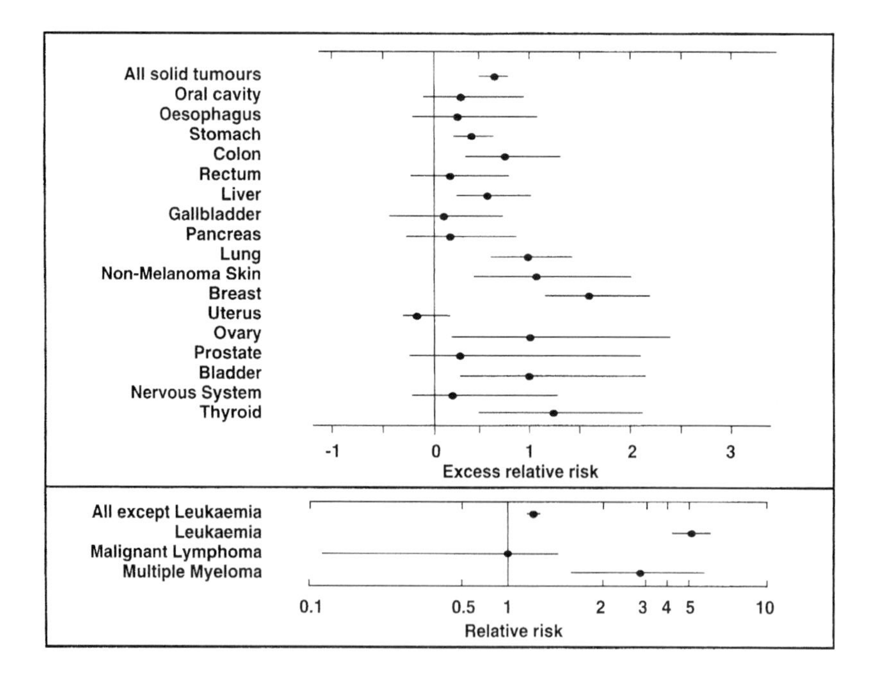

Figure 18.7. Estimated excess relative risks at 1 Sv (RBE 10) with 95% confidence intervals[23].

Table 18.16. Evidence from ABCC/RERF studies for late health-related effects of ATB irradiation.

Disease	Association with ATB irradiation		
	Strong[¶]	Weak[§]	None[‖]
ATB survivors excluding in utero *survivors*			
Cancers	• Leukaemia except chronic lymphoid leukaemia and adult T-cell leukaemia • Female breast • Thyroid • Colon • Stomach • Lung • Ovary	• Oesophagus • Salivary glands • Liver • Urinary bladder • Nervous system • Multiple myeloma • Malignant lymphoma	• Chronic lymphoid leukaemia • Adult T-cell leukaemia • Pancreas • Gallbladder • Rectum • Uterus • Bone
Non-cancer diseases and conditions	• Radiation cataract • Hyperpara-thyroidism • Delays in growth and development of those exposed at young ages	• Cardiovascular mortality and all non-cardiovascular mortality at high doses of >1.5 Gy • Thyroid diseases • Chronic hepatitis and liver cirrhosis • Uterine myoma • Early onset of menopause	• Infertility • Glaucoma • Autoimmune diseases • Generalized premature aging • Senile cataracts
Immune competence	• Decrease in T-cell mediated responses • Changes in humoral immune response	• Susceptibility to viral infections	• Changes in natural immune response
Chromosomal aberrations	• Lymphocytes		
Somatic mutations	• Erythrocytes	• Lymphocytes	

Table 18.16. (Continued)

Disease	Association with ATB irradiation		
	Strong¶	Weak§	None‖
In utero survivors			
Cancers		• All solid tumours	• Leukaemia
Non-cancer diseases and conditions	• Microcephaly • Mental retardation • Delays in growth and development • Lower IQ and poor school performance		• Non-cancer mortality
Chromosomal aberrations		• Lymphocytes	

¶ Statistically significant results in one or more studies. Questions about potential biases are largely resolved. Risk clearly related to amount of exposure.
§ Borderline statistical significance or inconsistent results. More studies may be needed.
‖ No statistically significant effect observed. This may reflect a true lack of effect or result from an inadequate sample size.

timates for 1950–87, that is, two years longer than the estimates in figure 18.6. The risk estimates are based on the Life Span Study cohort of 93 696 ATB survivors which for 1950–87 account for 2778 000 person-years.

The Life Span Study has been ongoing since 1958 and earlier reports for 1950–85, with fewer cases and less follow-up, concluded that there is an excess relative risk of multiple myeloma[39].

Figure 18.7 shows the variation with organ dose in the estimated relative risk of various cancers[40] and table 18.18 gives the time lapse between the ATB explosions in 1945 and the year when an increase in the incidence of a given cancer was observed[40]. This time can be assumed to be longer than the actual latency period as an increase was suspected earlier, but could not be statistically proven.

Table 18.17. Excess risks for leukaemia, lymphoma and multiple myeloma[38].

Cancer	No. of cases 1950–87	Excess absolute risk (cases per 10^4 PY Sv)	Excess relative risk at 1 Sv
Leukaemia			
All types	290		
Acute lymphocytic		0.6	9.1
Acute myelogenous		1.1	3.3
Chronic myelocytic		0.9	6.2
Lymphoma			
Male + female	208		
Male		0.6	
Female		No evidence of any excess risk	
Multiple myeloma			
Male + female	62	No evidence of any excess risk	

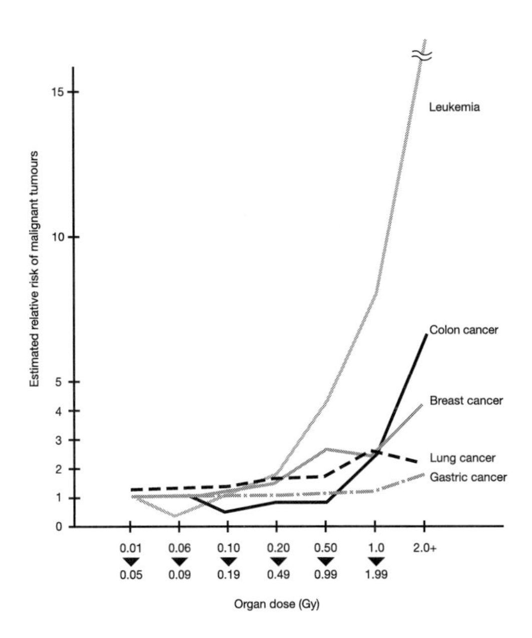

Figure 18.8. Estimated relative risk as a fuction of organ dose in Gy, for leukaemia and colon, breast, lung, and stomach cancers[40]. (Courtesy: RERF.)

Table 18.18. Year of development of cancers.

Cancer	Increase suspected (year)	Increase observed (year)
Leukaemia	1947	1950
Thyroid cancer	1951	mid-1954
Breast cancer	1955	1965
Lung cancer	1956	1965
Stomach cancer	1961	1975
Colon cancer	1961	1975

18.3.2 Liquidators and residents in contaminated territories

Based on the experience of the ATB survivors it was initially expected that there would be a significant increase in the incidence of leukaemia, particularly in the population of liquidators who received the highest doses.

No statistically significant increases of either leukaemia or of solid tumours other than thyroid cancer has yet been demonstrated[9,16]. Some increases in the incidence of leukaemia have been reported but these are small numbers and the results are difficult to interpret. This is mainly because of the differences in the intensity and method of follow-up between exposed populations and the general population with which they are compared. There is no doubt that there is improved cancer registration and this must be taken into account in the future for analysis of supposed increases in incidence.

Even so, not all studies have reported increases and, for example, that for a cohort of 4833 Estonian liquidators[41,42] found 144 deaths from all causes during 1986–93 compared to an expected number of 148, but no cases of leukaemia were observed. In a study[43] of 174 812 Ukranian liquidators, of whom more than 77% worked during 1986–87, a total of 86 cases of leukaemia were reported with figures of 13.35 per 10^5 for 1986 liquidators and 7.04 per 10^5 for 1987 liquidators. Data are not available for the Ukraine in the series[1,22] *Cancer Incidence in Five Continents* but are given for Belarus. The combined incidence per 10^5 males for all leukaemias is 10.3 per 10^5.

From prediction modelling[9], tables 18.19–18.21, the total lifetime numbers of excess solid cancer and leukaemia cases are small compared to the expected background number: for the cohort of 200 000 liquidators the increase in solid cancers is some 2000 compared to a background of 41 500. This excess would be difficult to detect epidemiologically. This is also true for the cohort of 6800 000 residents in contaminated territories for whom

Table 18.19. Predictions for a lifetime (95 years) of background and excess deaths from solid cancers in populations exposed as a result of the Chernobyl accident[9] (AF = attributable fraction).

Population	Population size	Average dose (mSv)	Background no. of solid cancer deaths	Predicted excess no. of solid cancer deaths	
			No. (%)	No. (%)	AF
Liquidators 1986–87	200 000	100	41 500 (21)	2000 (1)	5
Evacuees from the 30 km zone	135 000	10	21 500 (16)	150 (0.1)	0.1
Residents of SCZs	270 000	50	43 500 (16)	1500 (0.5)	3
Residents of other contaminated territories	6800 000	7	800 000 (16)	4600 (0.05)	0.6

Table 18.20. Predictions for the first 10 years of background and excess deaths from leukaemia in populations exposed as a result of the Chernobyl accident[9].

Population	Population size	Average dose (mSv)	Background no. of leukaemia deaths	Predicted excess no. of leukaemia deaths	
			No. (%)	No. (%)	AF
Liquidators 1986–87	200 000	100	40 (0.02)	150 (0.08)	79
Evacuees from the 30 km zone	135 000	10	65 (0.05)	5 (0.004)	7
Residents of SCZs	270 000	50	130 (0.05)	60 (0.02)	32
Residents of other contaminated territories	6800 000	7	3300 (0.05)	190 (0.003)	5.5

Table 18.21. Predictions for a lifetime (95 years) of background and excess deaths from leukaemia in populations exposed as a result of the Chernobyl accident[9].

Population	Population size	Average dose (mSv)	Background no. of leukaemia deaths	Predicted excess no. of leukaemia deaths	
			No. (%)	No. (%)	AF
Liquidators 1986–87	200 000	100	800 (0.4)	200 (0.1)	20
Evacuees from the 30 km zone	135 000	10	500 (0.3)	10 (0.01)	2
Residents of SCZs	270 000	50	1000 (0.3)	100 (0.04)	9
Residents of other contaminated territories	6800 000	7	24 000 (0.3)	370 (0.01)	1.5

the predicted excess solid cancers is some 4600 against a background of 800 000. The numbers are much smaller for leukaemia, tables 18.20–18.21.

Chapter 19

The Legasov Testament

Introduction

In 1986 Academician Valery Legasov was the First Deputy Director of the Kurchatov Institute of Atomic Energy, Moscow. He had been involved in the planning and design of the RBMK nuclear reactors of the type installed at Chernobyl and was expected to succeed the octogenarian Academician Anatoli Alexandrov as the next Director of the Kurchatov Institute. All the indications in the nuclear energy establishment of the Soviet Union were that Legasov, who in 1986 was only 50 years of age, would gain the very highest honours.

The Chernobyl accident ruined these prospects, in large part because Legasov finally began to speak out about the problems in the USSR which had contributed to the catastrophe: instead of keeping quiet and voicing only the Communist Party line. This was not known in the West at the time, and the cause of his death two years after Chernobyl gave rise to speculation, including leukaemia and suicide, but nothing was known for certain as the contents of the *Testament* published only by *Pravda* on 20 May 1988 using the title *My duty is to tell about this....*, could not be widely read as it was not in English.

For the first time, it is published in full in English[1] for this chapter. In addition, the comments of his widow eight years afterwards, published in the 1 June 1996 issue of *Trud*, have also been translated into English. Together these statements represent a valuable historical account of the events not only at Chernobyl, but also of those which befell a leading Soviet nuclear scientist who was the Head of the Soviet Delegation in August 1986 at the first international post-accident meeting held at the IAEA, Vienna.

[1] I am very grateful to Dr Igor Obodovskii for locating the issues of *Pravda* and *Trud* in the National Library of Ukraine, Kiev, and for providing me with the English translations. I have added some subsection titles for ease of reference and also the footnotes.

Vladimir Gubaryev, the first journalist to reach the site of the accident, and the Science Editor of *Pravda* who in 1986 had written the play[2] *Sarcophagus* which used hypothetical characters including a physician (based on Angelina Guskova), an American professor of surgery (based on Robert Gale) and the Director of a Nuclear Power Station (based on the Chernobyl NPP Director) in order to criticize the Soviet authorities[3]. Some of the comments are a damning indictment on the Soviet system and in 1986, even with glasnost and perestroika, it is surprising that *Sarcophagus* was allowed to be published.

In 1986, Gubaryev had also asked Legasov to write some notes on his Chernobyl experiences for *Pravda* and following Legasov's death, these notes have become known as his *Testament*, and were published in the 20 May 1988 issue with an introduction by Gubaryev which concluded with the following paragraph.

> It is difficult to explain and to understand the suicide of Valery Alexeevich Legasov: He committed suicide when he was a man of mature age and at the peak of his life. This tragedy should be a lesson for all of us and a reproach to those for whom a quiet life with materialistic benefits matters above all else.

[2] Gubaryev V 1987 *Sarcophagus* translated by Glenny M (Harmondsworth: Penguin).

[3] Some of the criticisms of NPP construction and organization included the following. 'Now that they've got to deal with the accident damage, they'll never find out what went wrong with the reactor. The construction boys were chivvied out of that foundation pit as if they were jet-propelled, so that the reactor could be handed over ahead of schedule. Under that reactor there's not only concrete blocks: if you were to poke around down there, you'd find a couple of excavators too. And all because of the bonuses they'd get if it were done before the original delivery date'.

The play also included claims that the authorities 'economized on the firemen's protective clothing' by not providing any and also economized by using sub-standard material for the roof of the turbine hall. It was claimed that the material was highly inflammable, 'the roof went up like gunpowder and melted', and was identical with that used 12 years earlier on factory roofs in Bokhara and on the Baikal–Amur railway line: 'both places burned to a cinder in 5–6 minutes'. The roofing material had been forbidden for use in industrial buildings but the excuse was that even so 'there was a lot of it in stock'.

From the NPP Director there was the following justification. 'Go and see the other nuclear power stations: take a good look. Were we any worse than them? No, and I can tell you this with authority because I've seen many of them. We were better. We've won the Ministry's efficiency award three times and our output has always been up to target. We're not the ones to blame. But can you explain to me why the quality of nuclear power station equipment has got steadily worse over the last 10 years? Why we are given obsolete instruments and spare parts? And, finally why our requests for the repair of those very same faulty switches, for instance, take three months to reach the Ministry and the replies take three months to come back? ... Have you heard what happens when managers of the big Moscow stores are selected? When he is selected and confirmed, they say to him: "We know you'll be on the fiddle, and we won't touch you, but don't take it amiss if we put you in jail one of these days: it's all part of the system. Every year we have to take one store manager to Court and make an example of him to pacify public opinion. So keep on fiddling until it's your turn to be this year's scapegoat".'

19.1 My duty is to tell about this...

I never thought that just having reached 50 years of age I would write my memoirs. But such a scale of events took place, with the involvement of people with contradictory interests, and with so many different explanations put forward, that to a certain degree it is my duty to record what I know and how I saw events.

26 April 1986 was a Saturday and I remember that I could not make up my mind what to do. Saturday is my usual day for visiting my Department at the University, but I could also attend a meeting of the Heads of the Local Administration at the University or perhaps cancel all work and have a rest day and go somewhere with my wife Margarita Mikhailovna. I followed my habit of many years of concentrating on work and went to the Local Administration meeting.

Before the meeting started I heard about an accident at the Chernobyl NPP from a Chief of one of the State Department Subdivisions to which our Institute is subordinate. He informed me quietly enough, though with a certain amount of irritation. Then we heard a report during the Agenda of the meeting.

Speaking frankly, the report was like most reports, very dull. We were always accustomed in our Department to reporting that everything is brilliant, all the indices of our work are very good and we are accomplishing with success all the planned tasks. Reports have the character of a victory communiqué singing an anthem to atomic energy. The speaker stated that 'they have made some mess, there was an accident, but this would not stop the development of atomic energy'.

About 12 noon a break was announced and I went to the office of the Scientific Secretary. There I learnt that a Government Commission had been established and that I had been included in the Commission which was to arrive at Vnukovo airport at 4pm.

I immediately went to my Institute and tried to find anyone who was a reactor specialist. With great difficulty I managed to find Alexandr Kalugin, the Chief of the Department which developed and controlled NPPs containing RBMK reactors. It was RBMKs that were installed at Chernobyl. Kalugin already knew about the accident because a signal 1-2-3-4 had been received from the Chernobyl NPP that night. This signal indicated an accident situation existed which required emergency measures to counter Nuclear, Radiation, Fire and Explosion effects. There was no higher state of emergency than this 1-2-3-4. I immediately returned home as did my wife from her work and I told her that I was leaving on a business trip, the situation was unknown to me and I did not know for how long I would stay.

At Vnukovo airport I was told that Boris Scherbina, Deputy Head of the Council of Ministers of the USSR and Head of the Bureau on Fuel and

Energy, had been appointed Chief of the Government Commission. We were to fly to Kiev and then travel to the Chernobyl NPP by cars. During the flight we were all very anxious and I described to Boris Scherbina the 1979 accident at Three Mile Island in the USA and emphasized that the causes of that accident had nothing in common with the Chernobyl accident because of significant differences in reactor power plant design. We passed the hour long flight making guesses.

In Kiev when we left our aircraft the first thing we saw was a long cavalcade of governmental black cars and an anxious group of Ukranian government officials. They had no precise information but said that the situation was very bad. As we approached the NPP by car I could not have imagined that we were approaching a situation of what I term planetary scale, in that it will never be forgotten in world history: as, for instance, the eruption of the volcano which destroyed the city of Pompeii[4].

19.1.1 Arrival at the power station

Although the NPP is called Chernobyl it is situated 18 km from the town of the same name. The town, which is at the centre of Chernobyl District, is very green and rural and we saw only people going about their everyday life. But in the town of Pripyat we already felt the anxiety when arriving at the town's Communist Party Headquarters located in a central square. The Heads of the local Authorities reported that at the fourth reactor during implementation of a turbine test when it was allowed free running down, two sequential explosions had taken place. The reactor premises were destroyed, several hundred people had received radiation injuries, two people had been killed and the remainder of the shift of workers on duty at the time of the accident were hospitalized in hospitals.

The Government Commission at its first meeting immediately despatched its members, by group, for various tasks. I headed the group which had the goal of developing measures to localize the accident.

When we approached the NPP we were surprised by the colour of the sky: at 8–10 km before the NPP a crimson glow[5] was already seen. It is common knowledge that an NPP is a very clean construction and that nothing visible can be seen from its ventilation chimneys. In contrast, this looked like a metallurgical plant or a large chemical complex above which was this enormous crimson glow which expanded to fill half the sky.

It was immediately clear to us that the managers of the NPP and the managers from the Ministry of Energy who were also present, generally behaved in a contradictory manner in that they were brave but had no

[4] The eruption of the Mount Etna volcano on 23 November AD 79 which destroyed the cities of both Pompeii and Herculaneum.

[5] This strange colour for the sky has already been described in section 3.1.2 by an eyewitness on the Polish–Byelorussian border.

idea what to do. Although the workers were willing to undertake any task, these managers had no emergency instructions to give them and the Government Commission which arrived at the NPP at 8pm on 26 April had to do everything. The operators of the first and second reactors had not left their working place and neither had those of the third reactor: despite the fact that the third and fourth reactors were in the same building structure.

The first instruction given by the Commission was to shut down the third reactor. The first and second reactors continued to function despite the fact that their premises, such as the control room, was already contaminated at a high level of radiation. This was due to the intake ventilation not being switched off immediately following the explosion and therefore contaminated air circulated through the ceiling air vents.

Boris Scherbina immediately called for Chemical Troops[6] which arrived quickly enough under the command of General Pikalov, and also Helicopter Troops[7] with pilots headed by General Antoshkin. The initial helicopter flight was to assess the condition of the fourth reactor and it was immediately seen that the reactor was completely destroyed. The upper biological shield was now almost vertical and this must have taken enormous explosive power. The upper part of the reactor hall was completely destroyed and graphite block pieces were scattered on the roof of the turbine hall. We could see a *white pillar* several hundred metres high consisting of burning products constantly flying from the crater of the reactor. Inside the reactor premises individual spots of deep crimson luminescence were seen: but it was difficult to determine what was exactly the cause of this luminescence.

Until the evening of 26 April all possible methods were tested of extinguishing the fires using water but they achieved nothing except to cause an increase in the level of water vapour and to flood the rooms and corridors of the reactor premises adjacent to the site of the fourth reactor.

By the end of the first night firemen had very effectively eliminated the series of individual fires in the turbine hall. It was variously reported that some firemen received higher radiation doses than necessary because they remained at observation points waiting for new fires to commence. This is not true but the firemen's bravery was still exceptional as the turbine hall contained much spilled oil and oxygen in the generators and these might have caused not only fires but also explosions which could have destroyed the third reactor. The first accurate information on the locations of possible extensions of the accident were provided by these firemen.

It then became clear that there was a very powerful flow of radioactive aerosols being emitted from the crater of the fourth reactor. As the graphite burned particles carried with them a high radioactive content and this

[6] General Pikalov's eyewitness account of the decontamination work of the Chemical Forces is given in section 12.1.6.

[7] See sections 3.1.7 and 3.5 for the experiences of the helicopter pilots.

presented us with a very complex task. The normal speed of graphite burning is about 1 tonne per hour and since about 2500 tonnes of graphite were in the fourth reactor this could constitute 250 hours of graphite burn-up which could by aerosols be distributed over a wide territory.

The radiation situation only allowed us to take action from the air from a height of not less than 200 metres above the reactor and there was no equipment capable of terminating by standard means the graphite burn-up. For a discussion of non-standard possibilities we were in constant telephone links with specialists from the Kurchatov Institute and the Ministry of Energy. Next day, we also received by telegram various suggestions from overseas on how we could stop the graphite burning. Eventually we decided upon the use of two materials to stabilize the temperature: lead and dolomite.

19.1.2 Evacuation of Pripyat

An even more urgent problem being solved by the Government Commission was that of the immediate future of the town of Pripyat. In the evening of 26 April the radiation situation in the town was more or less satisfactory in that dose rates were within the range of milliroentgen per hour to tens of milliroentgen per hour. This was certainly not healthy, but it did allow us some time to consider possible solutions.

The medical authorities were limited by the USSR Radiation Regulations according to which evacuation could only begin if there were a risk of members of the civilian population receiving a dose of 25 *biological roentgen*[8] per man. If there were a risk of receiving 75 biological roentgen during a period of stay in the affected zone then evacuation would become compulsory. For a risk in the range 25–75 the USSR Regulations stated that the decision should be left to the local authorities.

Physicists insisted that there should be compulsory evacuation of Pripyat because they had a presentiment that the radiation situation would rapidly worsen and Boris Scherbina and the Government Commission accepted this advice. Evacuation was planned for the next day, 27 April, but unfortunately this information was disseminated mainly orally among the population. This was by visits to houses and distributing notices but it became evident that not all had learned of this decision to evacuate. This was because on the morning of Sunday 27 April in the streets of Pripyat it was possible to see mothers pushing children in prams and children playing in the streets.

The official notification of the compulsory evacuation was given at

[8] The term *biological roentgen* is what is normally called a rem, a roentgen-equivalent-man, see section 1.1.1 where dose terminology is defined. 1 Sv = 100 rem. Table 1.1 gives ICRP recommended dose limits (1991) for occupationally exposed persons and for the public. These are not the same as action levels for evacuation of a population.

11am and up to 2pm all the necessary vehicles were obtained and the travel routes planned. The evacuation was conducted efficiently although under extraordinary conditions but there were also some errors. For example a large group of the population approached the Government Commission to receive permission to evacuate in their own cars: there were several thousand cars in the town. After some considerations, permission was given. However, this was an error because many cars were contaminated and it was only later that dosimetric check points were established. Nobody from the civilian population who did not stay at the NPP itself, about 50 000 people, received any serious detriment to health.

19.1.3 Dosimetric control

The next measures to be undertaken were to establish more accurate dosimetric control. This was undertaken by the State Committee on Hydrometeorology, the Chemical Forces, the staff of the NPP and by physicists who were to study the isotopic composition of the contamination. Military dosimetric services worked very well and we received the most accurate data on isotopic composition from a laboratory established on the affected territory. It was clear during the first days that the situation was often changing because of the changes in movement of air masses and of radioactive dust formation.

19.1.4 NPP workers and managers

Some of my personal impressions of that time are as follows. Of the personnel of the NPP we met, all were ready for any actions under any conditions but the managers of the NPP and of the Ministry of Energy had no understanding about the necessary sequence of actions to be taken, and of how to plan and organize. They had no guidelines written earlier and were incapable of making any decisions on the spot and therefore all work had to be managed by the members of the Government Commission..

19.1.5 Lack of equipment and facilities

Embarrassment was seen even in relatively minor items. I remember that when the Commission stayed in Pripyat there were insufficient numbers of protective respirators, of individual TLD dosimeters and of pencil-type ionization chamber personnel monitors.

Also, there was at the NPP no automatic system for environmental measurement of radiation within a range of several kilometres. This was why we had to enlist huge numbers of personnel for reconnaissance operations. There were no radio-controlled aircraft equipped with dosimetric instruments and therefore we required a large number of helicopter pilots for measurement and reconnaissance purposes.

A culture of elementary hygiene was also absent, at least during the first days. On 27, 28 and 29 April in Pripyat, premises were already very dirty, but when food products were supplied—sausages, cucumbers, bottles of Pepsi cola—everything was simply placed in rooms and people took it without their hands necessarily being clean. Only after several days were special refectories set up with appropriate sanitary and hygenic conditions.

19.1.6 Visit of Prime Minister Ryzhkov

On 2 May when the Government Commission was located in the town of Chernobyl, Nikolai Ryzhkov the Prime Minister of the USSR, and Igor Ligachev, the Secretary of the Central Committee of the Communist Party, came to the zone. From our reports they understood that it was a major accident which would have long-term consequences and that major works had to be achieved.

We explained the situation as we understood it and an operational group was established, headed by the Prime Minister, and practically all the industrial expertise of the USSR was engaged. From that moment the Government Commission became only a managerial mechanism of this Operational Group of the Political Bureau of the Central Committee of the Communist Party. I do not know of any major or minor decision which was not supervised by this Operational Group. In my opinion this was a correct organizational procedure.

19.1.7 The Army

The work of the military was very wasteful and seemed to be going round in circles[9] . The Chemical Troops first of all had to deal with reconnaissance and then the determination of which territories required decontamination. The army was authorized to work on the territory of the NPP itself, within the 30 km zone on decontamination of villages, settlements and roads. They also implemented an enormous amount of decontamination work in Pripyat, I never witnessed any case of a Soviet Army specialist or civilian attempting to avoid such difficult and dangerous work. I said to people that I would like to work with those who wanted voluntarily to help me and there were no cases of anyone refusing.

19.1.8 The information service

It appeared that in spite of the fact that we have *Atomenergoizdat* as a publisher for nuclear energy literature, publishing houses for medical literature, and the *Znaniye* (Knowledge) Society, there was no literature

[9] Land and forests and buildings which were decontaminated were later contaminated again because of the movements of the wind and the redistribution of the radioactive dust, thus rendering the initial attempts at decontamination useless.

available which could be instantly distributed to the population to give an explanation of radiation doses and which dose levels were relatively safe and which required immediate evacuation. There was also no literature giving advice to the population on how to measure radioactivity and what actions to take with vegetables, fruit, etc. There were many extensive books for the specialist but no pamphlets or brochures for the general public.

19.1.9 Training and education

Maybe there will never be an appropriate time to express my personal impressions on the history and development of atomic energy, and my entry into this subject, and certainly it has been very seldom that any of us have spoken sincerely on this subject.

I graduated from the Engineering, Physics and Chemistry Faculty of the D I Mendeleyev Chemical and Technological Institute in Moscow. This faculty trained specialists, mainly researchers, who were to work in the technological field of the atomic industry. For example, training on the separation of isotopes, handling of radioactive substances, extraction of uranium from its ore, production of nuclear fuel elements, processing of spent nuclear fuel and extraction of useful products from this fuel, and radioactive waste disposal. This included the use of radioactive sources not only in industry but also in medicine.

I then took my Diploma Course at the Kurchatov Institute where my study was in the field of processing nuclear fuel. Academician I K Kikoin wanted to include me in the Postgraduate Course but together with my friends I decided that we wanted to work at an Atomic Industry Plant so as to obtain practical skills in the field in which we would be working for many years.

I then left the Kurchatov Institute for Siberia and took part in commissioning a radiochemical plant. This was a very vivid and interesting period of my life and I worked at the plant for two years, before returning to the Kurchatov for my Postgraduate Course.

I developed a number of technological processes for my Candidate's and Doctor's theses and was elected to the Academy of Sciences of the USSR and my scientific work was awarded a State Prize. I have managed to attract to my research group many interesting young people with a good education. They are developing this field of chemical physics and I am sure their work will be important for the future.

My work was evidently noticed because I became Deputy Director of the Kurchatov Institute. My work included supervision of work in chemical physics, in radiochemistry and in the use of nuclear and plasma sources for technological purposes. When Anatoli Alexandrov was elected as the President of the Academy of Sciences he recommended me as First Deputy Director of the Institute.

19.1.10 Problems in the development of atomic energy

It was interesting for me to discuss what fraction of atomic energy should be present in the Soviet energy system and by what technology this should be achieved. We organized systems studies for different types of NPP and whether they should produce only electrical energy or also other outputs such as hydrogen, which could be used an energy carriers. Because of the safety problems we also compared the possible risks of atomic energy with the risks of other energy systems.

At the Scientific and Technical Council of the Institute we often discussed conceptual problems for the development of atomic energy, but very seldom discussed technical aspects: quality of different types of reactor and quality of the nuclear fuel. Nevertheless I had information available to me that convinced me that not all was as OK as it seemed to be in the development of atomic energy.

It could be seen by the naked eye that our equipment did not have major conceptual differences from foreign equipment: even excelling them in some details. But they had no good control systems or diagnostic systems. The American specialist Rasmussen[10] made an analysis of NPP safety where he methodically searched for all possible sources of breakdowns leading to accidents. He then systemized what he found and made probability evaluations of each event, and, for example, evaluated the probability of an external release of radioactivity. We learned about this from the foreign literature but I have not seen any Soviet teams of scientists who have made such detailed probabilistic studies.

Dr V A Sidorenko was the most active scientist in this field and he really knew the state of all aspects of an NPP including the quality of equipment: which led to unpleasant events from time to time. His efforts were mainly directed at coping with such events, first by organizational measures and second by a system of improved documentation to be kept at NPPs and with the designers. He was very anxious about the establishment of Supervisory Authorities controlling the situation.

Sidorenko and like-minded persons were also very worried about the quality of the equipment supplied to NPPs and also about the lack of proper training and education of the personnel designing, constructing and operating NPPs. Unfortunately Sidorenko did not receive adequate support and each of his documents were approved only with great difficulty. Generally, though, worries about the safety of NPPs were considered to be far-fetched,

[10] On 4 August 1972 the United States Atomic Energy Commission gave permission to commence a major Reactor Safety Study, to be directed by Professor Norman Rasmussen of the Massachusetts Institute of Technology. The US$3 million study, was carried out by AEC laboratories and contractors including Battelle, Oak Ridge, Brookhaven and Lawrence Livermore. Some of the early results were included in WASH-1250, *The Safety of Nuclear Power Reactors (Light Water Cooled) and Related Facilities*, published by the AEC in July 1973.

and the highly qualified specialists who also had the greatest responsibility, were convinced that the problems of safety could be exclusively solved by specialists having qualifications and by accurately instructing personnel.

Increasingly more funds were being spent on objectives which had no direct relationship to the previously mentioned atomic energy problems. These included expanding the capacity for the production of nuclear fuel elements and for studying metallurgical physics. The costs involved in building constructions, for example, were huge. Also, scientific organizations which were formerly the most powerful in the USSR became weakened, the level of supply of modern equipment decreased, people still in positions of responsibility passed into old age, and new approaches were not approved.

I saw everything but it was difficult for me to intervene in this process from a professional point of view. Declarations about the problems were viewed negatively by the authorities and were considered to be unprofessional. There was an entire generation of nuclear engineers who appeared who knew their work but had no critical attitude towards the equipment, believing that the systems under which they worked guaranteed safety. I had many doubts and considered that we had to try a new approach.

I took many risks. During my professional life I have managed 10 projects involving scientific and research work. Five of these failed and caused a detriment to State funds of 25 million roubles. They all failed because the initial planning was not correct. They were attractive and interesting scientifically but the necessary materials were unavailable and there was no proper organization: which, for instance, could develop a new compressor or a new heat-exchanger. As a result, initially attractive goals during the design development stage appeared to be expensive and cumbersome and were not approved for implementation. I am currently afraid that two further projects of the 10 will have the same fate. However, the remaining three projects appear to be very successful and one of these on which we spent 17 million roubles brings us an annual profit which completely covers the 25 million rouble loss mentioned earlier.

19.1.11 RBMK reactors

I was not interested in a traditional non-nuclear reactor construction because at that time I did not consider that there was any degree of danger associated with the construction. Although I had a feeling of anxiety, comparison with foreign equipment allowed me to conclude that though there were many problems connected with the safety of existing equipment the risk with a nuclear NPP is less than with a traditional design. This was because with the latter design there is an enormous release into the atmosphere of carcinogenic substances.

As for the RBMK reactor, reactor specialists considered that this was a bad one. Bad not because of safety considerations but because of economic reasons: high consumption of fuel and high capital expenditure. As a chemist I was worried about the amount of graphite, zirconium and water which was used in the design of the RBMK. I also considered that the design of the reactor was unusual and inefficient from the point of view of protective systems which should come into operation under extreme conditions. Only an operator could put in control rods in the case of an accident: either automatically because of the indications of a detector, or manually. The mechanical parts could function either well or poorly and there was no other protective system which was independent of the operator. I have heard that specialists made proposals to the designer about changes in these protective systems but these were either rejected or developed very slowly.

19.1.12 Organization and responsibility

I would like to express my personal point of view which I am sure is not shared by my colleagues and causes disputes among them. The problem is that in our aviation and other branches of industry there is the phenomenon of a *scientific manager* and a *designer*. When we speak about, for example, the design of an aircraft, there should be a single *owner*, who is both designer and scientific manager and all the authority and responsibility should be concentrated in his single pair of hands. This is obvious to me.

When the use of atomic energy began in the USSR everything was reasonable because it was a new branch of science: nuclear physics. Scientific management was limited and the problems of the basic design of equipment was given to designers. A scientific manager was responsible for physical correctness and for safety but the designer was the person who was responsible for the implementation, albeit with constant consultations with physicists.

This division of responsibility was justified at the beginning of the development of atomic energy, but when design organizations grew in size, divisions of responsibility were not helpful for certain types of equipment such as a reactor. Multiple Councils existed both within a single establishment and between establishments, creating a situation of collective responsibility where it is impossible to know who, if anyone, has ultimate responsibility.

I believe that this situation is not correct and leads to confusion and a complete absence of personal responsibility for the quality of equipment which in turn, as shown at the Chernobyl NPP, can lead to very significant irresponsibility.

19.1.13 Equipment faults

Prime Minister Ryzkhov in a speech on 14 July said that he considered that the Chernobyl NPP accident did not occur by chance and that it was inevitable. I was impressed by his words and considered them to be correct. I recollected, for example, a case at an NPP when instead of making a proper electrical weld in the main pipe system, a worker was careless with this task in order to save time. This could have caused a terrible accident: a fracture of the main pipe system, and an accident with the VVER[11] with melting of the active zone of the reactor. This time the flaw was detected by an operator and after the interrogation it was determined that the pipe system had been inadequately welded although in the documentation all the necessary signatures were in place: of the electrical welding worker and the gamma spectroscopist who inspected the weld that did not exist at all.

Frequent flaws existed in equipment such as broken slide-valves, broken channels in RBMK reactors. This took place every year. Talks about the necessity for simulators spanned 10 years, and talks about the establishment of a system of diagnostics of equipment conditions spanned five years: with nothing achieved. I remember that the number of quality control engineers and other associated personnel decreased year by year. When we visited the building construction of NPPs we were surprised at those with a high responsibility accepting slipshod work.

After Ryzkhov's speech I began to study these problems more closely[12], taking a more positive attitude, and speaking out. This caused exceptional bursts of indignation and I was accused of being illiterate and not minding my own business and that it was impossible to compare one type of reactor with another. This was a very difficult situation.

19.1.14 Like a samovar

On the evening of the Chernobyl accident the number of enterprises had increased rapidly, in which production of different elements of NPP equipment was authorized. The *Atommash* enterprise was one of these and employed very many young people with the quality of specialists being

[11] VVER is the designation for a pressurized water reactor, as distinct from the graphite moderated RBMK.

[12] Some six weeks earlier, 2 June 1986, Legasov published in *Pravda* the Soviet Government line as follows. 'I am deeply convinced that atomic stations are the pinnacle of power engineering's achievements ... They are not only economically advantageous in normal operation in comparison with thermal stations, and they are not only cleaner, they are also preparing the base for the next spurt in technology. The future of civilization is unthinkable without the peaceful use of atomic energy ... An accident has occurred that was considered improbable. Therefore, lessons must be drawn from it: technical, organizational and psychological lessons. People died, and the material and moral damage was immense, but I am convinced that atomic power engineering will emerge even more reliable from this ordeal.'

very poor. This was confirmed during my visits to NPPs. After I had visited Chernobyl NPP I came to the conclusion that the accident was the inevitable apotheosis of the economic system which had been developed in the USSR over many decades. Neglect by the scientific management and the designers was everywhere with no attention being paid to the condition of instruments or of equipment. One of the NPP Directors said directly to me 'Why are you so anxious? An atomic reactor is like a samovar, it is more simple than a thermal power plant, we have experienced personnel and nothing will happen'.

19.1.15 Impossible to find a culprit

When one considers the chain of events leading up to the Chernobyl accident, why one person behaved in such a way and why another person behaved in another etc, it is impossible to find a single culprit, a single initiator of events, because it was like a closed circle[13]. Operators made

[13] The headlines on the back page of *Pravda*, 1 August 1987, read *Severe Lessons of Chernobyl: Trial of Culprits in the Accident at the Atomic Energy Station has Finished.* The essentials of the report were as follows. The charges against the accused were a severe indictment of indiscipline and irresponsibility in the professional obligations of those involved as well as a serious lesson to us all. The trial was completed on 29 July and was held in Chernobyl town in what was formerly The House of Culture, continuing for more than three weeks under Judge Raimond Brize of the Supreme Court. The former Director of the NPP, Bryukhanov, was considered the main culprit. He was not reliable in carrying out regulations and safety instructions. Fomin, who was the former Chief Engineer and Deputy Director, and his Deputy Chief Engineer, Djatlov, were also accused. All three, together with the former Chief of the Reactor Room, Kovalenko, did not have the required discussions and did not make an analysis of the planned experiment and made no additional measurements to ensure safety. The former Shift Chief, Rogoshkin, knew the situation but did not take any action as he did not want to become involved. He also did not monitor the experiment and when he received information about the accident he failed to activate the system for informing the NPP personnel. The former State Inspector of Gosatom and Energonadzor of the USSR, Laushkin, did not ensure that all the safety instructions and regulations were carried out at the NPP. The maximum sentences allowed were 10 years and the following sentences were passed:

Viktor Bryukhanov age 51: 10 years
He had alerted the Kiev Regional Civil Defence only three and a half hours after the accident and then only to declare that it was only a fire on the roof and that it would be put out. On 10 September 1991 he petitioned for early release, and this was successful. He was the last of those convicted to remain in prison.

Nikolai Fomin age 50: 10 years
Before the trial started attempted suicide by slashing his wrists with glass from his spectacles. Released early from prison because of his mental state. Went to work at the Kalinin NPP where his mental state was still said to be fragile.

Anatoly Djatlov age 57: 10 years
Early release because of failing health.

Boris Rogoshkin age 52: 5 years
Returned to work at Chernobyl after early release.

mistakes because in any case they wanted to complete an experiment. The experimental plan was of poor quality and was not confirmed by the specialists who were required to give it confirmation.

I have a record of the telephone calls of the operators just before the accident and it is terrible to read them. One operator calls to another: 'In the programme it is written what to do and then many points are crossed out, what should I do?' Another operator thinks for a few seconds and replies: 'Do those things which are crossed out'.

Somebody compiled the plan, someone crossed something out, someone signed it, someone confirmed it, Also, when the representatives of the Atomic Energy Inspectorate visited the NPP they were not aware of anything about that experiment.

19.1.16 Victory Day

On 9 May it seemed to us that the fourth reactor had finished burning. It was quiet externally and we wanted on that Victory Day[14] to have a holiday ourselves. Unfortunately a small but bright crimson spot was detected inside the reactor: indicating the presence of a very high temperature. It was difficult to determine exactly what was burning, but our holiday was spoilt and we decided to deposit an additional 80 tonnes of lead into the reactor's crater. The crimson luminescence disappeared and we celebrated Victory Day on 10 May.

19.1.17 Euphoria and tragedy

Even during these difficult days we had a sense of euphoria. This was a paradox and was due to the fact that we were participants in the clean-up operations of such a tragic event. Tragedy was the background under which everything took place. Somehow, our joy was established by the way in which everybody worked: for example, how quickly they responded to our requests, and how quickly engineering calculations were made. We

Alexander Kovalenko age 45: 3 years
Returned to work at Chernobyl after early release.

Yuri Laushkin age 50: 2 years
Died of stomach cancer soon after early release.

It also emerged during the Court proceedings summing-up that there was an atmosphere of lack of control and lack of responsibility at the NPP, and that people had played cards and dominoes and written letters at work, It also emerged that other accidents had only narrowly been avoided at the plant, notably in 1982 and 1985 when elementary safety rules had been ignored and supervisors had not been alerted.

[14] Victory Day on 9 May each year is the anniversary of the end of the Second World War, known in the USSR as the Great Patriotic War, and this day is always an official holiday in the Soviet Union.

«Мой долг рассказать об этом...»

Из записок академика В. Легасова

Figure 19.1. Valery Legasov: *Pravda* 20 May 1988. (Courtesy: Pravda.)

began to make the first calculations for the erection of a cupola[15] over the destroyed reactor.

19.2 Defenceless Victor—Margarita Legasov's title of her reminiscences

This title in *Trud* was followed by a quotation by Valery Legasov

> *There are two colour photos hanging in my office at work. One of them is of a Nuclear Atomic Plant, the other of storks. These photos hang near each other as a reminder of the close relationship between life, nature and technology, letting one know beforehand of the fragility of life, about the necessity to keep it. I recalled these photos when I worked in Chernobyl eliminating the consequences of the accident at the NPP. Really, could storks in the future, living on the earth, feel themselves to be safe with modern industry? Is such a peaceful coexistence possible? And if possible, then what should be done to achieve this?*

It was not until 10 years after the accident and eight years after Valery Legasov's death that his widow published a short memoir in *Trud* that unequivocally confirmed that her husband had committed suicide on 27 April 1988. They had first met when students in the same institute and together worked at a students' building construction project in what were termed in the USSR as the *virgin lands*[16]. Under the title *Defenceless Victor* she described her memories of Legasov's troubled times at Chernobyl and the period afterwards when he was, to a certain extent, ostracized by the establishment. She also includes interesting comments on what life was like for a senior scientist and his family in the Soviet system: very different from the experiences of Western scientists.

Last year we at last completed erection of a gravestone on his grave. This was with thanks to my son and daughter and a few supporters and colleagues of the Academician who helped to cover the expenses. That day when the sculptor invited me to his workshop and showed me the completed work, Valery returned home in the form of his bronze sculpture. He often had to travel away on business trips, we tried to be patient and wait for his return, but on 27 April 1988 he was transported away, already lifeless, forever.

[15] The term 'Sarcophagus' had not yet been introduced.

[16] The development of virgin and long-fallow lands in Kazakhstan in order to create the opportunity for the cultivation of grain. In the 1950s such work by volunteer students were considered by the State to be of great importance.

On Saturday 26 April 1986, Valery left for an ordinary business meeting where he learned about the Chernobyl NPP accident and that evening he was already 2 km away from the destroyed reactor. Life seemingly continued but terrible forebodings did not allow us to relax and stop worrying about his health. After 27 April our acquaintances began to say that badly irradiated victims of the accident had begun to be transported to Moscow to Hospital No. 6. Nobody could tell me when he would return.

On the morning of 5 May about 8am there was a ring at the door bell and Valery entered in a borrowed suit of clothes and carrying a polythene bag with belongings rather than his normal case. He was very thin, with a dark face, red eyes and the palms of his hands were tanned black. He only had time to wash, change, breakfast and ask about his two grandchildren before he had to leave at 10am for a meeting. There was no time to tell us what was the state of events at Chernobyl. Then at lunchtime one of his assistants telephoned and said that Boris Scherbina wanted him again at Chernobyl.

It was only when he returned home later that he was able to tell us that he had personally entered the most dangerous areas in the fourth reactor and how shaken he was at the criminal carelessness displayed at the NPP before the explosion.

He next returned home on 13 May and it seemed to us that the biggest difficulties were in the past: but we soon understood that we were mistaken. By summer Valery was already in poor health, suffering from frequent headaches, chronic insomnia, nausea and stomach illness. It was difficult to recognize the earlier Valery in this morally depressed man. He was taken many times for medical investigation to Hospital No. 6 of the atomic establishment. Heart insufficiency, serious leukocytosis, problems with his myelocytes and bone marrow were diagnosed, as well as neurosis. But no official diagnosis was made of radiation syndrome, although I had no doubt that it was so.

He became an Academician at the early age of 45 but some of the leading figures of Soviet science called him 'A boy from the chemical suburbs'. However, he was interesting to work with and liked jokes, being famous as an amusing raconteur, although everyone knew that science was the principal interest of his life. His private family life was unknown to his colleagues.

For five years, 1964–69, we lived in a flat of 22 square metres at Nizhegorodskaya Street. Though we could use only communal transportation we often made trips together with our two little children to Kuskovo, Ostankino and Arkangelskoye. In Tsaritsino we enjoyed ski holidays. It now seems that these were the happiest times of our lives.

Valery was a car enthusiast for the last 10 years of his life and loved driving at very high speeds. He had always wanted a private car and his first, which was also his last, was a GAZ-25 Volga which we bought in

1977 for 9500 roubles when he was a Candidate Member of the Academy of Sciences. The initial capital for the purchase was his quota from his State Prize received for his achievements in the field of chemistry.

We usually celebrated New Year in the circle of our family, sometimes in a rest house. One of these days a pure bred chau chau puppy appeared in our family and it was assumed that it was my New Year's gift. Ma Lu-Thomas, as she was called, would recognize only Valery as his owner and loved being in our car. She was inseparable from him and died just after Valery's death. He was also an adoring grandfather to Misha and Valerik and invented little poems for them and played charades.

As a boy he received a musical education and for many years was interested in listening and understanding classical music: Grieg, Sibelius, Shostakovich and Prokofiev. He was also fond of Schnitke. Over the years we bought tickets for many concerts in the Tschaikovsky Concert Hall of the Musical Conservatoire. Valery's last concert was in Lithuania in the summer of 1987: for flute and organ. Little did I know that soon afterwards Valery would make a first attempt to commit suicide. He swallowed a handful of Triptizol tablets but that time the physicians managed to save him.

In one Soviet TV programme is was said that Academician Legasov was a sincere believer. It is not so. From autumn 1987 he began to read the Bible and thought much about what he read. He was not baptised a Christian, but respected religion even though he was brought up an atheist.

He considered that the East was weak and during his business trips he tried to see as much as possible of culture. He very much wanted to visit one of the sacred Islamic places, the mausoleum of Hoja Ahmed Iasavi, and the monument erected in honour of the ancient Turkish poet who lived in the twelfth century and was an advocate of Sufism. We visited the ancient city of Yami and worshipped at the grave of the philosopher, and Valery often recalled his verses:

Having met a man of another faith
Don't be evil to him
The God does not like people
With a cruel heart...
After their death punishment
Waits for them...

On his return from the Chernobyl NPP Valery told very sparingly, with tears in his eyes, about the unpreparedness for the accident. Those days nobody could precisely estimate the number of victims, but Legasov understood better than others, the lack of necessary means of health protection: pure water, food products, iodine prophylaxis.

In August 1986 Valery Legasov presented a report to IAEA experts at a meeting in Vienna, about the causes and the consequences of the accident. His five-hour report was very well received and he returned home triumphal. But soon his mood changed. During the last two years after the accident he suffered great psychological trauma and his inner strength was broken.

Twice he was nominated for a high award from the State, and twice the nomination was cancelled. He received a suggestion that he might take up a position with the IAEA in the field of nuclear technology: again, obstacles appeared. There was also the planned nomination for Director of a Research Centre on the Problems of Industrial and Nuclear Safety: this came to nothing. His election as a Member of the French Academy of Sciences was apparently assured and although we went to Paris on 4 February 1988, his last business trip, he did not receive Membership. Also, just after his Paris trip he was hospitalized with acute leukocytosis, pneumonia and severe neurosis.

Chernobyl was not only a tragedy of international importance but it was also the personal tragedy of the gifted scientist Valery Legasov.

Figure 19.2. This 6 cm diameter medal is inscribed *TO PARTICIPANT OF THE ACCIDENT 1986 CHERNOBYL ATOMIC ENERGY STATION*. On the back of the medal there is a symbolism for peace showing a dove flying over the power station. (Courtesy: H Mayer, IAEA.)

Chapter 20

Under the Star of Chernobyl

Introduction

Under the direction of Dr Rotislav Omelyashko, the Ukranian Ministry for Emergency Situations and on Affairs of Population Protection against Consequences of the Chernobyl NPP Accident is responsible for cultural expeditions within the 30 km zone to document, and as far as possible to preserve, the heritage of this area which is within the Polissya district of the Ukraine, such as local handicrafts and the fabric of the churches, including icons. In 1996 a limited number of copies were printed of a photographic album presenting some of the results of these expeditions. The album was entitled *Under The Star of Chernobyl* and this chapter includes a small selection from this Ukranian Ministry collection of photographs accompanied by a commentary on important aspects of the history and culture of the Polissya region.

20.1 Origin of the names Polissya and Chernobyl

Chernobyl (in Ukranian *Chornobyl*) is a small ancient town in the northern part of the present day Kiev Region, a region known as Kiev *Polissya* (in Russian this is *Polessie* or *Poles'ye*) where the word Polissya is the Ukranian name for the forest region in northern Ukraine and southern Belarus. *Lis* (in Russian *les*) means forest or wood, and *po* is a preposition/prefix meaning on or in, depending on the context. Thus Polissya may be translated as woodland or forest land: its vegetation is represented mostly by mixed forests of Scots pine on sandy hills, meadows, swamps and bogs in river valleys.

After the accident, many people recalled the lines from the Bible in the Book of Revelations, chapter 8, verses 10–11 which says in the New King James version: 'Then the third angel sounded and a great star fell from heaven, burning like a torch, and it fell on a third of the rivers and on the springs of water. The name of the star is Wormwood. A third of the waters became wormwood, and many men died from the water because it was bitter'.

Why were these lines recalled? In the Ukranian and Russian translations of the New Testament the star is called *Zvezda Polyn* (Russian) or *Zirka Polyn* (Ukranian). The word *Zirka* means Star and *Polyn* is the vernacular name of many species of the genus *Artemisia*, which is a herb which is widespread in many regions of the world. Most species of the genus contain bitter substances, and thus the name is sometimes used as a poetic synonym for bitterness and sorrow.

In Ukranian the word Chornobyl usually means a particular species of *Artemisia*, namely *A. vulgaris*, known in English as mugwort. However, true wormwood is the bitter wormwood which is *A. absinthium* and this also occurs in the area, but is never called chornobyl by the local population: they usually use the generic term *polyn*. Two other well known examples of aromatic plants of the genus Artemisia are sagebrush (*A. tridentata*) which is a shrub often found in North America, and tarragon (*A. dracunculus*) which is used for seasoning. Legend has it that as the serpent slithered out of the Garden of Eden, wormwood first sprang up in the impressions on the ground left by its tail. Another legend is that in the beginning it was called *Parthenis absinthium*, but Artemis, the Greek goddess of chastity, benefited from it so much that she named it after herself, *Artemisia absinthium*. The Latin meaning of *absinthium* is *to desist from*.

Although it is one of the most bitter herbs known, it has for centuries been a major ingredient of aperitifs and herb wines. Both absinthe and vermouth get their names from this plant, with vermouth being an 18th century French variation of the German *wermut*, itself the origin of the English name wormwood. In the previous centuries when many of the population were superstitious, wormwood was hung by the door to keep away evil spirits and to deter night-time visitations by goblins. It was also made a constituent of ink to stop mice eating old letters and was strewn on floors of rooms to prevent fleas. It is also believed to be the herb that William Shakespeare had in mind when Oberon lifted the spell from Titania with 'the juice of Dian's bud', Artemis being known to the Romans as Diane or Diana.

20.2 History

The region of Polissya contains evidence of early palaeolithic sites dating back to some 35–100 thousand years and during the later palaeolithic period some 10–25 thousand years ago the inhabitants would have been hunting mammoths. When the glaciers retreated the modern landscape started to form, creating the sand dunes which are special to Polissya.

Herodotus in the 6th century BC recorded that Polissya was an inaccessible fortress against enemy forays and hid the Nevrs tribe when the Scythians invaded. Herodotus also described the forest inhabitants in the following terms. 'Evidently these people are wizards and it is confirmed that once every year each Nevr for several days turns into a wolf and then changes back into a human being.' The Nevrs were an ancient tribe, who were not Slavs, and they inhabited Polissya in the early iron age.

The Slavic tribe of Drevlyans with their lands of Drevlanchina are mentioned in early chronicles and it is stated to be in the river valley among the rivers Irpen, Sluch and Pripyat. It is recorded that Prince Oleg, a Varangian by origin, subjugated Kiev and Drevlanchina during the years 883–885 but after his death the Drevlyans received their independence only to be invaded again when in 945 they rioted and killed the then ruler, Prince Igor. His widow, Olga, avenged the death of her husband by burning down the capital of Drevlyanschina and making the local Prince Mal and his daughter Malusha her slaves. Later, Malusha became the mother of the Great Prince of Kiev, Vladimir, who ruled in the years 980–1015.

The lands known as Kievan Rus obtained this name from the river Ross which flows near Kiev but after the foundation of Moscow as the centre of ancient Russian lands, power moved to what was originally called Moscow Rus, and only later, Russia. Historically, Kievan Rus was the land of the Eastern Slavs and the *cradle of three nations*: Russians, Ukranians and Belorussians. The Dnieper river was a natural waterway for journeys by the Varingians, who were Vikings from the north, and by the Greeks. The natural caves on the banks of the Dnieper were used in the earliest times to store trade goods and later these caves became part of the Kiev-Perchersk Lavra Monastery which was founded in 1061 and named after the old Russian word for cave, *perchera*.

Kiev as a city was founded some 1500 years ago: officially this date was celebrated in 1982 but historians consider the anniversary to be any year between 1975 and 2025. According to legend it was founded by three brothers, Kiy, Schtek, Khoriv and their sister Lybed, and it was called Kiev in honour of the eldest brother. Ancient chronicles first mention 'the Russian land' in connection with Kiev in 852 a decade the Varingians led by Askold and Dir captured the city. For many centuries the lands of the Eastern Slavs were separated into small kingdoms ruled by local princes but eventually they were unified by a Grand Prince of Kiev and from the late 10th century for well over 200 years the Kievan state dominated all Russian cities and principalities.

Christianity was introduced in 988 during the reign of Grand Prince Vladimir Svyatoslavych who declared it to be the state religion and in that year the whole population of Kiev was baptised in the waters of the Dnieper. Vladimir's statue can be seen in figure 20.1 facing the city of Kiev across the Dnieper. However, the first Christian in Kievan Rus was

Figure 20.1. Statue of Prince Vladimir overlooking the river Dnieper and the city of Kiev. (Courtesy: TASS.)

his grandmother Grand Princess Olga who was baptised in Byzantium. This was before the schism between the Catholic and Orthodox churches which took place in 1054 and is why in the cathedral of St Sophia in Kiev it is possible to see a fresco of Pope Clement. The city's main street is still called the Kreshchatik from the word *kreshcheniye* meaning baptism.

The Pechenegs tribe who lived on the steppes of what is now modern southern Ukraine regularly terrorized the city of Kiev and even northern Ukraine: Polissya. The were defeated in battle by the son of Prince Vladimir, Yaroslav the Wise, who in celebration of his victory built the cathedral of St Sophia and a series of Golden Gates in order to make his capital city the equal of Constantinople. At this time the population of Kiev was 50 000 in comparison with the 5000 of Paris and the 2000 of London. Moscow was founded more than a century later. Yaroslav at the age of 50 married a Scandinavian Princess Irina who was only 16 and they had four sons and four daughters of whom three became queens: of Norway, Hungary and France. One of Yaroslav's great grandsons was Prince Yuri Dolgoruki who is recognized as the founder of Moscow.

Many of the early monasteries and churches were burnt to the ground in 1240 as the Tatar–Mongol hordes led by Khan Batu, grandson of Ghengis Khan, swept through Russia, and even two centuries later travellers would write of Kiev as a 'dead city'. For the four centuries following the Tatar–Mongol yoke Kiev came under the sway of first the Grand Duchy of Lithua-

Figure 20.2. Example of a modern spinning wheel, abandoned in one of the deserted villages within the 30 km zone. (Courtesy: R Omelyashko.)

nia and then the Kingdom of Poland, but in 1654 they joined forces with the state of Muscovy and sufficiently recovered to win religious, political and economic independence. However, this renaissance and autonomy was taken away during the reigns of Peter the Great in the 18th century, and of Catherine the Great and Ukraine became only a province of Russia: called Small Russia to distinguish it from Great Russia, that is, Moscow. The lands of Kiev became border lands of the Moscow kingdom: perhaps the origin of the name Ukraine is from the Russian word *krai* which means *border*.

Then in the 19th century even the Ukranian language was prohibited in official life and in 1918 after the Bolshevik revolution there was an attempt to resurrect the state of Ukraine, as the Ukranian Peoples' Republic, but this failed and the Ukraine became one of the 15 republics within the USSR. Only with the declaration of Sovereignty of Ukraine from 16 July 1990 and the decision of the Ukranian parliament on 24 August 1991, confirmed by referendum on 1 December 1991, did Ukraine become again an independent sovereign state after a lapse of several centuries.

20.3 Culture

By the 10th century technical developments in handicraft making had occurred in Polissya which were in advance of those in many other areas.

Figure 20.3. Log beehive in the village of Glinka in Chernobyl district. (Courtesy: R Omelyashko.)

These included working the local brown Ovruch shale into spindles for weaving, figure 20.2, and these found their way into adjacent European countries and Polissya became a basic supplier of flax fibres.

This shale was also worked for other purposes, including the making of sarcophaguses, decorations for women and small icons. In addition, Ovruch quartzite was widely used in building construction for cornices and mosaic floors. Before the adoption of Christianity it was used for stone idols and later for Christian crosses. Pieces made by Polissya workmen are to be found in Kiev in the church of St Cyril and in the cathedral of St Sophia. Natural sources of multicoloured amber were also considered to be valuable and these have survived in examples of local necklaces and of amulets which were used for talismans.

The area also has a tradition of bee keeping and in ancient times forest nests of bees were kept in natural hollows within tree trunks. Surprisingly these log bee hives were still to be found in the 1980s in the exclusion zone, figure 20.3, particularly in Poznan in the Rivno region of Polissya, with the local peasants using instruments to obtain the honey which were similar to those which existed in the 10th to 12th centuries.

For centuries the inhabitants of Polissya, as well as bee keeping, also processed furs and skins, spun broadcloth, made wicker baskets, collected medicinal herbs and also produced wood-tar. The latter was due to the wide availability of oak and birch trees and this led also to a lucrative trade

Figure 20.4. Sleigh from the village of Maryanivka in Polissya district. (Courtesy: R Omelyashko.)

in charcoal. This was an essential raw material for the local production of glass (the silicates were obtained from the sand dunes), soap and dyes. In particular, Polissya manufactured a red dye called *chervets* using as a colour base a local insect. This became very valuable because it did not fade in the sun or rain and because of this property it was used by the Kiev Princes and by the Cossacks to colour their war banners with the special streamers: the *gonfalons*. The word *chervets* is derived from the Russian *cherv*, meaning *worm*, and is the name of a genus of insects called Coccus which are used to produce the carmine dye called cochineal. It is also recorded that when the Polish noble families were landlords they levied a tax which consisted of 'spoonful of *chervets* from each household'.

At the end of the 17th century there was intensive felling of trees, related to the widening of channels in the Dnieper and Oginsky river system, which connected with the smaller rivers of the Pripyat basin, so that river trade could extend to the Baltic. This affected not only the forests of Polissya but also many over the whole of Europe. Polissya timbers were sent as far away as Gdansk, Köningsberg (now Kaliningrad) and the Black Sea ports and by the time of the 19th century the oak forests of Polissya had been reduced by half. Wood working expertize is also evident in many everyday usage artefacts such as sleighs, figure 20.4, and oxen yokes.

For many centuries the houses of Polissya were made of wood with walls from rough hewn squared cross-section logs and roofs covered with

Figure 20.5. The Golden Gates were the principal entry to the old city of Kiev, and obtained their name from their similarity to the Constantinople Golden Gates which also served as the main entrance into the city. Their construction in Kiev is mentioned in chronicles of the year 1037. Over the passage is the golden domed Church of the Annunciation. (Photograph: R F Mould.)

shingle, wood chips or thatch. A feature of Polissya houses was the massive often ornamented longitudinal girder, the so-called *girder-father* which was imbued with magic qualities. The craftsmen of Polissya are considered to be specialist architects and builders for their times, and when the Golden Gate in Kiev was being reconstructed, figure 20.5, ethnographers linked its architectural design with those of Polissya, particularly the Drevlyan houses. This was the so-called six angled architecture which was the basis of typical Ukranian church cupolas, but this had been developed from earlier six angled wooden walls and six angled roofs of houses and grain storage buildings in Polissya.

Iron tools such as axes, hammers, pincers, scythes, sickles and iron ploughs were also made in Polissya as iron smelting was practical because of rich local sources of natural materials: bog iron ore, hematite and swamp ores. Archeologists have discovered such iron instruments dating from the 10th–12th centuries. In the 19th century the work of smiths extended to the production of decorative ironwork such as seen in figure 20.6.

Village fairs were a feature of the region and they always included sales of glass and pottery objects, and in some cases special grey or black ceramic ware. These were made in what were termed *pottery villages* which

Figure 20.6. Decorative ironwork with a geometrical design from the village of Zamoshnya in Chernobyl district. (Courtesy: R Omelyashko.)

specialized in this work, and in which the villages of Lubyanka, Mlachivka, Gavrylivka and Plakhtyanka were particularly famous.

The church of St Eliah in Chernobyl town has already been mentioned in section 6.5, with figures 6.9 and 6.11 showing Ukranian icons. Figure 20.7 is an aerial view of this church which is taken from cover of the book *Under the Star of Chernobyl* and figure 20.8 shows one of the wall paintings in the church. Figure 20.9 is of another church in Polissya and is included because of its typical wooden design for this area.

In Ukraine and also in Russia, St Eliah is often compared with the pagan gods of thunder: the Scandinavian Thor, eastern Slavic Perun, Greek Zeus or Roman Jupiter. This was because, after converting to Christianity, some of the functions of the pagan gods were in folk tradition transferred to Christian saints. Perhaps because Elijah rose to heaven in a chariot of fire, figure 6.9, superstition had it that when there is heavy rain with thunder and lightning it means that St Eliah is travelling across the sky in his fiery chariot.

In the Rivno region a pagan ritual still exists for showing respect to ancestors in that families visit the burial place and cry collectively. Such burial sites are often surrounded by a wooden fence and within this area is placed a small wooden log house. Grave crosses are also often draped with specially embroidered ritual towels called *rushniks*, examples of which are seen in figure 20.10. There are traditional days for remembering and for

Figure 20.7. Aeriel view of the church of St Eliah in Chernobyl town. (Courtesy: R Omelyashko.)

Figure 20.8. Wall painting in the church of St Eliah. (Courtesy: R Omelyashko.)

Figure 20.9. 17th century wooden church in the village of Tolsty Les. It was totally destroyed by an accidental fire on 24 April 1996. (Courtesy: R Omelyashko.)

Figure 20.10. Ritual embroidered towels placed to honour the dead. (Courtesy: R Omelyashko.)

Figure 20.11. An evacuated family from Polissya being welcomed to their new home. The plate of bread and salt is held in a rushnik and the girl on the left is wearing a traditional Ukranian blouse. (Courtesy: TASS.)

praying for the dead, usually in Spring after Easter. The people gather in the cemetery bringing vodka, wine and food and have what is in effect a funeral feast. On leaving the cemetery they place some food on the graves which is then taken and eaten by beggars. This is in the hope that the beggars will also respect the memory of those buried. Harvest festivals also have paganism as their origin and in many villages in Polissya this Orthodox feast is called among the peasants, the *Apple Saviour*. The women stand in line with baskets of applies and the priest, preceded by a large cross, blesses the apples and the people.

These *rushniks* have great ritual significance in the villages of Polissya and they serve several purposes as well as those described above. Apart from the obvious use for drying wet hands or dishes, they are also used to hold hot cooking pots to avoid being burnt, they are used as tablecloths with smaller versions being used as table napkins, as curtains for windows, and to place around icons which are usually hung in the corner of a house. They are also used to decorate icons in churches. In addition, according to Ukranian tradition a Ukranian girl before marriage should embroider a collection of *rushniks* for her future home and when a young man proposes marriage and the girl agrees she must give him one of her *rushniks* and wrap it around his hands.

Another tradition in the Ukraine, in Russia and in some other Slavic

Figure 20.12. Girls in local dress at the turn of the 20th century, from the village of Orane in Ivankiv district. The traditional folk dress for women of Polissya are kercheifs and aprons. (Courtesy: R Omelyashko.)

countries is that honoured guests are met with a gift of bread and salt and this is wrapped in a rushnik. This tradition is seen in figure 20.11 which shows an evacuated family from Chervone Polissya state farm being welcomed to their future home in September 1986 in the newly built village of Ternopolskoye in Makarov district, consisting of 150 homesteads which were constructed in 50 days.

Each region in Ukraine has its own traditional designs for rushniks, shirts for men and blouses for women and examples are shown in figures 20.12–20.14. In general, with an embroidered apron and blouse a married woman would wear a kerchief whereas young girls would wear their hair tied in ribbon or wear a crown made from artificial flowers with multicoloured ribbons, such as seen in figure 20.11.

Figure 20.13. Apron design from the village of Levkovichi in Ovruch district. (Courtesy: R Omelyashko.)

Figure 20.14. Shirt design from the village of Pokaliv in Ovruch district. (Courtesy: R Omelyashko.)

Figure 20.15. Collection of Ukrainian dolls which all fit within each other, inside father with the balalaika. From a street seller in the main street of Kiev, the Kreschatik, 1998.

Glossary

ABCC: Atomic Bomb Casualty Commission, the forerunner of the RERF.

ACs: Automatic control absorbing rods of the RBMK-1000.

ACS: American Cancer Society.

Acute effects: Early effects.

Aetiology: The study or the theory of the factors that cause disease and the method of their introduction into the host. The cause(s) or origin of a disease or disorder.

AF: Attributable fraction, *see section 17.3.*

Agranulocytosis: Symptom complex characterized by marked decrease in the number of granulocytes and by lesions of the throat and other mucous membranes, of the GI tract and of the skin.

AHS: Adult Health Study of atomic bomb survivors, *see section 16.1.*

AJCC: American Joint Committee for Cancer Staging.

Allogenic TABM: Donor or homologous transplantation of allogenic bone marrow.

Anaemia: A reduction below normal in the number of erythrocytes per cm^3, in the quantity of haemoglobin or in the volume of packed red cells per 100 ml of blood which occurs when the equilibrium between blood loss and blood production is disturbed.

Angiopathy: Any disease of the vessels.

Aplasia: Lack of development of an organ or tissue, or of the cellular products from an organ or tissue.

ARS: Acute radiation syndrome, *see section 1.2.1.*

Artemisia: Botanical genus for a series of aromatic herbs which include wormwood which in the Ukranian language is called Chornobyl and in the Russian language Chernobyl.

Arteriosclerosis: A group of diseases characterized by thickening and loss of elasticity of arterial walls.

ASTRO: American Society for Therapeutic Radiology and Oncology.

ATB: Atomic bomb.

Atherosclerosis: An extremely common form of arteriosclerosis, in which deposits of yellowish plaques (atheroma) containing cholesterol, lipoid material and lipophages are formed within the intima and inner media of large and medium arteries.

^{198}Au: Symbol for the radionuclide of gold with mass number 198.

BEIR V: National Academy of Sciences, National Research Council, Committee V on Biological Effects of Ionizing Radiation, USA.

Benign: Not malignant.

Bfs: Bundesamtes für Strahlenschutz und der Strahlenschutzkommission.

BMT: Bone marrow transplantation.

Bq: Becquerel, the SI unit of radiation activity.

Carcinogenesis: Production of carcinoma.

Cataract: An opacity, partial or complete, of one or both eyes, on or in the lens or capsule: especially an opacity impairing vision or causing blindness.

Centromere: The constricted portion of the chromosome at which the chromatids are joined and by which the chromosome is attached to the spindle during cell division: see also Dicentric.

Chromosome: *see section 9.5.2.*

Chronic effects: Late effects.

Ci: Curie, a unit of activity.

CI: Confidence interval, *see section 17.5.*

CNS: Central nervous system.

CRIs: Combined radiation injuries.

^{137}Cs: Symbol for the radionuclide of caesium with mass number 137.

Cytogenetics: *see section 9.5.2.*

Cytoplasm: The protoplasm of a cell exclusive of that of the nucleus.

DDREF: Reduction factor for extrapolating from high dose and high dose rate (e.g. ATB) to low dose and low dose rate (e.g. Chernobyl).

Deletions: In genetics, usually a chromosome aberration in which a portion of the chromosome is lost.

Dentin: The hard portion of the tooth surrounding the pulp, covered by enamel on the crown and cementum on the root, which is harder and denser than bone but softer than enamel.

Deuterium: Radionuclide of hydrogen with one neutron and one proton in the nucleus. Symbol is usually D but can be ^{2}H. The symbol for heavy water is D_2O. Stable, i.e. non-radioactive, hydrogen has only a single proton in its nucleus.

Df: Decontamination factor, *see section 12.1.7.*

Dicentric: In genetics, a structurally abnormal chromosome with two centromeres.

dl: decilitre = 10^{-1} litre.

DNA: Deoxyribonucleic acid, *see section 9.5.2.*

DU: Depleted uranium.

EAW: Emergency accident worker in the clean-up operations or healthcare delivery at Chernobyl.

EBRD: European Bank for Reconstruction and Development.

EDR: External dose rate reduction factor, *see section 12.1.7.*

Eosin: A rose coloured stain or dye, typically the sodium salt of tetrabromfluorescein.

Eosinophil: A structure, cell or histological element readily stained by eosin.

EPR: Electron paramagnetic resonance, *see section 9.4.*

ERR: Excess relative risk, alternative terminology is ESR, *see section 17.1.*

Erythrocyte: One of the elements found in peripheral blood, also called red blood cell or corpuscle.

ESR: Electron spin resonance, alternative terminology is EPR which is now more commonly used, *see section 9.4.*

Etiology: see Aetiology.

EURT: East Urals Radioactive Trace, *see figure 4.6.*

FAO: Food and Agriculture Organization of the United Nations.

FISH: Fluorescent *in situ* hybridization, *see section 9.5.4.*

Follicle: A sac or pouch-like depression or cavity, e.g. follicular carcinoma of the thyroid.

FRG: Federal Republic of Germany.

Gaussian distribution: An alternative name for the normal distribution, *see section 17.5.*

Goitre: Enlargement of the thyroid gland.

Granulocyte: Any cell containing granules, especially a leucocyte containing granules in its cytoplasm.

Granulocytopenia: Alternative term to agranulocytosis.

GW(e): Gigawatts of electrical energy.

Gy: Gray, the SI unit of absorbed dose.

GyEq: Gray equivalent, *see section 4.10* for doses from gamma rays and neutrons combined.

^{3}H: Symbol for the radionuclide of hydrogen with two neutrons and one proton in the nucleus: its special name is tritium.

ha: Hectare. 1 ha = 10 000 m^2 = 2.4711 acres.

Heamopoiesis: Blood cell production.

Haemoglobin: The oxygen-carrying pigment of the erythrocytes, formed by the developing erythrocyte in bone marrow.

HICARE: Hiroshima International Council for Healthcare of the Radiation Exposed.

Hiroshima ATB: This ATB, called *Little Boy*, employed a gun barrel in which a *bullet* of ^{235}U was fired against a target of ^{235}U. The fusion of these two sub-critical masses of ^{235}U produced a critical mass that caused the uranium nucleii to undergo rapid fission. The instantaneous energy released produced a detonation in which the gun barrel was vapourised. The heat and explosive shock wave that followed was responsible for most of the deaths. Fewer than 10% of those people within 0.5 km of the hypocentre survived. At distances of 2 km some 80–90% survived. The ATB was dropped on 6 August 1945 with a yield equivalent to 12 500 tonnes of TNT and exploded at a height of 580 m. By the end of 1945 the death toll was estimated at 140 000.

Histiocytosis: Condition marked by the abnormal appearance of histiocytes (macrophages) in the blood.

Histiocytosis X: A generic term embracing eosinophilic granuloma, Letterer–Siwe disease and Hand–Schiller–Christian disease.

Hypertension: Persistently high arterial blood pressure.

Hyperthyroidism: Condition of excessive functional activity of the thyroid and excess secretion of thyroid hormones marked by goitre, tachycardia or atrial fibrillation and other signs.

Hypothyroidism: Deficiency of thyroid activity.

^{131}I: Symbol for the radionuclide of iodine with mass number 131.

IAEA: International Atomic Energy Agency.

IARC: International Agency for Research on Cancer, WHO, also termed UICC, Union International Contre le Cancer.

ICD No.: International Classification of Disease numbers for specific cancer sites, *see table 18.1.*

ICD-10: ICD revision 10.

ICRP: International Commission on Radiological Protection.

ICRU: International Commission on Radiological Units.

INSAG: International Nuclear Safety Advisory Group.

IPHECA: International Programme on the Health Effects of the Chernobyl Accident. This is a WHO programme.

K2/R4: Khmelnitsky 2 and Rovno 4 proposed nuclear reactors in the Ukraine.

KGAE: State Committee for the Utilization of Atomic Energy, USSR.

LACs: Local automatic control absorbing rods of the RBMK-1000.

LD_{50}: Median lethal dose.

Lens: The transparent biconvex body of the eye situated between the posterior chamber and the vitreous body.

LEP: Local emergency protection absorbing rods of the RBMK-1000.

LET: Linear energy transfer.

Leucocyte: *see section 9.5.2.*

Leucopenia: *see section 9.5.2.*

Leukocytosis: Transient increase in the number of leucocytes in the blood.

Liquidator: Clean-up or health care delivery worker at Chernobyl 1986–89, *see also* EAW.

LSS: Life span study of atomic bomb survivors, *see section 16.1.*

Lymph: A transparent, slightly yellow liquid or alkaline reaction, found in the lymphatic vessels and derived from the tissue fluids.

Lymphocyte: *see section 9.5.2.*

Lymphoid: Resembling or pertaining to lymph of tissue of the lymphoid system.

m: metre.

Malignant tumour or malignancy: Cancer.

Mass number: Number of protons and neutrons in the nucleus of an atom.

MCP: Main circulating pump in Unit No. 4.

Metastasis or metastatic cancer: Secondary spread of cancer from the primary site.

Morbidity: The incidence or prevalence of a disease or of all diseases in a population.

MPD: Maximum permissible dose.

MR: Mortality ratio, *see section 17.4.*

MRR RAMS: Medical Radiological Research Centre of the Russian Academy of Medical Sciences, Obninsk.

MW(e): Megawatts of electrical energy.

Myelodysplasia: Defective development of any part of the spinal cord.

Myelocyte: A precursor in the granulocytic series. Any cell of the grey matter of the nervous system.

Myeloid: Pertaining to, derived from or resembling bone marrow. Pertaining to the spinal cord. Having the appearance of myelocytes but not derived from bone marrow.

Nagasaki ATB: This ATB, called *Fat Man*, was an implosive device that employed a sphere of ^{239}Pu that was compressed uniformly by enormous force by an outer mantle of conventional high explosive. The ATB was dropped on 9 August 1945 and exploded at a height of 503 m with an estimated force of 22 000 tonnes of TNT. By the end of 1945 the death toll was estimated at 70 000.

NCRP: National Council on Radiation and Measurements, USA.

NEA: Nuclear Energy Agency of OECD.

NIST: National Institute of Standards and Technology, Gaithersburg, Maryland, USA.

NPP: Chernobyl Nuclear Power Plant, or nuclear power station.

NRC: Nuclear Regulatory Commission in the USA.

Oblast: A large territorial and administrative division. Each country (Ukraine, Belarus, Russia) consists of a number of oblasts.

OECD: Organization for Economic Cooperation and Development.

Oncogene: A gene found in the chromosomes of tumour cells whose activation is associated with the initial and continuing conversion of normal cells into cancer cells.

Oncogenesis: Production or causation of tumours.

ORM: Operational reactivity margin.

$P < 0.05$: Statistically significant at the $P = 0.05$ level of probability. $P > 0.05$ is not significant.

Papilla: A small nipple-shaped projection, elevation or structure, as in papillary carcinoma of the thyroid.

PBSCT: Peripheral blood stem cell transplantation.

Perinatal: Occurring in the period shortly before and after birth.

Plasma: *see section 9.5.2.*

Platelet: A disc shaped structure, 2–4 μm in diameter, found in the blood of all mammals and chiefly known for its role in blood coagulation.

Polissya: Forest region in northern Ukraine and southern Belarus which includes the Chernobyl NPP and the 30 km exclusion zone.

Protoplasm: The translucent polyphasic colloid with water as the continuous phase that makes up the essential material of all plant and animal cells: *see also Cytoplasm.*

Pu: Symbol for plutonium, used when referring to the radionuclides ^{239}Pu, ^{240}Pu and ^{241}Pu.

PY: Person-years, *see section 17.2.*

R or r: Roentgen, the unit of exposure.

R4: *see* K2/R4.

Ra: Symbol for radium.

rad: Unit of radiation dose, 1 cGy = 1 rad.

Radionuclide: An alternative term for a radioactive isotope.

Rayon: A smaller territorial and administrative division than an oblast. Each oblast consists of a number of rayons.

RBE: Relative biological effectiveness.

RBMK: The name RBMK-1000 and RBMK-1500 are those for designs of Soviet nuclear reactors. The initials RBMK is an abbreviation in Russian of terms meaning reactor of high output, multichannel type.

REAC/TS: US Department of Energy, Radiation Emergency Assistance Center and Training Site, Oak Ridge.

Reciprocal translocation: *see section 9.5.2.*

RERF: Radiation Effects Research Foundation, Hiroshima.

Retina: The innermost of the three tunics of the eyeball, surrounding the vitreous body and continuous with the optic nerve.

Retinopathy: Inflammation of the retina or degenerative non-inflammatory conditions of the retina.

RNMDR: Russian National Medical Dosimetric Registry.

RR: Relative risk, *see section 17.2.*

s: second.

Sclerosis: Induration, or hardening, especially hardening of a part from inflammation and in diseases of the interstitial substance.

SE: Standard error, *see section 17.5.*

SEER programme: Surveillance, Epidemiology and End Results progam in the USA. The nine participating cancer registries include the five states of Connecticut, Iowa, New Mexico, Utah and Hawaii and four metropolitan areas: San Francisco Bay area in California, the Detroit metropolitan area in Michigan, the Atlanta, Georgia metropolitan area and the Seattle, Washington area. This represents about 10% of the total population of the USA.

Semi-natural environment: An environment with characteristics intermediate between those of managed agricultural land and those of the natural environment.

SI: Systéme Internationale. This is a system of units of measurement.

Solid tumours: All malignant tumours excluding tumours of the blood and blood-forming organs, plus brain and central nervous system tumours of benign and uncertain behaviour. This definition is used by RERF when analysing follow-up of atomic bomb survivors.

^{90}Sr: Symbol for the radionuclide of strontium with a mass number 90.

Sv: Sievert, the SI unit of dose equivalent.

T_4: Symbol for thyroxine.

TABM: Transplantation of allogenic bone marrow.

THELC: Transplantation of human embryonic liver cells.

Thrombocytopenia: Decrease in the number of blood platelets.

Thyroidectomy: Removal of the thyroid gland.

Thyroxine: A crystalline iodine-containing hormone.

TMI: Three Mile Island.

Translocation: *see section 9.5.2.*

Tritium: *see* ^{3}H.

TSH: Thyroid stimulating hormone.

U: Symbol for uranium used when describing the radionuclides ^{235}U and ^{238}U.

UICC: *See IARC.*

UNSCEAR: United Nations Scientific Committee on the Effects of Atomic Radiation.

UV: Ultraviolet.

VNIIAES: All-Union Scientific Research Institute for Nuclear Power Plant Operation.

WHO: World Health Organization.

Cartoon of the author paying his bill at the Hotel Minsk, Gorky Street, Moscow. (*New Scientist* 31 January 1984.)

References

Chapter 1

[1] Mould R F 1983 *A Century of X-Rays and Radioactivity in Medicine* (Bristol: Institute of Physics Publishing)

[2] International Commission on Radiological Protection 1991 *Recommendations of the International Commission on Radiological Protection* ICRP Publication No. 60 in Annals of the ICRP (New York: Pergamon)

[3] Robeau D G 1996 Environmental impact assessment *Proc. Int. Conf. One Decade After Chernobyl, Summing up the Consequences of the Accident* Vienna 8–12 April 1996 (Vienna: IAEA) pp 73–6

[4] United Nations Scientific Committee on the Effects of Atomic Radiation (UNSCEAR) 1993 *Sources, Effects and Risks of Ionizing Radiation* 1993 Report to the General Assembly (New York: United Nations)

[5] USSR State Committee for the Utilization of Atomic Energy 1986 *The Accident at the Chernobyl Power Plant and its Consequences* Reports compiled for the International Atomic Energy Agency's (IAEA) Experts' Meeting 25–29 August 1986 (Vienna: IAEA)

[6a] Souchkevitch G N 1999 Classification and terminology of radiation injuries *Int. J. Radiation Medicine* **1** 14–20

[6b] Bily D O and Obodovskii I A 1999 *Personal communication*

[7] Guskova A 1987 *Lessons of Chernobyl* Presentation Institute of Biology Seminar 11 April 1987 London

[8] Kovalenko A N ed 1998 *Acute Radiation Syndrome: Medical Consequences of the Chernobyl Catastrophe* (Kiev: Ivan Fedorov)

[9] International Atomic Energy Agency 1998 *Diagnosis and Treatment of Radiation Injuries* Safety reports series no. 2 (Vienna: IAEA)

[10] Court L and Dollo R 1999 *Accident de Tokai-Mura* Memorandum 15 October 1999 Electricité de France

[11] Clarke R 1999 Control of low-level radiation exposure: time for a change? *J. Radiol. Prot.* **19** 107-15

[12] Jaworski Z 1995 Stimulating effects of ionizing radiation: new issues for regulatory policy *Regulatory Toxicology and Pharmacology* **22** 172–9

[13] Rossi H H and Zaider M 1997 Radiogenic lung cancer: the effects of low doses of low linear energy transfer (LET) radiation *Radiat. Environ. Biophys.* **36** 85–8

[14] Waligorski M P R 1997 On the present paradigm of radiation protection: a track structure perspective *Nukleonika* **42** 889–94
[15] Rossi H H 1999 Risks from less than 10 millisievert *Radiation Protection Dosimetry* **83** 277–9
[16] Jaworski Z 1999 Radiation risks and ethics *Physics Today* **52** 24–9

Chapter 2

[1] Patterson W C 1986 *Nuclear Power* (Harmondsworth: Penguin Books)
[2] USSR State Committee for the Utilization of Atomic Energy 1986 *The Accident at the Chernobyl Power Plant and its Consequences* Reports compiled for the International Atomic Energy Agency's (IAEA) Experts' Meeting 25–29 August 1986 (Vienna: IAEA)
[3] International Nuclear Safety Advisory Group 1986 *Summary Report on the Post-Accident Review Meeting on the Chernobyl Accident* Safety series No. 75-INSAG-1 (Vienna: IAEA)
[4] United Nations Scientific Committee on the Effects of Atomic Radiation (UNSCEAR) 1988 *Sources, Effects and Risks of Ionizing Radiation* 1988 Report to the General Assembly (New York: United Nations)
[5] Lederman L 1996 Safety of RBMK reactors: setting the technical framework *IAEA Bulletin* **38/1** 10–16
[6] International Nuclear Safety Advisory Group 1992 *The Chernobyl Accident: Updating of INSAG-1* Safety series No. 75-INSAG-7 (Vienna: IAEA)

Chapter 3

[1] Marchenko T, Smolyar I and Torbin V 1996 *Community Development Centres for Social and Psychological Rehabilitation in Belarus, Russia and Ukraine* (Paris: UNESCO Chernobyl Programme)
[2] *Pravda* 8 May 1986
[3] Towpik E and Mould R F eds 1998 *Maria Sklodowska-Curie Memorial Issue of the Polish Oncological Journal Nowotwory* (Warsaw: Nowotwory)
[4a] *Izvestia* 19 May 1986
[4b] Mettler F A 1999 *Personal communication*
[5] Scherbak Y M 1996 Ten years of the Chornobyl era *Scientific American* **April** 44–9
[6] Scherbak Y M 1989 *Chernobyl: a Documentary Story* (London: Macmillan Press)
[7] *Wochenspiegel Lübeck* 22 June 1990
[8] Wendhausen H 1996 *Personal communication*
[9] *Washington Post* national weekly edn 26 June–2 July 1995
[10] Velikhov E 1993 *The aftermath of Chernobyl: developments in Russia, Ukraine and Belarus* Presentation British Pugwash Group public discussion meeting Royal Society London 3 December

[11a] USSR State Committee for the Utilization of Atomic Energy 1986 *The Accident at the Chernobyl Power Plant and its Consequences* Reports compiled for the International Atomic Energy Agency's (IAEA) Experts' Meeting 25–29 August 1986 (Vienna: IAEA)

[11b] International Nuclear Safety Advisory Group 1986 *Summary Report on the Post-Accident Review Meeting on the Chernobyl Accident* Safety series No. 75-INSAG-1 (Vienna: IAEA)

[11c] International Nuclear Safety Advisory Group 1992 *The Chernobyl Accident: Updating of INSAG-1* Safety series No. 75-INSAG-7 (Vienna: IAEA)

[12] Reason J 1990 *Human Error* (New York: Cambridge University Press)

[13] *Pravda* 20 May 1986

[14] Meshkati N 1991 Human factors in large-scale technological systems accidents: Three Mile Island, Bhopal and Chernobyl *Industrial Crisis Quarterly* **5** 131–54

[15] Munipov V M 1990 Human engineering analysis of the Chernobyl accident *Unpublished manuscript* (Moscow: VNIITE)

[16] Ukraine MinChernobyl, Academy of Sciences of Ukraine, *Description of the Ukritiye 1992 Encasement and Requirements for its Conversion* Tender document for competition work to make a design and to find a technological solution for the conversion of the Chernobyl Nuclear Plant Ukritiye encasement Russian and English languages (Kiev: Naukova Dumka)

[17] Belyayev S T, Borovoy A A and Bouzouloukov I 1991 Technical management on the Chernobyl site, status and future of the Sarcophagus Societe Francaise d'Energie Nucleaire Proc. *International Conference on Nuclear Accidents and the Future of Energy* Paris 15–17 April 1991 (Paris: SFEN) pp 26–46

[18] Briffa E ed 1991 *Suicide mission to Chernobyl* BBC TV Horizon in association with WGBH of Boston, distributed by Films for the Humanities and Sciences (Princeton: NOVA)

[19] Lyabakh M ed 1996 *And the Name of the Star is Chornobyl* (Kiev: Chernobylinterinform)

[20] Medvedev Z 1990 The Kyshtym nuclear accident *The Legacy of Chernobyl* (Oxford: Blackwell) p 279

[21] Svensson H 1988 The Chernobyl accident, impact on Western Europe 6th Klaus Breuer lecture *Radiotherapy and Oncology* **12** 1–13

[22] *Personal communication* 1998

[23] *Daily Mail* London 30 April 1986

[24] *Daily Mirror* London 30 April 1986

[25] *New York Post* May 1986

[26] TASS Rome 15 May 1986

[27] *Il Meridiano* Trieste May 1986

[28] *The Sunday Times* London 11 May 1986

[29] *Time* 12 May 1986

[30] Mould R F 1988 *Chernobyl the Real Story* (Oxford: Pergamon) 1992 Japanese edn (Nigata-shi: Nishimura) p 45

Chapter 4

[1] United Nations Scientific Committee on the Effects of Atomic Radiation (UNSCEAR) 1988 *Sources, Effects and Risks of Ionizing Radiation* 1988 Report to the General Assembly (New York: United Nations)

[2] International Nuclear Safety Advisory Group 1986 *Summary Report on the Post-Accident Review Meeting on the Chernobyl Accident* Safety series No. 75-INSAG-1 (Vienna: IAEA)

[3] Ukranian Council of Ukranian Social Services Chornobyl Commission 1987 *Chernobyl Commission Report* (Kiev: Chernobyl Commission)

[4] United Nations Scientific Committee on the Effects of Atomic Radiation (UNSCEAR) 1993 *Sources, Effects and Risks of Ionizing Radiation* 1993 Report to the General Assembly (New York: United Nations)

[5] Ilyin L A, Balonov M I and Bukldakov L A 1990 Radiocontamination patterns and possible health consequences of the accident at the Chernobyl nuclear power station *J. Radiation Protection* **10** 3–29

[6] Committee on Radiation Protection and Public Health, Nuclear Energy Agency for Economic Co-operation and Development 1995 *Chernobyl ten years on, radiological and health impact, an appraisal by the NEA* (Paris: OECD)

[7] Williams E D, Becker D, Dimidchik E F, Nagataki S, Finchera A and Tronko N D 1996 Effects on the thyroid in populations exposed to radiation as a result of the Chernobyl accident *Proc. Int. Conf. One Decade After Chernobyl, Summing up the Consequences of the Accident* Vienna 8–12 April 1996 (Vienna: IAEA) pp 207–30

[8] Dreicer M, Aarkrog A, Alexakhin R, Anspaugh L, Arkhipov N P and Johansson K I 1996 Consequences of the Chernobyl accident for the natural and human environments *Proc. Int. Conf. One Decade After Chernobyl, Summing up the Consequences of the Accident* Vienna 8–12 April 1996 (Vienna: IAEA) pp 319–66

[9] Borovoy A 1992 Characteristics of the nuclear fuel of power unit no. 4 of Chernobyl NPP Kryshev I I ed *Radioecological Consequences of the Chernobyl Accident* (Moscow: Nuclear Society International) pp 9–20

[10] Likhtarev I A, Kovgan L N, Repin V S, Los I P, Chumak V V, Novak D N, Sobolev B G, Kairo I A, Chepurnoy N I, Perevosnikov O N and Litvinets L A 1996 Dosimetric support of the International Programme on the Health Effects of the Chernobyl Accident (IPHECA) pilot project, main results and problems *World Health Statistics Quarterly* **49/1** 40–51

[11] Bennett R G 1996 Assessment by UNSCEAR of worldwide doses from the Chernobyl accident *One Decade After Chernobyl, Summing up the Consequences of the Accident* Proc. International Conference Vienna 8–12 April 1996 (Vienna: IAEA) pp 117–26

[12] World Health Organization 1995 *Int. Conf. Health Consequences of the Chernobyl and Other Radiological Accidents* Geneva 20–23 November 1995 Fact sheets

[13] Shigematsu I, Ito C, Kamada N, Akiyama M and Sasaki H 1993 *A-Bomb Radiation Effects Digest* (Tokyo: Bunkodo)

[14] Ito C 1996 Atomic bomb damage and health care for A-bomb survivors in Japan Shigematsu I ed *Proc. Int. Symp. in Commemoration of the 50th Year of the Atomic Bombing The Present State and Perspective of Health Care fir the Radiation Exposed Case Studies in the World and the Contribution of Hiroshima* (Hiroshima: Hiroshima International Council for Health Care of the Radiation Exposed) pp 73–81

[15] Smithsonian Institution *Science in American Life* exhibit 1999

[16] Adamson G 1981 *We All Live on Three Mile Island The Case Against Nuclear Power* (Sydney: Pathfinder) pp 42–3

[17] *Reuters* 3 November 1999 US Court says Three Mile Island suits can proceed

[18] *US News* 29 March 1999 When the world stopped: 20 years after Three Mile Island the debate still rages

[19] Gonzalez A J 1998 Radioactive residues of the cold war period a radiological legacy *IAEA Bulletin* **40/4** 2–11

[20a] Institut für Strahlenhygiene BfS Germany 1995 Altlasten aus den sowjetischen Kernwaffentests: Das Polygon von Semipalatinsk *Strahlenschutz Praxis* **3** 19–20

[20b] International Atomic Energy Agency 1998 *Radiological Conditions at the Semipalatinsk Test Site, Kazakhstan: Preliminary Assessment and Recommendations for Further Study* IAEA Radiological Assessment Report Series (Vienna: IAEA)

[20c] International Atomic Energy Agency 2000 *The Polygon* Videotape in preparation (Vienna: IAEA)

[21] Bradley D J, Frank C W and Mikerin Y 1996 Nuclear contamination from weapons complexes in the former Soviet Union and the United States *Physics Today* **49/4** 40–5

[22] Akleyev A V and Lyubchansky E R 1994 Environmental and medical effects of nuclear weapon production in the southern Urals *The Science of the Environment* **142/1** 9–18

[23] Burkhart W and Kellerer A M 1994 *Preface* Radiation exposure in the southern Urals *The Science of the Environment* special issue **142/1**

[24] Kosenko M M 1996 Cancer mortality in the exposed population of the Techa river area *World Health Statistics Quarterly* **49/1** 17–21

[25] IAEA Emergency Response Centre 1999 Information advisory concerning incident in Japan

[26] IAEA Press Release 1 October 1999 Accident at the Tokaimura fuel conversion plant

[27] International Atomic Energy Agency 1998 *Diagnosis and Treatment of Radiation Injuries Safety* reports series no. 2 (Vienna: IAEA)

[28] Court L and Dollo R 1999 *Accident de Tokai-Mura* Memorandum 15 October 1999 Electricité de France

[29] Turai I 1999 *Personal communication*

[30a] International Atomic Energy Agency and OECD 1999 *Echelle Internationale des Événements Nucléaires* (Vienna: IAEA and Paris: OECD)

[30b] International Atomic Energy Agency 1999 *Report on the Preliminary Fact Finding Mission Following the Accident at the Nuclear Fuel Processing Facility in Tokaimura Japan* (Vienna: IAEA)

Chapter 5

[1] Scherbak Y 1987 Chernobyl documentary story *Unost (Youth) Magazine* in Russian

[2] Romanenko A Y 1998 Organizational arrangements efficiency for minimising the consequences of the Chernobyl disaster *2nd Int. Conf. Long-term Health Consequences of the Chernobyl Disaster* Kiev Presentation of the Conference Vice-President 1–6 June

[3] Afanasyev D E 1998 *Personal communication*

[4] Shapiro A 1998 *Personal communication*

[5] Cade S 1957 Radiation induced cancer in man *Brit. J. Radiology* **30** 393–402

[6] Mould R F 1983 *Cancer Statistics* (Bristol: Adam Hilger) pp 81–3

[7] Williams E D, Becker D, Dimidchik E P, Nagataki S, Pinchera A and Tronko N D 1996 Effects on the thyroid in populations exposed to radiation as a result of the Chernobyl accident *Proc. Int. Conf. One Decade After Chernobyl, Summing up the Consequences of the Accident* Vienna 8–12 April 1996 (Vienna: IAEA) pp 207–38

[8] Hall P and Holm L E 1998 Radiation associated thyroid cancer facts and fiction *Acta Oncologia* **37** 325–30

[9] Belcher E H and Vetter H eds 1971 *Radioisotopes in Medical Diagnosis* (London: Butterworths) pp 606–8

[10] Souchkevitch G N, Tsyb A F, Repacholi M N and Mould R F 1996 *Health Consequences of the Chernobyl Accident, Results of the IPHECA Pilot Projects and Related National Programmes* Scientific report (Geneva: World Health Organization)

[11] Kaul A 1998 Dosimetric support of epidemiological researches Summary report *2nd Int. Conf. Long-term Health Consequences of the Chernobyl Disaster* Kiev 1–6 June 1998

[12] USSR State Committee for the Utilization of Atomic Energy 1986 *The Accident at the Chernobyl Power Plant and its Consequences* Reports compiled for the International Atomic Energy Agency's (IAEA) Experts' Meeting 25–29 August 1986 (Vienna: IAEA)

[13] Geiger H J 1986 The accident at Chernobyl and the medical response *JAMA* **256** 609–12

[14] Shchepin O 1987 *Report of the Soviet First Deputy Health Minister* Moscow 15 January

[15] Guscova A 1987 *Report to the ICRP Meeting* Como 7–12 September

[16] Kutkov V A, Gusev I A and Dementiev S I 1996 Internal exposure of the staff involved in the cleanup after the accident at the Chernobyl nuclear power plant in 1986 *World Health Statistics Quarterly* **49/1** 62–6

[17] IAEA Advisory Group Meeting 1987 *Medical Handling of Skin Lesions Following High Level Accidental Irradiation* Paris 28 September–2 October

[18a] Kassabian M 1907 *Röntgen Rays and Electro-Therapeutics* (Philadelphia: Lippincott)

[18b] Mould R F 1983 *A Century of X-Rays and Radioactivity in Medicine* (Bristol: Institute of Physics Publishing)

[19] Borden W C 1900 *The Use of the Röntgen Ray by the Medical Department of the United States Army in the War with Spain* House of Representatives 56th Congress 1st session Document no. 729 (Washington, DC: US Government Printing Office)

[20] Beck C 1904 *Röntgen Ray Diagnosis and Therapy* (New York: Appleton and Lange)

[21] Araki T and Morotani Y (Mayors of the cities of Hiroshima and Nagasaki) 1976 *Appeal to the Secretary General of the United Nations* (Hiroshima: Hiroshima Peace Culture Foundation)

[22] Araki T and Morotani Y (Mayors of the cities of Hiroshima and Nagasaki) 1978 *Hiroshima and Nagasaki a Photographic Record of a Historical Event* (Hiroshima: Hiroshima Peace Culture Foundation)

[23] Hersey J 1984 *Hiroshima* (London: Penguin)

[24] Chisholm A 1985 *Faces of Hiroshima* (London: Cape)

[25] Marchenko T, Smolyar I and Torbin V 1996 *Community Development Centres for Social and Psychological Rehabilitation in Belarus, Russia and Ukraine* (Paris: UNESCO Chernobyl Programme)

[26] Telyatnikov L 1987 *Personal communication*

[27] Souchkevitch G N, Burkhart W, Stepanenko V F, Mould R F, Keningsberg J E and Likhtarev I A 1995 *World Health Organization IPHECA Dosimetry Project Meeting* Salzgitter Munich September

[28] Kreisel W, Tsyb A, Krishenko N, Bobyleva O, Napalkov N P, Kjellström T, Schmidt R and Souchkevitch G 1996 WHO updating report on the WHO conference on 'Health consequences of Chernobyl and other radiological accidents' including results of the IPHECA programme *Proc. Int. Conf. One Decade After Chernobyl, Summing up the Consequences of the Accident* Vienna 8–12 April 1996 (Vienna: IAEA) pp 85–99

[29] Guskova A 1987 *Lessons of Chernobyl* Presentation Institute of Biology Seminar 11 April 1987 London

[30] Wagemaker G, Guskova A K, Bebeshko V G, Griffiths N M and Krishenko NA 1996 Clinically observed effects in individuals exposed to radiation as a result of the Chernobyl accident *Proc. Int. Conf. One Decade After Chernobyl, Summing up the Consequences of the Accident* Vienna 8–12 April 1996 (Vienna: IAEA) pp 173–204

[31] Vriesendorp H M, Wagemaker G and van Bekkum D W 1981 Engraftment of alllogenic bone marrow *Transplant Proc.* **13** 643

[32] Wagemaker G, Heidt P J, Merchav S and van Bekkum D W 1982 Abrogation of histocompatibility barriers to bone marrow transplantation in rhesus monkeys Baum S J, Ledney G D and Thierfelder S eds *Experimental Haematology Today* (Basel: Karger) pp 111–18

[33] Guskova A and World Health Organization 1995 *Int. Conf. Health Consequences of the Chernobyl and Other Radiological Accidents* Geneva 20–23 November 1995 Press release WHO/84

[34] Wagemaker G and Bebeshko V G eds 1996 *Diagnosis and Treatment of Patients with Acute Radiation Syndromes* European Commission, Belarus, the Russian Federation and Ukraine joint study project no. 3 Final report EUR 16535 (Brussels: European Commission)

[35] Till J E and McCulloch E A 1961 A direct measurement of the radiation sensitivity of normal bone marrow cells *Radiation Research* **14** 213–22
[36] Hendry J H and Howard A 1971 The response of haemopoietic colony forming units to single and split doses of gamma-rays of D-T neutrons *Int. J. Radiation Biology* **19** 51–64
[37] Vriesendorp H M and van Bekkum D W 1980 Role of total body irradiation in conditioning for bone marrow transplantation *Immunology, Bone Marrow Transplantation* 345–64
[38] Wielenga J J, van Gils F C J M and Wagemaker G 1989 The radiosensitivity of primate haemopoietic stem cells based on *in vivo* measurements *Int. J. Radiation Biology* **55** 1041
[39] Wagemaker G 1995 Heterogeneity of the radiation sensitivty of immature haemopoietic stem cells subsets *Stem Cells* **13** 257–60
[40] Meijne E I M, van der Winden-van Groenwegen A J M, Ploemacher R E, Vos O Davids J A G and Huiskamp R 1991 The effects of X-radiation on hematopoietic stem cell compartments in the mouse *Experimental Hematology* **19** 617–23
[41] Ploemacher R E, van Os R, van Beurden C A J and Down J D 1992 Murine hemopoietic stem cells with long-term engraftment and marrow repopulating ability are less radiosensitive to gamma radiation than are spleen colony forming cells *Int. J. Radiation Biology* **61** 489–99
[42] Down J D, Boudewun A, van Os R, Thames H D and Ploemacher R E 1995 Variations in radiation sensitivity and repair among different hemopoietic stem cell subsets following fractionated irradiation *Blood* **86** 122–37
[43] Baird M C, Hendry J H and Testa N G 1990 Radiosensitivity increases with differentiation status of murine hemopoietic progenitor cells selected using enriched marrow subpopulations and recombinant growth factors *Radiation Research* **123** 292–8
[44] Hammer A and Lyndon N 1987 *Hammer: Witness to History* (London: Simon and Schuster)
[45] Epstein E J 1997 *Dossier: a Biography of Armand Hammer* (New York: Random House)
[46] Gale R P and Hauser T 1989 *Final Warning* (New York: Warner Books)
[47] Hopewell J 1987 Personal communication *Conf. Medical Handling of Skin Lesions Following High Level Accidental Irradiation* Paris 28 September–2 October
[48] Ganja K 1998 Medical organization in Chernobyl and its environs. Presentation in Chernobyl town during a visit at the *2nd Int. Conf. Long-term Health Consequences of the Chernobyl Disaster* Kiev 1–6 June

Chapter 6

[1] Souchkevitch G N, Tsyb A F, Repacholi M N and Mould R F eds 1996 *Health Consequences of the Chernobyl Accident, Results of the IPHECA Pilot Projects and Related National Programmes* Scientific report (Geneva: World Health Organization)

[2] Nyagu A I, Souchkevitch G N, Loganovsky K N, Yuryev K L and Afansyev D F eds 1998 *Proc. 2nd Int. Conf. Long-term Health Consequences of the Chernobyl Disaster* Kiev 1–6 June

[3] Minenko V F, Drozdovich V V and Tret'yakevich 1996 Methodological approaches to calculation of annual effective dose for the population of Belarus *Bulletin of the All-Russian Medical and Dosimetric State Registry* **7** 246–52

[4] Savkin M N, Titov A V and Lebedev A N 1996 Distribution of individual and collective exposure doses for the population of Belarus in the first year after the Chernobyl accident *Bulletin of the All-Russian Medical and Dosimetric State Registry* **7** 87–113

[5] Sivintsev Y V and Kachalov V eds 1992 *Five Hard Years* in Russian (Moscow: Publishing House)

[6] Ukranian Civil Defense 1990 *The Chernobyl Accident, Events Facts Numbers April 1986 through March 1990* Guide of the Ukranian Civil Defense (Kiev: Ukranian Civil Defense)

[7] Jensen P H 1996 One decade after Chernobyl: environmental impact assessments *Proc. Int. Conf. One Decade After Chernobyl, Summing up the Consequences of the Accident* Vienna 8–12 April 1996 (Vienna: IAEA) pp 77-83

[8] Zufarov V 1988 *Chernobyl Record* in Russian (Moscow: Planeta)

[9] Marchuk Yu 1996 National statement *Proc. Int. Conf. One Decade After Chernobyl, Summing up the Consequences of the Accident* Proc. International Conference Vienna 8–12 April 1996 (Vienna: IAEA) pp 57–61

[10] International Atomic Energy Agency 1996 Conf. Summary *Proc. Int. Conf. One Decade After Chernobyl, Summing up the Consequences of the Accident* Vienna 8–12 April 1996 (Vienna: IAEA) p 12

[11] Ukranian Council of Ukranian Social Services Chernobyl Commission 1987 *Chernobyl Commission Report* (Kiev: Chernobyl Commission)

[12] Marchenko T, Smolyar I and Torbin V 1996 *Community Development Centres for Social and Psychological Rehabilitation in Belarus, Russia and Ukraine* (Paris: UNESCO Chernobyl Programme)

[13] Lyashko A P 1987 Meeting of the Council of Ministers of the Ukraine

[14] Voznyak V Y 1996 Social, economic, institutional and political impact *Proc. Int. Conf. One Decade After Chernobyl, Summing up the Consequences of the Accident* Vienna 8–12 April 1996 (Vienna: IAEA) pp 369–452

[15] Mozgovaia T 1999 The time has stopped: Atlantis of Polesye leaves for eternity *Segodnya (Today)* 23 December 1999 pp 8–9

[16] World Health Organization 1995 *Int. Conf. Health Consequences of the Chernobyl and Other Radiological Accidents* Geneva 20–23 November Press release WHO/84

[17] Korol N 1998 *Personal communication* Studies by the Laboratory for Population Investigations of Health Effects of Chernobyl, Kiev

Chapter 7

[1] Ukraine MinChernobyl, Academy of Sciences of Ukraine 1992 *Description of the Ukritiye Encasement and Requirements for its Conversion* Tender document for competition work to make a design and to find a technological solution for the conversion of the Chernobyl Nuclear Plant Ukritiye encasement Russian and English languages (Kiev: Naukova Dumka)
[2] USSR State Committee for the Utilization of Atomic Energy 1986 *The Accident at the Chernobyl Power Plant and its Consequences* Reports compiled for the International Atomic Energy Agency's (IAEA) Experts' Meeting 25–29 August 1986 (Vienna: IAEA)
[3] Soviet Television 1986 *The Warning* Videotape supplied by the Soviet Embassy London
[4] Velikhov E 1993 *The aftermath of Chernobyl: developments in Russia, Ukraine and Belarus* Presentation British Pugwash Group public discussion meeting Royal Society London 3 December
[5] Briffa E ed 1991 *Suicide mission to Chernobyl* BBC TV Horizon in association with WGBH of Boston, distributed by Films for the Humanities and Sciences (Princeton, NJ: NOVA)
[6] Zufarov V 1989 *Personal communication* TASS photographer
[7] Belyayev S T, Borovoy A A and Bouzouloukov I 1991 Technical management on the Chernobyl site, status and future of the Sarcophagus Societe Francaise d'Energie Nucleaire *Proc. Int. Conf. Nuclear Accidents and the Future of Energy* Paris 15–17 April 1991 (Paris: SFEN) pp 26–46

Chapter 8

[1] Char N I and Csik B J 1987 Nuclear power development history and outlook *IAEA Bulletin* **29/3** 19–23
[2] Smithsonian Institution 1999 *Science in American Life* exhibit
[3] International Nuclear Safety Advisory Group 1986 *Summary Report on the Post-Accident Review Meeting on the Chernobyl Accident* Safety series No. 75-INSAG-1 (Vienna: IAEA)
[4] Wedekind L H ed 1997 Energy and environment the drive for safer cleaner development *IAEA Bulletin* **39/3** 26–40
[5a] International Atomic Energy Agency 1988 *The Radiological Accident in Goiânia* (Vienna: IAEA)
[5b] Gonzàlez A J 1999 Strengthening the safety of radiation sources and the security of radioactive materials: timely action *IAEA Bulletin* **41** No. 3 2–17, see also IAEA/WHO 1998 *Planning the Medical Response to Radiological Accidents*, IAEA Safety Report Series No. 4 (Vienna: IAEA)
[5c] International Atomic Energy Agency 1990 *The Radiological Accident in San Salvador* (Vienna: IAEA)
[5d] International Atomic Energy Agency 1993 *The Radiological Accident in Soreq* (Vienna: IAEA)
[5e] International Atomic Energy Agency 1996 *The Radiological Accident at the Irradiation Facility in Nesvizh* (Vienna: IAEA)

[5f] International Atomic Energy Agency 1998 *The Radiological Accident in Tammiku* (Vienna: IAEA)

[5g] International Atomic Energy Agency 1998 *Accidental Overexposure of Radiotherapy Patients in San Jose, Costa Rica* (Vienna: IAEA)

[5h] International Atomic Energy Agency 2000 *The Radiological Accident in Turkey* in preparation (Vienna: IAEA)

[5j] International Atomic Energy Agency 1988 *An Electron Accelerator Accident in Hanoi, Vietnam* (Vienna: IAEA)

[5k] International Atomic Energy Agency 1998 *The Radiological Accident in the Reprocessing Plant at Tomsk* (Vienna: IAEA)

[6] Trichopoulos D, Zavitsanos X, Koutis C, Drogari P, Proukakis C and Petridou E 1987 The victims of Chernobyl in Greece, induced abortions after the accident *Brit. Med. J.* **295** 1100

[7] Warnecke E 1994 Disposal of radioactive waste, a completing overview *Kerntechnik* **59** 64–71

[8] Juhn P E and Kupitz J 1996 Nuclear power beyond Chernobyl: a changing international perspective *IAEA Bulletin* **38/1** 2–9

[9] Guinnessy P 1999 Germany set to abandon nuclear power *Physics World* **12** 9

[10] Marchuk Yu 1996 National statement *Proc. Int. Conf. One Decade After Chernobyl, Summing up the Consequences of the Accident* Vienna 8–12 April 1996 (Vienna: IAEA) pp 57–61

[11] Surrey J (chairman) 1997 *Economic assessment of the Khmelnitsky 2 and Rovno 4 nuclear reactors in Ukraine* Report to the European Bank for Reconstruction and Development, the European Commission and the US Agency for International Development by an international panel of experts (Brighton: Science and Technology Policy Research Unit University of Sussex)

[12] Stone and Webster Management Consultants Inc 1998 *Least Cost Electric Power System Development Analysis* Report to the European Bank for Reconstruction and Development (Eaglewood: Stone andWebster)

[13] Bradford P, MacKerron G, Surrey J and Thomas S 1998 *The Case for Completing the K2/R4 Nuclear Plants in the Ukraine: a critique of the Stone and Webster report of May 1998* Report to the Austrian Energy Agency (Brighton: Science and Technology Policy Research Unit University of Sussex)

[14] Velikhov E 1993 *The aftermath of Chernobyl: developments in Russia, Ukraine and Belarus* Presentation British Pugwash Group public discussion meeting Royal Society London 3 December

[15] Briffa E ed 1991 *Suicide mission to Chernobyl* BBC TV Horizon in association with WGBH of Boston, distributed by Films for the Humanities and Sciences (Princeton, NJ: NOVA)

[16] Ukraine MinChernobyl, Academy of Sciences of Ukraine 1992 *Description of the Ukritiye Encasement and Requirements for its Conversion* Tender document for competition work to make a design and to find a technological solution for the conversion of the Chernobyl Nuclear Plant Ukritiye encasement Russian and English languages (Kiev: Naukova Dumka)

[17] Scherbak Y M 1996 Ten years of the Chernobyl era *Scientific American* April 44–9

[18] 1998 *Personal communications* Kiev and Chernobyl, June

[19] Merkel A 1996 Conference opening address *Proc. Int. Conf. One Decade After Chernobyl, Summing up the Consequences of the Accident* Vienna 8–12 April 1996 (Vienna: IAEA) pp 39–43

[20] Lukashenko A G 1996 National statement *Proc. Int. Conf. One Decade After Chernobyl, Summing up the Consequences of the Accident* Vienna 8–12 April 1996 (Vienna: IAEA) pp 47–51

[21] Mould R F, Swaine F and Irwin S 1999 Breast radiation injury litigation *Brit. J. Radiology* **72** 925

[22] Ritch J B 1999 Nuclear green perspectives on science, diplomacy and atoms for peace *IAEA Bulletin* **41** No. 2 2–7

[23] McGowan P 1978 Markov killed by 'cancer gun' *London Evening Standard* 24 November 1978

[24] Hodt H J, Sinclair W K and Smithers D W 1952 A gun for interstitial implantation of radioactive gold grains *Brit. J. Radiology* **25** 419

[25] Mould R F 1978 The golden gun *London Evening Standard* 4 December 1978

[26] Cade S 1929 *Radium Treatment of Cancer* (London: Churchill)

[27] Marshall E 1984 Juarez: an unprecedented radiation accident *Science* **223** 1152

[28] Jaskowski J ed 1992 *Katastrofa w Czarnobylu a Polska* (Gdansk: Wydawnictowo Gdanskie)

Chapter 9

[1] Shigematsu I, Ito C, Kamada N, Akiyama M and Sasaki H 1993 *A-Bomb Radiation Effects Digest* (Tokyo: Bunkodo)

[2] Ilyin L A 1995 *Chernobyl Myth and Reality* English edn (Moscow: Megapolis)

[3] Ivanov V K, Rastopchin E M, Gorsky A I and Ryvkin V B 1998 *Health Physics* **74** 309–15

[4] Souchkevitch G N, Tsyb A F, Repacholi M N and Mould R F eds 1996 *Health Consequences of the Chernobyl Accident, Results of the IPHECA Pilot Projects and Related National Programmes* Scientific report (Geneva: World Health Organization)

[5] Bailiff I K and Stepanenko V eds 1996 *Retrospective Dosimetry and Dose Reconstruction* European Commission, Belarus, the Russian Federation and Ukraine experimental collaboration project no. 10 Final report EUR 16540 (Brussels: European Commission)

[6] Likhtarev I A, Kovgan L N, Repin V S, Los I P, Chumak V V, Novak D N, Sobolev B G, Kairo I A, Chepurnoy N I, Perevosnikov O N and Litvinets L A 1996 Dosimetric support of the International Programme on the Health Effects of the Chernobyl Accident (IPHECA) pilot project, main results and problems *World Health Statistics Quarterly* **49/1** 40–51

[7] Romanyukha A A, Ignat ev E A, Degteva M O, Kozheurov V P, Wieser A and Jacob P 1996 Radiation doses from Ural region *Nature* **381** 199–200

[8] Romanyukha A A, Degteva M O, Kozheurov V P, Wieser A, Jacob P, Ignatiev E A and Vorobiova M I 1996 Pilot study of the Urals population by tooth electron paramagnetic resonance dosimetry *Radiat. Environ. Biophys.* **35** 305–10

[9] Romanyukha A A, Desrosiers M, Seltzer S, Ignatiev E A, Ivanov D, Dayankin S, Degteva M O, Wieser A and Jacob P 2000 Ultrahigh doses reconstructed by EPR for Techa riverside population highly exposed by ^{90}Sr as key for developing new dose reconstruction methodology *Proc. 10th Int. Congress of the International Radiation Protection Association (IRPA-10) July 2000 Hiroshima* Abstract

[10] Romanyukha A A 1999 *Personal communication*

[11] Romanyukha A A, Ignatiev E A, Ivanov D V and Vasilyev A G 1999 The distance effect on the individual exposures evaluated from the Soviet nuclear bomb test at Totskoye test site in 1954 *Radiation Protection Dosimetry* **86** 53–8

[12] Desrosiers M F and Romanyukha A A 1998 Technical aspects of the electron paramagnetic resonance method for tooth enamel dosimetry *Biomarkers: Medical and Workplace Applications* (Washington, DC: Joseph Henry Press) pp 53–64

[13] Desrosiers M F 1991 *In vivo* assessment of radiation exposure *Health Physics* **61** 859–61

[14] Schauer D A, Coursey B M, Dick C E, McLaughlin W L, Puhl J M, Desrosiers M F and Jacobson A D 1993 A radiation accident at an industrial accelerator facility *Health Physics* **65** 131–40

[15] Schauer D A, Desrosiers M F, Kuppusamy P and Zweier J L 1996 Radiation dosimetry of an accidental overexposure using EPR spectrometry and imaging of human bone *Appl. Radiat. Isot.* **47** 1345–50

[16] Desrosiers M F 1999 *Personal communication*

[17] International Atomic Energy Agency 1990 *The Radiological Accident in San Salvador* (Vienna: IAEA)

[18] Ikeya M 1993 *New Applications of Electron Spin Resonance: Dating Dosimetry and Microscopy* (Singapore: World Scientific)

[19] Nakamure N, Miyazawa C, Akiyama M, Sawada S and Awa A A 1996 Biodosimetry: chromosome aberration in lymphocytes and electron paramagnetic resonance in tooth enamel from atomic bomb survivors *World Health Statistics Quarterly* **49/1** 67–71

[20] Awa A 1990 *Personal communication* 1996 Unpublished data Radiation Effects Research Foundation Hiroshima

[21] Shigematsu I 1993 *Radiation Effects Research Foundation a Brief Description* (Hiroshima: RERF)

[22] Guskova A 1987 *Lessons of Chernobyl* Presentation Institute of Biology Seminar 11 April 1987 London

[23] Lloyd D C and Sevan'kaev A V eds 1996 *Biological Dosimetry for Persons Irradiated by the Chernobyl Accident* European Commission, Belarus, the Russian Federation and Ukraine experimental collaboration project no. 6 Final report EUR 16532 (Brussels: European Commission)

[24] Bauchinger M, Schmid E, Zitzelsberger H, Braselmann H and Nahrstedt U 1993 Radiation induced chromosome aberrations analysed by two-colour fluorescence *in situ* hybridization with composite whole chromosome specific DNA probes and a pancentromeric DNA probe *Int. J. Radiation Biology* **64** 179–84

Chapter 10

[1] International Commission on Radiological Protection 1991 *Recommendations of the International Commission on Radiological Protection* ICRP Publication No. 60 in Annals of the ICRP (New York: Pergamon)

[2] Souchkevitch G N, Tsyb A F, Repacholi M N and Mould R F eds 1996 *Health Consequences of the Chernobyl Accident, Results of the IPHECA Pilot Projects and Related National Programmes* Scientific report (Geneva: World Health Organization)

[3] United Nations Scientific Committee on the Effects of Atomic Radiation (UNSCEAR) 1993 *Sources, Effects and Risks of Ionizing Radiation* 1993 Report to the General Assembly (New York: United Nations)

[4] Linsley G 1996 Environmental impact of radioactive releases: addressing global issues *IAEA Bulletin* **38/1** 36–8

[5] Jensen P H 1996 One decade after Chernobyl: environmental impact assessments *Proc. Int. Conf. One Decade After Chernobyl, Summing up the Consequences of the Accident* Vienna 8–12 April 1996 (Vienna: IAEA) pp 77–83

[6] Radiation Safety Control Center National Laboratory for High Energy Physics 1996 *Radiation and Life* (Tsukuba: RSCC NLHEP)

[7] United Nations Scientific Committee on the Effects of Atomic Radiation (UNSCEAR) 1988 *Sources, Effects and Risks of Ionizing Radiation* 1988 Report to the General Assembly (New York: United Nations)

[8] Bennett R G 1996 Assessment by UNSCEAR of worldwide doses from the Chernobyl accident *Proc. Int. Conf. One Decade After Chernobyl, Summing up the Consequences of the Accident* Vienna 8–12 April 1996 (Vienna: IAEA) pp 117–26

[9] Kaul A 1998 Dosimetric support of epidemiological researches Summary report *2nd Int. Conf. Long-term Health Consequences of the Chernobyl Disaster* Kiev 1–6 June 1998

[10] Belyayev S T and Demin V F 1991 Long-term Chernobyl consequences, countermeasures and their effectiveness Societe Francaise d'Energie Nucleaire *Proc. Int. Conf. Nuclear Accidents and the Future of Energy* Paris 15–17 April 1991 (Paris: SFEN) pp 208–26

[11] Cardis E, Anspaugh L, Ivanov V K, Likhtarev I A, Mabuchi K, Okeanov A E and Prisyazhniuk A E 1996 Estimated long-term health effects of the Chernobyl accident *Proc. Int. Conf. One Decade After Chernobyl, Summing up the Consequences of the Accident* Vienna 8–12 April 1996 (Vienna: IAEA) pp 241–71

[12] Buzunov V A, Strapko N P, Pirogova E A, Krasnikova L I, Bugayev V N, Korol N A, Treskunova T V, Ledoschuk B A, Gudzenko N A, Bomko E I, Bobyleva O A and Kartushin G I 1996 Epidemiological survey of the medical consequences of the Chernobyl accident in Ukraine *World Health Statistics Quarterly* **49/1** 4–6

[13] Ivanov V K and Tsyb A F 1996 The Chernobyl accident and radiation risks: dynamics of epidemiological rates according to the data in the national registry *World Health Statistics Quarterly* **49/1** 22–8

[14] Keningsberg J E, Minenko V F and Buglova E E 1996 Radiation effects on the population of Belarus after the Chernobyl accident and the prediction of stochastic effects *World Health Statistics Quarterly* **49/1** 58–61

[15] Ilyin L A 1995 *Chernobyl Myth and Reality* English edn (Moscow: Megapolis)

[16] Ilyin L A, Krjuchkov V P and Osanov D P 1995 Exposure levels in 1986–1987 for persons involved in recovery operations following the Chernobyl accident and dosimetric data verification *J. Radiation Biology Radioecology* in Russian **35** 803–28

Chapter 11

[1] Kaul A ed 1986 *Umweltradioactivität und Strahlenexposition in SüdBayern durch den Tschernobyl-Unfall* report 16/86 (München/Neuherberg: Institut für Strahlenschutz der Gesellschaft für Strahlen- und Umweltforchung)

[2] Nuclear Energy Agency Committee on Radiation Protection and Public Health 1995 *Chernobyl Ten Years On Radiological and Health Impact* (Paris: OECD)

[3] Robeau D G 1996 Environmental impact assessment *Proc. Int. Conf. One Decade After Chernobyl, Summing up the Consequences of the Accident* Vienna 8–12 April 1996 (Vienna: IAEA) pp 73–6

[4] International Atomic Energy Agency 1996 Summary of the conference results *Proc. Int. Conf. One Decade After Chernobyl, Summing up the Consequences of the Accident* Vienna 8–12 April 1996 (Vienna: IAEA) pp 3–17

[5] United Nations Scientific Committee on the Effects of Atomic Radiation (UNSCEAR) 1988 *Sources, Effects and Risks of Ionizing Radiation* 1988 Report to the General Assembly (New York: United Nations)

[6] Bennett B G 1996 Assessment by UNSCEAR of worldwide doses from the Chernobyl accident Bayer A, Kaul A and Reinders C eds *Zehn Jahre nach Tschernobyl eine Balanz* Proc. Seminar des Bundesamtes für Strahlenschutz und der Strahlenschutzkommission München 6–7 March (Stuttgart: Gustav Fischer) pp 52–64

[7] World Health Organization 1986 *Updated Summary of Data Situation with Regard to Activity Measurements 12 June 1986* (Geneva: WHO)

[8] USSR State Committee for the Utilization of Atomic Energy 1986 *The Accident at the Chernobyl Power Plant and its Consequences* Reports compiled for the IAEA Experts' Meeting 25–29 August 1986 (Vienna: IAEA)

[9] Institut de Protection et de Sûreté Nucléaire 1996 Status in the exclusion zone *Ecological Bulletin* Kiev, no. 3

[10] Jensen P H 1996 One decade after Chernobyl: environmental impact assessments *Proc. Int. Conf. One Decade After Chernobyl, Summing up the Consequences of the Accident* Vienna 8–12 April (Vienna: IAEA) pp 77–83

[11a] Lomakin M D 1998 *Personal communication* Chernobyl Center on Nuclear Safety, Radioactive Waste and Radioecology, Kiev

[11b] Sawicki L K 1999 Wilki w krainic smierci (Wolves in the land of death) *Lowiec Polski (Polish Hunter)* Issue No. 1842 November 27–29

[12] Voice E 1995 Postcard from Pripyat *Physics World* April 1995 **84**

[13] Svensson H 1988 The Chernobyl accident, impact on Western Europe 6th Klaus Breuer lecture *Radiotherapy and Oncology* **12** 1–13

[14] Commission Fédérale pour la Protection Atomique/Chimique 1995 *10 ans Après Tchernobyl une Contribution Suisse* (Berne: COPAC)

[15] Kryshev I I 1991 Radioactive contamination and radioecological consequences of the Chernobyl accident Societe Francaise d'Energie Nucleaire *Proc. Int. Conf. Nuclear Accidents and the Future of Energy* Paris 15–17 April (Paris: SFEN) pp 167–77

[16] Strand P, Howard B and Averin V eds 1996 *Transfer of Radionuclides to Animals, their Comparative Importance under Different Agricultural Ecosystems and Appropriate Countermeasures* European Commission, Belarus, the Russian Federation and Ukraine, Experimental collaboration project no. 9 Final report EUR 165394 (Brussels: European Commission)

[17] Souchkevitch G N, Tsyb A F, Repacholi M N and Mould R F 1996 *Health Consequences of the Chernobyl Accident, Results of the IPHECA Pilot Projects and Related National Programmes* Scientific report (Geneva: World Health Organization)

[18] Lochard J and Beleyaev S eds 1996 *Decision Aiding System for the Management of Post-Accident Situations* European Commission, Belarus, the Russian Federation and Ukraine, Joint study project no. 2 Final report EUR 16534 (Brussels: European Commission)

[19] World Health Organization 1995 International conference *Health Consequences of the Chernobyl and Other Radiological Accidents* Geneva 20–23 November 1995 Fact sheets

[20] Johnston K 1987 British sheep still contaminated by Chernobyl fallout *Nature* **328** 661

[21] World Health Organization 1988 Guidelines for application after widespread radioactive contamination resulting from a major radiation accident *Derived Intervention Levels for Radionuclides in Food* (Geneva: WHO)

Chapter 12

[1] United Nations Scientific Committee on the Effects of Atomic Radiation (UNSCEAR) 1988 *Sources, Effects and Risks of Ionizing Radiation* 1988 Report to the General Assembly (New York: United Nations)

[2] Ilyin L A, Guskova A K, Kiruyshkin V I and Kosenko M M 1986 *Soviet Manual for the Medical Treatment of Radiation Victims* in Russian (Moscow: Energoatomizdat)

[3] Velikhov E 1993 *The Aftermath of Chernobyl: Developments in Russia, Ukraine and Belarus* Presentation British Pugwash Group public discussion meeting Royal Society London 3 December

[4] USSR State Committee for the Utilization of Atomic Energy 1986 *The Accident at the Chernobyl Power Plant and its Consequences* Reports compiled for the IAEA Experts' Meeting 25–29 August 1986 (Vienna: IAEA)

[5] Briffa E ed 1991 *Suicide mission to Chernobyl* BBC TV Horizon in association with WGBH of Boston, distributed by Films for the Humanities and Sciences (Princeton, NJ: NOVA)

[6] Marchenko T, Smolyar I and Torbin V 1996 *Community Development Centres for Social and Psychological Rehabilitation in Belarus, Russia and Ukraine* (Paris: UNESCO Chernobyl Programme)

[7] Kozubov G M and Taskaeva A I eds 1990 *Radiation Impact on Coniferous Forests in the Chernobyl Area after the Accident* (Moscow: Syktykar)

[8] Dreicer M, Aarkrog A, Alexakhin R, Anspaugh L, Arkhipov N P and Johansson K I 1996 Consequences of the Chernobyl accident for the natural and human environments *Proc. Int. Conf. One Decade After Chernobyl, Summing up the Consequences of the Accident* Vienna 8–12 April 1996 (Vienna: IAEA) pp 319–66

[9] Pikalov V 1987 Harsh lessons of the Chernobyl disaster *Military Bulletin* **14** no. 8

[10] Hubert P, Annisomova L, Antsipov G, Ramsaev V and Sobotovitch V eds 1996 *Strategies of decontamination* European Commission, Belarus, the Russian Federation and Ukraine experimental collaboration project no. 4 Final report EUR 16530 (Brussels: European Commission)

Chapter 13

[1] USSR State Committee for the Utilization of Atomic Energy 1986 *The Accident at the Chernobyl Power Plant and its Consequences* Reports compiled for the IAEA Experts' Meeting 25–29 August 1986 (Vienna: IAEA)

[2] Tkachenko A 1987 *In the Fields and Farms of the Ukraine in 1987* (Moscow: Novosti)

[3] Robeau D G 1996 Environmental impact assessment *Proc. Int. Conf. One Decade After Chernobyl, Summing up the Consequences of the Accident* Vienna 8–12 April 1996 (Vienna: IAEA) pp 73–6

[4] Marchuk Yu 1996 National statement *Proc. Int. Conf. One Decade After Chernobyl, Summing up the Consequences of the Accident* Vienna 8–12 April 1996 (Vienna: IAEA) pp 57–61

[5] Povinec P, Fowler S and Baxter M 1996 Chernobyl and the marine environment: the radiological impact in context *IAEA Bulletin* **38/1** 18–22

[6] Sansone U and Voitsekhovitch O eds 1996 *Modelling and Study of the Mechanisms of the Transfer of Radioactive Material from Terrestrial Ecosystems to and in Water Bodies around Chernobyl* European Commission, Belarus, the Russian Federation and Ukraine experimental collaboration project no. 3 Final report EUR 16529 (Brussels: European Commission)

[7] Svensson H 1988 The Chernobyl accident, impact on Western Europe 6th Klaus Breuer lecture *Radiotherapy and Oncology* **12** 1–13

Chapter 14

[1] United Nations Scientific Committee on the Effects of Atomic Radiation (UNSCEAR) 1988 *Sources, Effects and Risks of Ionizing Radiation* 1988 Report to the General Assembly (New York: United Nations)

[2] Izrael Y A, De Cort M, Jones A R, Nazarov I M, Fridman S G, Kvasnikova E V, Stukin E D, Kelly G N, Matveenko I I, Pokumelko Y M, Tabatchnyi L Y and Tasturov Y 1996 The atlas of ^{137}Cs contamination of Europe after the Chernobyl accident in Karaoglou A, Desmet G, Kelly G N and Menzel H G eds *Proc. 1st Int. Conf. The Radiological Consequences of the Chernobyl Accident* Minsk 18–22 March EUR 16544 (Brussels: European Commission) pp 1–10

[3] Shoigu S K 1996 National statement *Proc. Int. Conf. One Decade After Chernobyl, Summing up the Consequences of the Accident* Vienna 8–12 April 1996 (Vienna: IAEA) pp 53–6

[4] Lukashenko A G National statement *Proc. Int. Conf. One Decade After Chernobyl, Summing up the Consequences of the Accident* Vienna 8–12 April 1996 (Vienna: IAEA) pp 47–51

[5] Marchuk Yu 1996 National statement *Proc. Int. Conf. One Decade After Chernobyl, Summing up the Consequences of the Accident* Vienna 8–12 April 1996 (Vienna: IAEA) pp 57–61

[6] Kryshev I I 1991 Radioactive contamination and radioecological consequences of the Chernobyl accident Societe Francaise d'Energie Nucleaire *Proc. Int. Conf. Nuclear Accidents and the Future of Energy* Paris 15–17 April (Paris: SFEN) pp 167–77

[7] Institut de Protection et de Sûreté Nucléaire 1996 Status in the exclusion zone *Ecological Bulletin* Kiev, no. 3

[8] Robeau D G 1996 Environmental impact assessment *Proc. Int. Conf. One Decade After Chernobyl, Summing up the Consequences of the Accident* Vienna 8–12 April 1996 (Vienna: IAEA) pp 73–6

[9] Sobolev B, Likhtarev I, Kairo I, Tronko N, Oleynik V and Bogdanova T 1996 Radiation risk assessment of the thyroid cancer in Ukranian children exposed due to Chernobyl in Karaoglou A, Desmet G, Kelly G N and Menzel H G eds *Proc. 1st Int. Conf. The Radiological Consequences of the Chernobyl Accident* Minsk 18–22 March EUR 16544 (Brussels: European Commission) pp 741–8

[10] Souchkevitch G N, Tsyb A F, Repacholi M N and Mould R F eds 1996 *Health Consequences of the Chernobyl Accident, Results of the IPHECA Pilot Projects and Related National Programmes* Scientific report (Geneva: World Health Organization)

[11] Belyayev S T and Demin V F 1991 Long-term Chernobyl consequences, countermeasures and their effectiveness Societe Francaise d'Energie Nucleaire *Proc. Int. Conf. Nuclear Accidents and the Future of Energy* Paris 15–17 April 1991 (Paris: SFEN) pp 208–26

[12] Nuclear Energy Agency Committee on Radiation Protection and Public Health 1995 *Chernobyl Ten Years On Radiological and Health Impact* (Paris: OECD)

[13] Balonov M I 1993 Overview of doses to the Soviet population from the Chernobyl accident and the protective actions applied in Merwin S E and Balonov M I eds *The Chernobyl Papers* vol 1 (Richland: Research Enterprises) pp 23–45

[14] USSR State Committee on Hydrometeorology 1991 *Radiation Maps in the Territory of the European Part of the USSR as of December 1990. Densities of Area Contamination by* $^{137}Cs\,^{90}Sr\,^{239}Pu\,^{240}Pu$ (Minsk: SCH)

[15] International Advisory Committee IAEA 1991 *The International Chernobyl Project. Assessment of Radiological Consequences and Evaluation of Protective Measures* Technical Report and Surface Contamination Maps (Vienna: IAEA)

[16] Shestopalov V M ed 1996 *Atlas of Chernobyl Exclusion Zone* (Kiev: National Academy of Sciences of Ukraine)

Chapter 15

[1] Jaworski Z 1996 Chernobyl in Poland the first few days, ten years later Bayer A, Kaul A and Reinders C eds *Zehn Jahre nach Tschernobyl eine Balanz* Proc. Seminar des Bundesamtes für Strahlenschutz und der Strahlenschutzkommission München 6–7 March (Stuttgart: Gustav Fischer) pp 284–300

[2] United Nations Scientific Committee on the Effects of Atomic Radiation (UNSCEAR) 1988 *Sources, Effects and Risks of Ionizing Radiation* 1988 Report to the General Assembly (New York: United Nations)

[3] Bowonder B, Kasperson J X and Kasperson R E 1985 Avoiding future Bhopals *Environment* **27** 6–37

[4] Association of Physicians of Chernobyl and World Health Organization 1998 *2nd Int. Conf. Long-term Consequences of the Chernobyl Disaster* Kiev June 1–6

[5] World Health Organization 1995 *Int. Conf. Health Consequences of the Chernobyl and Other Radiological Accidents* Geneva 20–23 November 1995 Fact sheets

[6] Souchkevitch G N, Tsyb A F, Repacholi M N and Mould R F eds 1996 *Health Consequences of the Chernobyl Accident, Results of the IPHECA Pilot Projects and Related National Programmes* Scientific report (Geneva: World Health Organization)

[7] Buzunov V A, Strapko N P, Pirogova E A, Krasnikova L I, Bugayev V N, Korol N A, Treskunova T V, Ledoschuk B A, Gudzenko N A, Bomko E I, Bobyleva O A and Kartushin G I 1996 Epidemiological survey of the medical consequences of the Chernobyl accident in Ukraine *World Health Statistics Quarterly* **49/1** 4–6

[8] Marchuk Yu 1996 National statement *Proc. Int. Conf. One Decade After Chernobyl, Summing up the Consequences of the Accident* Vienna 8–12 April 1996 (Vienna: IAEA) pp 57–61

[9] Shoigu S K 1996 National statement *Proc. Int. Conf. One Decade After Chernobyl, Summing up the Consequences of the Accident* Vienna 8–12 April 1996 (Vienna: IAEA) pp 53–6

[10] Ivanov V K and Tsyb A F 1996 The Chernobyl accident and radiation risks: dynamics of epidemiological rates according to the data in the national registry *World Health Statistics Quarterly* **49/1** 22–8

[11] Kamarli Z and Abdulina A 1996 Health conditions among workers who participated in the cleanup of the Chernobyl accident *World Health Statistics Quarterly* **49/1** 29–31

[12a] Kesminiene A Z, Kurtinaitis J and Vilkeliene Z 1997 Thyroid nodularity aming Chernobyl cleanup workers from Lithuania *Acta Medica Lituanica* **2** 51–4

[12b] Meshkati N 1991 Human factors in large-scale technological systems accidents: Three Mile Island, Bhopal and Chernobyl *Industrial Crisis Quarterly* **5** 131–54

[13] Tekkel M, Rahu M and Veidbaum T 1997 The Estonian study of Chernobyl cleanup workers: design and questionnaire data *Radiation Research* **147** 641–52

[14] Rahu M, Tekkel M and Veidebaum T 1997 The Estonian study of Chernobyl cleanup workers II Incidence of cancer mortality *Radiation Research* **147** 653–7

[15] Mettler F A 1996 Overview of the International Chernobyl Project: health effects *Proc. Int. Conf. One Decade After Chernobyl, Summing up the Consequences of the Accident* Vienna 8–12 April 1996 (Vienna: IAEA) pp 67–72

[16] Marchenko T, Smolyar I and Torbin V 1996 *Community Development Centres for Social and Psychological Rehabilitation in Belarus, Russia and Ukraine* (Paris: UNESCO Chernobyl Programme)

[17] International Advisory Committee IAEA 1991 *The International Chernobyl Project. Assessment of Radiological Consequences and Evaluation of Protective Measures* Technical Report and Surface Contamination Maps (Vienna: IAEA)

[18] Lee T R 1996 Environmental stress reactions following the Chernobyl accident *Proc. Int. Conf. One Decade After Chernobyl, Summing up the Consequences of the Accident* Vienna 8–12 April 1996 (Vienna: IAEA) pp 283–310

[19a] Myllkangas M, Kumpusalo E, Salomaa S, Viinmäki H, Kumpusalo L, Kolmakov S, Ilchenko I, Zhukowski G and Nissinen A 1995 Perceived quality of life in a contaminated and non-contaminated village seven years after the accident World Health Organization *Int. Conf. Health Consequences of the Chernobyl and Other Radiological Accidents* Geneva 20–23 November

[19b] International Atomic Energy Agency 1996 *One Decade After Chernobyl: Environmental Impact and Prospects for the Future* Document IAEA/J1-CN-63 (Vienna: IAEA)

[20] Goldberg D and Hillier V F 1979 A scaled version of the General Health Questionnaire *Psychological Medicine* **9** 139–45

[21] Goldberg D and Williams P 1988 *A User's Guide to the General Health Questionnaire* (Windsor: NFER-Nelson)

[22] International Atomic Energy Agency 1996 Summary of the conference results *Proc. Int. Conf. One Decade After Chernobyl, Summing up the Consequences of the Accident* Vienna 8–12 April 1996 (Vienna: IAEA) pp 3–17

[23] Yamada M, Kodama K and Wong F L 1991 The long-term psychological sequelae of atomic bomb survivors in Hiroshima and Nagasaki Ricks R C, Berger M and O'Hara F M eds *The Medical Basis for Radiation Accident Preparedness Vol III The Psychological Perspective Proc. 3rd Int. REAC/TS Conf.* 5–7 December 1990 Oak Ridge (Amsterdam: Elsevier) pp 155–63

[24] Naklani Y, Honda S, Mine M, Tomonaga M, Tagawa M and Imamura Y 1996 The mental health of atomic bomb survivors Nagataki S and Yamashita S eds *Nagasaki Symposium Radiation and Human Health* (Amsterdam: Elsevier) pp 239–49

[25] Nyagu A I, Loganofsky K N and Loganofsky T K 1998 *Pre-natal Irradiation Effects on the Brain* (Kiev: Association of Physicians of Chernobyl)

[26] World Health Organization 1993 *The ICD-10 Classification of Mental and Behavioural Disorders. Diagnostic Criteria for Research* (Geneva: WHO)

Chapter 16

[1] Kodama K, Fujiwara S, Yamada M, Kasagi F, Shimizu Y and Shigematsu I 1996 Profiles of non-cancer diseases in atomic bomb survivors *World Health Statistics Quarterly* **49/1** 7–16

[2] Shigematsu I 1993 *Radiation Effects Research Foundation a Brief Description* (Hiroshima: RERF)

[3] Shimizu Y *et al* 1992 Studies of the mortality of A-bomb survivors 9. Mortality 1950–85 Part 3 Non-cancer mortality based on the revised doses DS86 *Radiation Research* **130** 249–66

[4] Wong F L *et al* 1993 Non-cancer disease incidence in the atomic bomb survivors *Radiation Research* **135** 418–30

[5] Jannoun L and Bloom H J G 1990 Long-term psychological effects in children treated for intracranial tumours *Int. J. Radiation Oncology Biology Physics* **18** 747–53

[6] Robinson L L, Nesbit M E, Sather H N, Meadows A T, Ortega J T and Hamoul G D 1984 Factors associated with IQ scores in long-term survivors of acute lymphoblastic leukaemia *Amer. J. Pediatric Hematology Oncology* **6** 115–21

[7] Kolotas C, Demetriou L, Daniel M, Schneider L, Göbel U, Schmitt G and Zamboglou N 2000 Long-term effects on intelligence of children treated for acute lymphoblastic leukemia *Cancer Investigation* at press

[8] United Nations Scientific Committee on the Effects of Atomic Radiation (UNSCEAR) 1993 *Sources, Effects and Risks of Ionizing Radiation* 1993 Report to the General Assembly (New York: United Nations)

[9] Hasegawa Y 1994 RERF epidemiological studies on health effects of exposure to atomic bomb radiation Paper for the 5th Coordination meeting of WHO collaboration centres in radiation emergency, medical preparedness and assistance (REMPAN) 5–8 December 1994 Paris pp 161–5

[10] Souchkevitch G N, Tsyb A F, Repacholi M N and Mould R F eds 1996 *Health Consequences of the Chernobyl Accident, Results of the IPHECA Pilot Projects and Related National Programmes* Scientific report (Geneva: World Health Organization)

[11] Loganovskaya T F 1998 Clinical and psychophysiological characteristics of children exposed to acute prenatal irradiation. Abstracts *2nd Int. Conf. Long-term Health Consequences of the Chernobyl Disaster* Kiev 1–6 June 1998 (Kiev: Association of Physicians of Chernobyl) p 281

[12] Bazyttchik S V and Astakova L 1996 Mental development of children exposed to ionizing radiation in utero and in infancy Nagataki S and Yamashita S eds *Nagasaki Symposium Radiation and Human Health* (Amsterdam: Elsevier) pp 97–102

[13] Wechsler D 1956 *Die Messung der Intelligenz Erwachsener* (Stuttgart: Hans Huberbern)

[14] TASS 1991 *Invalid new born children in Byelorussia*

[15] Petrova A M, Gnedko T, and Hansen H 1995 Reproductive health patterns in Belarus before and after the Chernobyl World Health Organization International conference *Health Consequences of the Chernobyl and Other Radiological Accidents* Geneva 20–23 November

[16] International Advisory Committee IAEA 1991 *The International Chernobyl Project. Assessment of Radiological Consequences and Evaluation of Protective Measures* Technical Report and Surface Contamination Maps (Vienna: IAEA)

[17] Alikhanyan Y 1987 *Babies born after the Chernobyl accident* (Moscow: Novosti Press)

[18] Ivanov V K and Tsyb A F 1996 The Chernobyl accident and radiation risks: dynamics of epidemiological rates according to the data in the national registry *World Health Statistics Quarterly* **49/1** 22–8

[19] Belyayev S T and Demin V F 1991 Long-term Chernobyl consequences, countermeasures and their effectiveness Societe Francaise d'Energie Nucleaire *Proc. Int. Conf. Nuclear Accidents and the Future of Energy* Paris 15–17 April 1991 (Paris: SFEN) pp 208–26

[20] Yamashita S and Shibata Y eds 1997 *Chernobyl a Decade Proc. 5th Chernobyl Sasakawa Medical Cooperation Symp.* 14–15 October 1996 Kiev (Amsterdam: Elsevier)

[21] Williams E D, Becker D, Dimidchik E P, Nagataki S, Pinchera A and Tronko N D 1996 Effects on the thyroid in populations exposed to radiation as a result of the Chernobyl accident *Proc. Int. Conf. One Decade After Chernobyl, Summing up the Consequences of the Accident* Vienna 8–12 April 1996 (Vienna: IAEA) pp 207–38

[22] Mettler F A and Upton A C 1995 *Medical Effects of Ionizing Radiation* 2nd edn (Philadelphia, PA: Saunders) p 232

[23] Dalley V M 1957 Cancer of the antrum and ethmoid: classification and treatment *Proc. Roy. Soc. Med.* **50** 533–4

[24] Carling E R, Windeyer B W and Smithers D W eds 1955 *British Practice in Radiotherapy* (London: Butterworth) p 225

[25a] Lederman M and Mould R F 1968 Radiation treatment of cancer of the pharynx *Brit. J. Radiology* **41** 251–74

[25b] Zierhuit D, Lohr F, Schraube P, Huber P, Wenz F, Haas R, Fehrentz D, Flentje M, Hunstein W and Wannenmacher M 2000 Cataract incidence after total-body irradiation *Int. J. Radiation Oncology Biology Physics* **46** 131–5

[26] Shigematsu I, Ito C, Kamada N, Akiyama M and Sasaki H 1993 *A-Bomb Radiation Effects Digest* (Tokyo: Bunkodo)

[27] International Commission on Radiological Protection 1991 *Recommendations of the International Commission on Radiological protection* ICRP Publication No. 60 in Annals of the ICRP (New York: Pergamon)

[28] Baranov A E, Guskova A K and Nadejina N M 1994 Experience of the Chernobyl accident: early and delayed clinical effects and biological markers of ionizing radiation. Presentation at *Int. Conf. Biological Markers of Ionizing Radiation* Ulm 25–29 September 1994

[29] Eglite A, Ozola G and Curbakova E 1998 Eye pathologies of Chernobyl clean-up workers Proc. Latvian Academy of Sciences **52** Suppl. *SECOTEX 97 Central and Eastern European Conf. Ecotoxicology and Environmental Safety* Jurmala Latvia 24–27 August 1997 pp 191–3

[30] Curbakova E, Farbtuha T, Zvagule T, Eglite M, Jekabsone I and Eglite A 1998 The health status of Chernobyl nuclear power plant accident liquidators in Latvia Proc. Latvian Academy of Sciences **52** Suppl. *SECOTEX 97 Central and Eastern European Conf. Ecotoxicology and Environmental Safety* Jurmala Latvia 24–27 August 1997 pp 187–91

[31] Bake M-A 1998 Biological monitoring of metals as indicators of pollution Proc. Latvian Academy of Sciences **52** Suppl. *SECOTEX 97 Central and Eastern European Conf. Ecotoxicology and Environmental Safety* Jurmala Latvia 24–27 August 1997 pp 24–32

[32] Orlikovs G A, Farbtuha T A, Seleznovs Y V and Kuzenko A V 1999 Role of lead in the pathogenesis of development of the post-radiation syndrome in the clean-up workers of the Chernobyl accident *Epidemiology* **10** Supp *Abstracts of the 11th Conf. Int. Soc. Environmental Epidemiology and 9th Conf. Int. Soc. Exposure Analysis* Athens, 5–8 September 1999, abstract 523SP

[33] Kaye B H 1995 *Science and the Detective, Selected Reading in Forensic Science* (Weinheim: VCH Verlagsgescllschaft mbH) p 247

[34] Luksiene B, Salavejus S and Druteikiene R 1998 Plutonium and ^{210}Lead as environmental pollutants Proc. Latvian Academy of Sciences **52** Suppl. *SECOTEX 97 Central and Eastern European Conf. Ecotoxicology and Environmental Safety* Jurmala Latvia 24–27 August 1997 pp 32–4

[35] Research Triangle Institute (Contract No. 205-93-0606) for US Department of Health and Human Services 1997 *Toxicological Profile for Uranium* Draft for Public Comment by 17 February 1998 (Atlanta: US Public Health Service Agency for Toxic Substances and Disease Registry)

[36] McDiarmid M A, Hooper F J, Squibb K and McPhaul K 1999 The utility of spot collection for urinary uranium determinations in depleted uranium exposed Gulf War veterans *Health Physics* **77** 261–4

Chapter 17

[1] Hirayama T, Waterhouse J A H and Fraumeni J F 1980 *Cancer risks by site* Union International Contre le Cancer Technical Report Series 41 (Geneva: UICC)

[2] Mould R F 1983 *A Century of X-Rays and Radioactivity in Medicine* (Bristol: Institute of Physics Publishing)

[3] National Council on Radiation Protection and Measurements 1987 *Induction of Thyroid Cancer by Ionizing Radiation* NCRP Report 80 (Bethesda: NCRP)

[4] Ivanov V K and Tsyb A F 1996 The Chernobyl accident and radiation risks: dynamics of epidemiological rates according to the data in the national registry *World Health Statistics Quarterly* **49/1** 22–8

[5] Kato H and Shimizu Y 1991 Cancer mortality risk among A-bomb survivors Kobayashi S, Uchiyama M and Matsudaira H eds *Health Effects of Atomic Radiation, Hiroshima-Nagasaki, Lucky Dragon, Techa River and Chernobyl* Proc. Japan-USSR seminar on radiation effects research 25–29 June 1990 Tokyo (Tokyo: National Institute of Radiological Sciences) pp 225–36

[6] Cardis E, Anspaugh L, Ivanov V K, Likhtarev I A, Mabuchi K, Okeanov A E and Prisyazhniuk A E 1996 Estimated long-term health effects of the Chernobyl accident *Proc. Int. Conf. One Decade After Chernobyl, Summing up the Consequences of the Accident* Vienna 8–12 April 1996 (Vienna: IAEA) pp 241–71

[7] National Academy of Sciences, National Research Council, Report of the Committee on the Biological Effects of Ionizing Radiations (BEIR) 1990 *Health Effects of Exposure to Low Levels of Ionizing Radiation (BEIR V)* (Washington, DC: National Academy of Sciences)

[8] Jablon S and Kato H 1972 Studies of the mortality of A-bomb survivors. Part 5 Radiation dose and mortality 1950–1970 *Radiation Research* **50** 649–98

[9] Edwards F M 1996 Models for estimating risk of radiation carcinogenesis Hendee W R and Edwards F M eds *Health Effects of Exposure to Low-Level Ionizing Radiation* (Bristol: Institute of Physics) pp 215–35

[10] United Nations Scientific Committee on the Effects of Atomic Radiation (UNSCEAR) 1988 *Sources, Effects and Risks of Ionizing Radiation* 1988 Report to the General Assembly (New York: United Nations)

[11] Souchkevitch G N, Tsyb A F, Repacholi M N and Mould R F ed 1996 *Health Consequences of the Chernobyl Accident, Results of the IPHECA Pilot Projects and Related National Programmes* Scientific report (Geneva: World Health Organization)

[12] Weisburger J H and Williams G M 1995 Causes of cancer Murphy G P, Lawrence W and Lenhard R E eds *American Society Textbook of Clinical Oncology* 2nd edn (Atlanta, GA: ACS) pp 10–39

[13] Brenner D J 1999 Radiation therapy: long-term risks, carcinogenic, hereditary and teratogenetic effects *41st Annual Scientific Meeting of ASTRO* San Antonio November 1999 Refresher course 303C (Fairfax: ASTRO)

Chapter 18

[1] Parkin D M, Whelan S L, Ferlay J, Raymond L and Young J eds 1997 *Cancer Incidence in Five Continents* vol VII IARC Scientific publications no. 143 (Lyon: International Agency for Research on Cancer)

[2a] Smans M, Muir C S and Boyle P eds 1992 *Atlas of Cancer Mortality in the European Economic Community* IARC Scientific publications no. 107 (Lyon: International Agency for Research on Cancer)

[2b] Shaha A R 1999 Thyroid cancer: natural history, management strategies and outcomes *41st Annual Scientific Meeting of ASTRO* San Antonio November 1999 Refresher course 109 (Fairfax: ASTRO)

[3] Berrino F, Sant M, Verdecchia A, Capocaccia R, Hakulinen T and Estève J eds 1995 *Survival of Cancer Patients in Europe The EUROCARE Study* IARC Scientific publications no. 132 (Lyon: International Agency for Research on Cancer)

[4] DeVita V T, Hellman S and Rosenberg S A eds 1989 *Cancer Principles and Practice of Oncology* 3rd edn (Philadelphia, PA: Lippincott) pp 1269–84

[5] Schneider A B, Shore-Freedman E, Ryo U Y *et al* 1985 Radiation induced tumours of the head and neck following childhood irradiation *Medicine* **64** 1

[6] Million R R and Cassisi N J eds 1994 *Management of Head and Neck Cancer a Multidisciplinary Approach* 2nd edn (Philadelphia, PA: Lippincott) p 788

[7] Ivanov V K, Tsyb A F, Gorsky A I, Maksyutov M A, Rastopchin E M, Konogorov A P, Biryukov A P, Matayash V A and Mould R F 1997 Thyroid cancer among liquidators of the Chernobyl accident *Brit. J. Radiology* **70** 937–41

[8] National Academy of Sciences, National Research Council, Report of the Committee on the Biological Effects of Ionizing Radiations (BEIR) 1990 *Health Effects of Exposure to Low Levels of Ionizing Radiation (BEIR V)* (Washington, DC: National Academy of Sciences)

[9] Cardis E, Anspaugh L, Ivanov V K, Likhtarev I A, Mabuchi K, Okeanov A E and Prisyazhniuk A E 1996 Estimated long-term health effects of the Chernobyl accident *Proc. Int. Conf. One Decade After Chernobyl, Summing up the Consequences of the Accident* Vienna 8–12 April 1996 (Vienna: IAEA) pp 241–71

[10] Kesminiene A Z, Kurtinaitis J and Vilkeliene Z 1997 Thyroid nodularity among Chernobyl cleanup workers from Lithuania *Acta Medica Lituanica* **2** 51–4

[11] Kazakov V S, Demidchik E P and Astakhova L N 1992 Thyroid cancer after Chernobyl *Nature* **359** 21

[12] Nikiforov Y and Gnepp D R 1994 Pediatric thyroid cancer after the Chernobyl disaster. Pathomorphologic study of 84 cases (1991–1992) from the Republic of Belarus *Cancer* **74** 748–64

[13] Pinchera A and Demidchik E P eds 1996 *Development of Optimal Treatment and Preventative Measures for Radiation-Induced Childhood Thyroid Cancer* European Commission, Belarus, the Russian Federation and Ukraine joint study project no. 4 Final report EUR 16536 (Brussels: European Commission)

[14] Wedekind L 1997 From dreams to new realities. Chernobyl's challenge *IAEA Bulletin* **39/3** 24–5

[15a] Mrochek A G 1998 Belarus population health status Chernobyl disaster long-term follow-up. Abstracts *2nd Int. Conf. Long-term Health Consequences of the Chernobyl Disaster* Kiev 1–6 June (Kiev: Association of Physicians of Chernobyl) p 91 and 1998 Personal communication

[15b] International Atomic Energy Agency 1996 *One Decade After Chernobyl: Environmental Impact and Prospects for the Future* Document IAEA/J1-CN-63 (Vienna: IAEA)

[16] Souchkevitch G N, Tsyb A F, Repacholi M N and Mould R F eds 1996 *Health Consequences of the Chernobyl Accident, Results of the IPHECA Pilot Projects and Related National Programmes* Scientific report (Geneva: World Health Organization)

[17] Baverstock K, Egloff B, Pinchera A, Ruchti C and Williams D 1992 Thyroid cancer after Chernobyl *Nature* **359** 21–2

[18] Tronko N, Epstein Y, Oleinik V, Bogdanova T, Likhtarev I, Gulko G, Kairo I and Sobolev B 1994 Thyroid gland in children after the Chernobyl accident (yesterday and today) Nagataki S ed *Nagasaki Symposium Radiation and Human Health* (Amsterdam: Elsevier) pp 31–46

[19] Williams E D, Becker D, Dimidchik E P, Nagataki S, Pinchera A and Tronko N D 1996 Effects on the thyroid in populations exposed to radiation as a result of the Chernobyl accident *Proc. Int. Conf. One Decade After Chernobyl, Summing up the Consequences of the Accident* Vienna 8–12 April 1996 (Vienna: IAEA) pp 207–38

[20] Likhtarev I A, Tronko N D, Gulko G M, Kairo I A, Sobolev B G and Chepurny N I 1994 Doses and consequences of irradiation of the thyroid of the Ukranian population *Reports of the Ukranian Academy of Sciences* **3** 164–6

[21] Likhtarev I A and Paretzke H G 1996 Exposure of different population groups of Ukraine after the Chernobyl accident and main health-risk assessments Bayer A, Kaul A and Reinders C eds *Zehn Jahre nach Tschernobyl eine Balanz* Proc. Seminar des Bundesamtes für Strahlenschutz und der Strahlenschutzkommission München 6–7 March (Stuttgart: Gustav Fischer) pp 65–85

[22] Parkin D M, Muir C S, Whelan S L, Gao Y T, Ferlay J and Powell J eds 1992 *Cancer Incidence in Five Continents* vol VI IARC Scientific publications no. 1203 (Lyon: International Agency for Research on Cancer)

[23] Thomson D E, Mabuchi K, Ron E, Soda M, Tokunaga M, Ochikubo S, Sugimoto S, Ikeda T, Terasaki M, Izumi S and Preston D L 1994 Cancer incidence in atomic bomb survivors Part II Solid tumours 1958–1987 *Radiation Research* **137** S17–67

[24] Hall P and Holm L E 1998 Radiation associated thyroid cancer facts and fiction *Acta Oncologia* **37** 325–30

[25] Kiikuni K 1994 Chernobyl Sasakawa health and medical cooperation Nagataki S eds *Nagasaki Symposium Radiation and Human Health* (Amsterdam: Elsevier) pp 155–60

[26] Sasakawa Memorial Health Foundation 1995 *Report on the 1994 Sasakawa Project Workshop* (Tokyo: Sasakawa Foundation)

[27] Yamashita S, Ito M, Namba H, Hara T, Ashizawa K, Nishikawa T, Hoshi M, Shibata Y, Nagataki S, Shigematsu I and Kiikuni K 1996 Nagataki S ed *Nagasaki Symposium Radiation and Human Health* (Amsterdam: Elsevier) pp 103–16

[28] Baverstock K F 1994 WHO International thyroid project Nagataki S ed *Nagasaki Symposium Radiation and Human Health* (Amsterdam: Elsevier) pp 127–31

[29] Duffy B J and Fitzgerald P J 1950 Thyroid cancer in childhood and adolescence a report on twenty-eight cases *Cancer* **3** 1018–32

[30] Wood J W, Tamagaki H, Neriishi S, Sato T, Sheldon W F, Archer P G, Hamilton H G and Johnson K G 1969 Thyroid carcinoma in atomic bomb survivors Hiroshima and Nagasaki *J. Epidemiology* **89** 4–10

[31] Ishimaru M and Ishimaru T 1975 Leukaemia and related disorders *Radiation Research* **16** S89–96

[32] Mettler F A and Upton A C 1995 *Medical Effects of Ionizing Radiation* 2nd edn (Philadelphia, PA: Saunders)

[33] McGregor D H, Land C E, Choi K *et al* 1977 Breast cancer incidence among atomic bomb survivors Hiroshima and Nagasaki 1950–1969 *J. National Cancer Institute* **59** 799–811

[34] Tokunaga M, Norman J E, Asano M *et al* 1979 Malignant breast tumours among atomic bomb survivors Hiroshima and Nagasaki 1950–1974 *J. National Cancer Institute* **62** 1347–59

[35] American Cancer Society 1998 *Cancer Facts and Figures* (Atlanta, GA: American Cancer Society)

[36] Shigematsu I 1993 *Radiation Effects Research Foundation a Brief Description* (Hiroshima: RERF)

[37] Hasegawa Y 1994 RERF epidemiological studies on health effects of exposure to atomic bomb radiation. Paper for the *5th Coordination Meeting of WHO Collaboration Centres in Radiation Emergency, Medical Preparedness and Assistance* (REMPAN) Paris 5–8 December pp 161–5

[38] Preston D L, Kusumi S, Tomonaga M, Izumi S, Ron E, Kuramoto A, Kamada N, Dohy H, Matsui T, Nonaka H, Thompson D E, Soda M and Mabuchi K 1994 Cancer incidence in atomic bomb survivors Part III Leukemia, lymphoma and multiple myeloma 1950–1987 *Radiation Research* **137** S68–97

[39] Shimizu Y, Kato H and Schull W J 1990 Life Span Study report 11 Part 2 Cancer mortality in the years 1950–85 based on the recently revised doses (DS86) *Radiation Research* **121** 120–41

[40] Shigematsu I, Ito C, Kamada N, Akiyama M and Sasaki H 1993 *A-Bomb Radiation Effects Digest* (Tokyo: Bunkodo)

[41] Tekkel M, Rahu M and Veidbaum T 1997 The Estonian study of Chernobyl cleanup workers: design and questionnaire data *Radiation Research* **147** 641–52

[42] Rahu M, Tekkel M and Veidebaum T 1997 The Estonian study of Chernobyl cleanup workers II Incidence of cancer mortality *Radiation Research* **147** 653–7

[43] Buzunov V, Omelyanetz N, Strapko N, Ledoshuck B, Krasnikova L and Kartushin G 1996 Chernobyl NPP accident consequences for the cleaning up participants in Ukraine, health status epidemiological study main results Karaoglou A, Desmet G, Kelly G N and Menzel H G eds *Proc. 1st Int. Conf. the Radiological Consequences of the Chernobyl Accident* Minsk 18–22 March EUR 16544 (Brussels: European Commission) pp 871–8

UNSCEAR reference in press

United Nations Scientific Committee on the Effects of Atomic Radiation (UNSCEAR) 2000 Report to the General Assembly Annex G *Exposures and Effects of the Chernobyl Accident* (New York: United Nations) to be published September 2000

Index

A

Absolute risk
  see Attributable fraction
  Defintion 250
  Formula for risk coefficient, thyroid cancer, NCRP 250
  Leukaemia, ATB survivors 251
Absorbed dose, *see* Dose
Absorbing rods, *see* Reactors, Control rods
Accidents, nuclear, other than Chernobyl
  IAEA International Nuclear Event scale 69
  see Kyshtym
  see Mayak Production Association
  see Techa river area
  see Three Mile Island
  see Tokaimura
  Windscale 69
Acre, hectare (ha) and km^2 units of area and their relationships 4
Activity, definition 4
Acute radiation syndrome (ARS)
  see Bone marrow transplantation (BMT)
  Breathing problems 73
  Characteristics of ARS 6, 29, 74
  see Combined radiation injury (CRI)
Acute radiation syndrome (ARS)
  Definitions of ARS 1st–4th degrees in terms of absorbed dose 6, 79, 80
  Dose-effect relationships 81
  European Commission report on diagnosis and treatment of patients with ARS 91
  Grouping of 56 patients by medical history of radiation burns 83
  Initial diagnosis at the accident site 79
  Lack of standardised terminology 6
  Latent phase of ARS 11
  Liquidators, radiation sickness, chromosome aberrations 167
  Lung doses of 21 liquidators who died of ARS 82
  Medical management and treatment 7, 8, 29, 30
  Quality of life study of liquidators living in Kiev 226
  see Radiation burns
  Sleep disturbances, lethargy, chronic tiredness 225
  Survival times of Chernobyl victims with ARS 6, 7
  Thyroid doses of 21 liquidators who died of ARS 82

Adolescents, *see* Children and adolescents
Aegean sea 205
Aetiology 321
Afghanistan war, comparison with flying at Chernobyl by helicopter pilot 45
Agranulocytosis 321
Agriculture, *see* Farming, contamination
Albania
  Measurements of ^{131}I in milk in May 1986 185
Alcohol, *see* Vodka and red wine
Alexandrov, Anatoli, Director of Kurchatov Institute 287
Alpha rays 1, 3
American satellite detection of accident not publicly reported 48
^{241}Americium
  Daughter product of ^{241}Pu 53
  Residual activity in the global environment after 70 years 57
  Time evolution of exposure pathways 56
Anaemia 321
Animals
  Evacuation of livestock 108
  in Polissya Ecological Reserve 186, 187, Plate VII
  see Wild animals, contamination
^{125}Antimony
  Half-life 63
  Worldwide activity release from atmospheric testing 63
Aplasia 321
Artemiaia 321
Asbestos fibre exposure and mesothelioma of the lung 257
Association of Physicians of Chernobyl 6
Atmospheric nuclear testing
  Long-term committed doses and main radionuclides 175
  see Nuclear weapons
Atomic bombs, *see* Hiroshima and Nagasaki
Atomic lake, Balapan, Semipalatinsk 62, 66
Attributable fraction
  Background and excess deaths, leukaemia, solid cancers, liquidators 251, 285, 286
  Definition 251
Austria
  Area (in 1000 km^2) contaminated by ^{137}Cs in the range 10–185 kBq/m^2 213
  Dose from natural background compared to Chernobyl dose 58
  First year effective dose from Chernobyl 176
  see International Atomic Energy Agency (IAEA)
  Legally abandoning nuclear power 137
  Lifetime dose from natural background 173

B

Background doses
  Annual effective doses 171, 172
  Cosmic rays 12, 171, 172
  Global average annual effective dose 171
  Lifetime dose from natural background, by country 173

from Natural background compared to Chernobyl dose, by country 58
Natural background sources 3, 171
^{40} Potassium (^{40}K) 171
Radon 171, 172
Terrestrial gamma rays 172
Thorium radionuclide series 172
Uranium radionuclide series 172
Balapan lake, Semipalatinsk 62, 66
Baltic sea 205
^{140}Barium
Core inventory 54
Half-life 54
Release from the accident 54
becquerel (Bq) as the SI unit of activity, definition 4
Belarus
Area (in 1000 km^2) contaminated by ^{137}Cs in the range 10–185 kBq/m^2 213
Belarus–Ukraine woodlands 16
Bryansk–Belarus *hot spot*, ^{137}Cs 211, 215, 216
Chamkov village, Gomel region 27
EPR dosimetry 159
see Environmental contamination
see Gomel
see Minsk
Population 16
see Radioactive plume
Belgium
Dose from natural background compared to Chernobyl dose 58
First year effective dose from Chernobyl 176
Lifetime dose from natural background 173
Measurements of ^{131}I in milk in May 1986 185
Bell of Chernobyl 169
Belokon, Valentin, first physician to arrive at the accident site 73, 74, 75
Bhopal, poison gas leak accident 225, 226
The Bible 115, 305, 307
Biological warfare, Novosibirsk 1979, biological/chemical warfare plant accident, 300 deaths 226
Birth abnormalities, *see* Reproductive health patterns
Black sea 16, 17, 119, 205, 313
Bobruisk 198, 204, 215
Bogdany village 231
Bone marrow transplantation (BMT)
Advantages of PBSCT over autologous bone marrow transplants 71
Case history, NPP worker 93
European Commission report on diagnosis and treatment of patients with ARS 91
Failures as reported by Guscova 91
Gale, Robert and *Hammer, Armand* 92
Graft rejection 91
Immune suppression comparison with kidney and heart transplants 90
Limitations 9, 72
Radiation damage to bone marrow 90
Risks 71, 90

see also Transplantation of human embryonic liver cells (THELC)
Bonuses for NPP workers linked to attempt to complete experiment which caused accident 32
Boron as neutron absorber 13
Brain damage and mental retardation, children treated for leukaemia and for brain tumours 238
Brain damage *in utero* and mental retardation
Atomic bomb survivors 237
Belarus children study 238
Highest risk period during gestation 237
IPHECA WHO project 238, 239
IQ measurements 237
Radiation induced IQ downwards shift per Sv 237
Ukranian children study 238
Breast cancer
Excess deaths, ATB survivors, function of attained age and age at time of exposure 256
Mean annual cancer incidence crude rates, Belarus and Hiroshima 262
Brest 238, 269, 270, 271, 274, 277, 276
British Nuclear Fuels (BNFL), Sellafield, quality control records falsified 234
Bryansk 105, 157, 198, 211, 215, 216, 274
Bulgaria
Area (in 1000 km^2) contaminated by ^{137}Cs, 10–185 kBq/m^2 213
Dose from natural background compared to Chernobyl dose 58
First year effective dose from Chernobyl 176
Measurements of ^{131}I in milk in May 1986 185
Poisoned umbrella murder 150

C

^{134}Caesium
Core inventory 54
Half-life 54
Pathways into growing plants 182
Releases from the accident 52, 54
Residual activity in the global environment after 70 years 57
Time evolution of exposure pathways 56
^{135}Caesium
Decay to ^{135}Xe 14
^{136}Caesium
Releases from the accident 54
^{137}Caesium
Absorption by the gastrointestinal tract 182
Area (in 1000 km^2) by country, contaminated by ^{137}Cs in the range 10–185 kBq/m^2 213
Atmospheric fallout from nuclear weapons testing 62, 63
see Contamination
Core inventory 54
see Decontamination
Deposition in east-central Europe, nuclear weapons testing and Chernobyl 61, 62

see Environmental contamination
Half-life 54, 63
Lappland, reindeer, grazing on lichen 184
Pathways into growing plants 182
Releases from the accident 5, 52, 54
Residual activity in the global environment after 70 years 57
Techa river contamination 65
Time evolution of exposure pathways 56
Worldwide activity release from atmospheric testing and Chernobyl 61, 63

Canada
and Radioactive plume 53

Cancer
man-m Attributable fraction, solid cancers, liquidators 251, 285
Bladder cancer 256
Brain damage, mental retardation, children treated for leukaemia and for brain tumours 238
see Breast cancer
of Digestive organs, radiation induced, Russian liquidators 257
EUROCARE study of survival of cancer patients in Europe 262
Excess relative risks, all solid cancers, ATB survivors 280
How does radiation cause cancer? 256–259
Incidence and mortality rates, crude, age-specific, definitions 260, 261

Cancer (continued)
Industrial chemicals causing cancer 256
see Leukaemia
Lung cancer and uranium miners 171
Mean annual cancer incidence crude rates, Belarus and Hiroshima 262
Mean annual cancer incidence crude rates, Nagasaki 263
Mesothelioma of the lung and exposure to asbestos fibres 257
Predictions, solid cancers, leukaemia 285, 286
see Radiation induced cancer
see Relative risk
of Respiratory organs, radiation induced, Russian liquidators 257
of the Scrotum 256
Skin cancer and x-rays 77, 257
see Thyroid cancer
Year of development, all cancers, atomic bomb survivors 284

^{14}Carbon
Atmospheric fallout from nuclear weapons testing 63
Half-life 63
Radiocarbon dating 164
Worldwide activity release from atmospheric testing 61, 63

Cardiovascular diseases 87, 236
Cartoons 72, 140, 142, 234
Cateract
in Atomic bomb survivors, relative risk 237, 246
Definition 322
ICRP recommended dose limits to the lens 4, 246

in Latvian liquidators 246, 247
Opthalmological examinations 246, 247
after Radiation therapy 246
Shielding of the eye during radiation therapy 246
after Total body irradiation 246
Causes of the accident
Broader problems which contributed to the accident 34
Conduct of the experiment 32
see Countdown to the accident by seconds and minutes
Emergency scram 38
Fuel channel rupture 38
Human error 32, 33
Man–machine interface 33
Inadequate safety measures 32, 33
see INSAG reports
see Legasov, Valery
see Press reports
Technological 32, 33
Void coefficient 39
^{141}Cerium
Core inventory 54
Half-life 54, 63
Release from the accident 54
Worldwide activity release from atmospheric testing and Chernobyl 63
^{144}Cerium
Atmospheric fallout from nuclear weapons testing 61
Core inventory 54
Half-life 54
Release from the accident 54
Time evolution of exposure pathways 56
Certificates, *see* Liquidator certificates
Chain reaction 12
Challenger, American space shuttle accident 225
Chemical warfare
Novosibirsk USSR 1979, biological/chemical warfare plant accident, 300 deaths 226
see Pikalov, Vladimir, general
Chelyabinsk 64, 68, 227, 259
Chelyabinsk tractor works 194
Chemical Forces, USSR 154, 198
Chernigov 110, 124, 157, 204, 215
Chernobyl
History and culture 308, 311
Origin of name 307
Chernobyl Medical Commission 96
Chernobyl Museum, Kiev city Fire Department
Exhibition of children's paintings 110, 111, Plate VI
Fire suits donated by Germany, United Kingdom, France 88
Chernobyl nuclear power plant (NPP)
Abandonment of Units No. 5 and 6 17
Aerial view of site, September 1986 Plate I
Arrival of firemen at the accident site 28
Burning core of the reactor 152, Plate IV
Closure 143, 144
Cooling pond 8, 17, 186, 188
Emergency scram activated too late 14
Lenin, statue outside the administration building 26

Power increase immediately before explosion 14
Power lines 46, 47
see Radioactive releases from the accident
see Sarcophagus
Satellite pictures 8, 48
Chernobyl RBMK
Biological shield blown out of position, 15° to vertical 14, 40, 45, 120
see Causes of the accident
Central reactor hall 18, 19, 22, 27, 38, 40, 43, 44, 120, 122, 126
Construction stages for Units No. 1–6 17
Control room 22, 23, 30
Cross-sectional views, before and after accident 19, 41, 43, 44
Design features 18–22
Electricity production for the years 1981–85 15
Emergency cooling system 120
Faults in the concept of the RBMK 33
Foundations of the NPP buildings 119
Fuel masses after the accident, *see* Lava
Fuel rod covers, blown 1 km into the sky 18
Geographical location 16
Geology of the NPP site 118
Graphite block debris on turbine hall roof 195
Isotopic composition of unloaded fuel, uranium, plutonium 24
Main circulation pumps 19, 20, 36, 40, 120, 122, 123
Meteorology of the NPP site 118
Chernobyl RBMK (continued)
Missing fuel which could not be initially located 20, 126, 128
No containment building 14
see Press reports
Principal specifications 22, 24
Radiation measurement devices, *including* ionization chambers 21
see also RBMK reactors
see Reactors
Roof, damage and contamination 17, 40, 42, 193, Plate I
Spent fuel storage pool 43
Turbine hall 17, 18, 22, 23, 27, 28, 40, 120, 122, 124, 132, 193, 194
Ventilation stack 17, 31, 40, 46, 120
Chernobyl town
Church of St Eliah 112, 315, 316
Evacuation of population 106
Medical centre 94
Origin of the name Chernobyl/Chornobyl 18, 308
Population *and* location from Chernobyl NPP 16
Chernobyl Union organization 88
Chernobylite 128, Plate V
Children and adolescents
Attitude changes of evacuees 116
Chromosome aberrations, Hiroshima and Nagasaki 168
Examination of school children, by haematologist, endocrinologist, opthalmologist 76

Exhibition of paintings visualizing their concept of the accident 110, 111, Plate VI
see Gomel
Initial health assessment, triage, healthy, probably healthy, sick 76
Lack of facilities for initial examinations, including water supply 76
see Thyroid cancer
Thyroid dose estimates 156
China
and Radioactive plume 53
see Semipalatinsk, bordering Russia and China
China syndrome 119, 135, 203
Chromosome 165
Chromosomal aberrations, dose estimation methods
for Atomic bomb survivors 164, 167
see Cytogenetics
Dose-response relationships 166
see Leucocyte
Liquidators, radiation sickness, chromosome aberrations 167
for Tokaimura workers 71
see Translocation
Underlying theory 164
Chukseyev, Dimitri 117
Churches
in the Communist era 113
Father Melody 113
St Eliah, Chernobyl town 112, 315, 316
St Michael, Krasnoye village 113
St Sophia cathedral, Kiev 312
of Tolsty Les in the 30 km zone 317
Cirrhosis, *see* Liver cirrhosis
Coffin bonuses, workers at Chernobyl NPP after the accident 88
Collective effective dose equivalent, collective dose, collective dose equivalent *see* Doses
Combined radiation injury (CRI)
Chemical, mechanical and thermal CRI 9
Definitions 9
Committed dose, committed tissue or organ equivalent dose, dose commitment, *see* Doses
Confidence interval and confidence limits 252, 253
Contamination
Conversion between units Ci/km^2 and kBq/m^2 4, 177
see Decontamination
see Environmental contamination
European Commission foodstuff recommended activity limits 192
Exposure pathways for humans 181
see Farming, contamination
see Fish, contamination
Food contamination *versus* food consumption for ^{137}Cs, graph 191
Lappland, reindeer, grazing on lichen, ^{137}Cs contamination 184
see Milk, contamination
Radiation warning sign on hard shoulder of the road in the 30 km zone 192
of Turbine hall floor 124

see Vegetables and fruit, contamination
Washout contamination measurements of grass 182
see Water contamination
see Wild animals, contamination
Control rods, *see* RBMK reactors *and* Reactors
Cooling pond
see Chernobyl nuclear power plant
Dimensions 206
Radiation burns of member of the public fishing in the pond at the time of the accident 7
Satellite picture 8
see Water contamination
Cooling slab to prevent ground water contamination 119, 120
Core inventory, Chernobyl RBMK 51, 54
Coronary heart disease
4 of 14 deaths by end of 1995 80
see Cardiovascular diseases
Costs
of Closing nuclear power plants 141
of Dismantling Three Mile Island-2 60
European Bank of Reconstruction and Development (EBRD) 144
of Housing, schools and hospitals, Ukraine 104
of Khmelnitzi and Rovno power plants (K2/R4), Ukraine 141, 144
see Liquidator certificates
Memorandum of Understanding 1995, G7 countries, Chernobyl NPP closure 143
Salaries and wages, Ukraine 145
of Second Sarcophagus 145
State compensation cash benefit payments, USSR 111
Ukranian economy 111
Ukranian investment in clean-up operations 143
Ukranian 2 Gryvna coin minted in 1996, *Bell of Chernobyl* 169
US\$–Ukranian Gryvna exchange rate 169
Countdown to the accident by seconds and minutes
25 April from 01:06 to 23:10 hours 34–35
26 April from 00:05 to 01:23:48 hours 35–38
Final exponential rise of power 39
Variation of thermal power with time 39
Criticality accident
Definition 70
see Tokaimura
the Cross-tree 111, 112, 113
Curie, Marie 1, 28
curie (Ci) as a unit of activity, definition 4
^{242}Curium
Core inventory 54
Half-life 54
Release from the accident 54
Cytogenetics 165
Czechoslovakia
Area (in 1000 km^2) contaminated by ^{137}Cs, 10–185 kBq/m^2 213

Dose from natural background compared to Chernobyl dose 58
First year effective dose from Chernobyl 176
Measurements of ^{131}I in milk in May 1986 185

D

Danube river 17, 205
Deaths of 28 NPP workers and firemen, survival times 81
Decontamination
Bio-robot 194
Burning of contaminated homes and farms 105, Plate IV
of the Cross-tree 112
Disposal pits for contaminated vehicles 195
Forests 196
Gamma dose rates 194
Gomel, forest fire 196
Graveyards for contaminated vehicles, including helicopters 195, Plate II, Plate III
Helicopters and crop spraying biplanes 196, 197
Major problems 199
Monitoring and decontamination of transport 195, 198, Plate II, Plate III
Post-accident decontamination studies 199–202
the Red forest 196, 197
Remote controlled robots 193
Sequence for the NPP area 194
Soil removal 194, 195
Turbine hall roof 194
Waste disposal burial sites 5, 195, 216
Denmark
Dose from natural background compared to Chernobyl dose 58
First year effective dose from Chernobyl 176
Lifetime dose from natural background 173
Dentin 323
Depleted uranium and Gulf war syndrome
Constituents of depleted uranium 164
Legal actions, USA, United Kingdom 164
Possible use of EPR tooth enamel dosimetry estimates 164
Desert Storm, Gulf war, Iraq 164
Desna river 188, 204
Deterministic effects, *see* Somatic effects
Deuterium 15, 60, 323
Diagnostic radiology, *see* medical diagnostic x-rays
Digestive tract damage 6
Dnieper river 16, 117, 119, 188, 204, 205
Dniester river 205
Donets basin 119
Doses
Absorbed dose 2
Annual dose equivalents due to medical examinations, Japan 174
Annual effective background dose 171
see Background doses
Collective effective dose equivalent 2
Collective effective doses, residents in contaminated territories, evacuees, liquidators 175–180

Doses (continued)
- Committed dose, dose commitment 3
- Contribution to total dose from different radionuclides to 10 May 1986 188
- Dose equivalent 3
- *see* Dose limits
- *see* Dosimetry
- Effective dose for 1996–2056 compared to 1986–95, and food intake 183
- Effective doses, geographical variation 171
- Errors in dose estimations 155
- Evaluation of dose from ingestion of foods 183
- First year effective dose from Chernobyl, by country 176
- Foetal brain irradiation, ATB survivors 237, 238
- Group dose 155
- ICRP recommended dose limits 4, 246
- Itinerary dose 155
- $LD^{50/30}$ for humans 3
- Lethal dose and median lethal dose (LD^{50}), definition 3
- Long-term committed doses from man-made sources 175
- *to* Lungs and thyroids of 21 liquidators who died of ARS 82
- Maximum permissible dose (MPD) concept, definition 3
- *at* Mayak 64
- *from* Natural background compared to Chernobyl dose, by country 58
- *see* Populations exposed
- *to* Servicemen liquidators 154
- Thyroid doses in adults 158, 210
- Thyroid doses from ^{131}I received by Ukranian population 210
- Tokaimura accident, Japan, dose estimates for the three severely irradiated workers 9
- Turbine hall roof dose rate 195
- Uterine absorbed ATB dose, small head circumference, severe mental retardation 237
- Weighted absorbed dose 3

Dose commitment, *see* Dose

Dose equivalent, *see* Dose

Dose limits
- ICRP recommended dose limits 3, 4, 170, 246
- Occupational workers, general public 4
- Permitted emergency dose, USSR 154
- Hands and feet 4
- Lens of eye 4
- Skin 4
- Radiation workers, annual, United Kingdom 3
- USSR Commission for Accidental Whole-Body Radiation Doses, limits set 1986–89 177

Dosimetry
- *see* Chromosomal aberrations, dose estimation methods
- *see* Doses
- *see* Electron paramagnetic resonance (EPR) dosimetry using tooth enamel
- Film badge monitoring 1, 153

Dosimetry (continued)
see Fluorescence *in situ* hybridization (FISH) dosimetry
Hiroshima and Nagasaki T65D and DS86 153
Ionization personnel dosimeters 153
Mobile dosimetry laboratories 75
see ^{24}Sodium
Thermoluminescence dosimetry (TLD) 1, 153
Thyroid dosimetry 155, 156, 157
see Units, radiation
Whole-body dosimetry 155
Down's syndrome 238
Drum bomb, Semipalatinsk 63, 65
Dytiatky control point at the border of the 30 km zone 220, 224

E

Earthquakes in the region of Chernobyl 119
East Urals Radioactive Trace (EURT), *see* Kyshtym *and* Techa river area
Economics *see* Costs
Ekaterinburg 159, 160
Electron paramagnetic resonance (EPR) dosimetry using tooth enamel
for Accidental overexposure, industrial facilities 162, 163
Back extrapolation to accident dose 163
Background spectra 159
for Belarus residents 159
Cautionary notes 163
for Chernobyl doses 158
Lowest measurable dose 159, 162
for Mayak workers 162
Methods, procedure, attributes 158, 159, 161
Underlying principles 158
Elena, see Chernobyl RBMK biological shield
Elephant's foot 118, 130, Plate V
Emergency accident worker (EAW), *see* Liquidator
Emergency scram, *see* Reactors
Engraving symbolizing Chernobyl, Soviet magazine *Unost* 180
Enriched uranium, *see* Uranium
Environmental contamination
Area (in 1000 km^2) by country, contaminated by ^{137}Cs in the range 10–185 kBq/m^2 213
Areas (km^2) contaminated, Belarus, Ukraine, Russia 207, 212
Atlas of Chernobyl Exclusion Zone, National Academy of Ukraine 214, 220
Beta contamination map, 30 km exclusion zone 220
Bryansk–Belarus *hot spot*, ^{137}Cs 211, 215, 216
^{137}Cs chosen as the reference radionuclide 207
^{137}Cs contamination in countries of Europe 211
^{137}Cs contamination in 60 km zone around NPP 214
Central *hot spot* around NPP, ^{137}Cs 211, 215, Plate VIII
Doses received by forest areas, effect on tree growth 222
Environmental contamination stages, few days, one month, years 208

Fertile arable land area (km^2) removed from economic use 207
Forest area contaminated in Ukraine 207
Hot spots 104
IAEA International Chernobyl Project 213, 216, 217, 218, 219, Plate VIII
see Individual radioactive isotopes, *including those of* Caesium, Iodine, Plutonium, Strontium
Kaluga–Tula–Orel *hot spot*, ^{137}Cs 211, 215, 217
on NPP site 51
^{239}Pu and ^{240}Pu contamination 213, 219
see Radioactive plume
see Radioactive releases
Residual activity in the global environment after 70 years 55, 57
Soil contamination, conclusions from studies 209
Sosnovyi Bor, reduction of ^{137}Cs activity with time 208
^{90}Sr contamination 212, 218
beyond the USSR 51
Waste disposal burial sites 5, 195, 216
see Water contamination
Yanov railway station 105, 220, 224

Epilation, *see* Radiation effects
Erythema, *see* Skin Erythema
Erythrocyte 324
Estonia
Area (in 1000 km^2) contaminated by ^{137}Cs in the range 10–185 kBq/m^2 213
Deaths by accident, poisoning, suicide, violence 227
Estonian Komsomal report on Estonian liquidators going on strike 227
Study of 4833 liquidators 227, 284
EUROCARE study of survival of cancer patients in Europe 262
European Commission 91, 143, 144, 168, 192, 199, 205, 210

Evacuation
Belarus 105
Chernobyl town 106
from a Collective farm, eyewitness account 109
see Exclusion zone
Housing, schools and hospitals, Ukraine 104
Livestock 108
Love Canal disaster, New York, USA, 1978 135
Number of persons evacuated 103, 104, 105
in the Period 1990–95 104
Pripyat 103, 105, 107
see Radiation phobia
Radii of zones 104
Russia 105
Slavutich 94,110
Time frame 105
Typical wooden built cottage 223
Ukraine 104
Villages 105
Voluntary resettlement 104, 107
Zeleny Mys 94, 110, 111

Evacuees
Average doses (mSv) received by those living 3–15 km from the NPP 3

Collective effective doses 178
called Fire flies and *Glow worms* 114
see Resettlement of evacuees
see Returnees
from Techa river area 67
Time evolution of important exposure pathways 56
Total collective effective dose equivalent (man-Sv), external irradiation alone 3
Exclusion zone, 30 km
Atlas of Chernobyl Exclusion Zone, National Academy of Ukraine 214, 220
Beta contamination map 220
Maps 16, 214, 219, 220
Number of radioactive waste sites 5
also termed Alienation zone 103, 104
see Zones
Exposure
Radiation exposure thresholds for selected somatic effects 10
Relationship to absorbed dose 2
SI unit of exposure, coulomb/kg of air 2
The roentgen as a unit of exposure 2
Eye disease
see Cataract
see Ocular diseases

F

Farming, contamination
Contaminated milk, Ukraine, collective and private farms 186
Private production plots 189
Subsistence farming in Ukraine, Belarus, Russia 189
see Techa river area
Finland
Area (in 1000 km^2) contaminated by ^{137}Cs in the range 10–185 kBq/m2 213
Dose from natural background compared to Chernobyl dose 58
First year effective dose from Chernobyl 176
Gulf of Finland, Loviisa, radionuclide content of algae 188
Lifetime dose from natural background 173
and Radioactive plume 50, 53, 55
Fires after the explosion
see Firemen
Polymerizable liquids and trisodium phosphate dropped by helicopters 45
Prevention of fire reaching Unit No. 3 39
Silicates, clay, dolomite and lead to put out the fire 31, 43, 52
view of Evgenii Velikhov, May 1986, inverse square law of radiation, ventilation stack 31
Firemen
see Acute radiation syndrome
see Chernobyl museum, Kiev city Fire Department
Chromosome aberrations 164, 165
Decontamination of skin of the neck 81, Plate III(a)
Design of fire helmets 81

1st degree ARS diagnosis and treatment 29
see Liquidator certificates
Memorial to the bravery of the firemen 88, 89
Nausea and vomiting thought to be due to food poisoning 29
No training and education on radiation protection procedures 29
Radiation burns 28, 29, 80, 83, Plate II, Plate III
Radiation doses 27
Second wave of beta-ray burns 81
see Telyatnikov, Leonid
Treatment and rehabilitation schedule 88
Unit No. 4 roof where several firemen received lethal radiation doses 18, 27
Fish, contamination
in 30 km zone 187
Algae, radionuclide content 188
in Estuary fisheries 182
in Freshwater lakes, Belarus 182
Lake fish in Swedish Lappland, perch, pike, white fish, salmon, trout 188
Perch, lake Lugano, Switzerland 186
Fission
Build-up of fission products which absorb neutrons 14
Decontamination of skin of the neck 81
Design of fire helmets 81
Energy transferred from fission reaction, RBMK 21
Induced 12, 13
and Nuclear weapons 60
Spontaneous 12
see Tokaimura
Yield from atmospheric testing of nuclear weapons 61
Fluorescence *in situ* hybridization (FISH) dosimetry
Chromosome painting 168
Combination with EPR 160
Two colour derivative chromosomes 168
Liquidators 168
Foetal brain irradiation, ATB survivors 237
Follow-up of the evacuees and those living on contaminated territories
see also Psychological problems
see Suicides and accidents
Follow-up of the liquidators
Afraid to have more children 87
Alcohol-related problems 87, 226
Arthritis 87
Cardiovascular problems 87
Case histories 93, 94
Chronic tiredness, lethargy 226
Depression 87
Emotional and sleep disturbances 87, 225
Gastrointestinal problems 87
Immune function illnesses 87
see Liquidator certificates
Medical and social rehabilitation problems 227
NPP worker who received a BMT 93

Premature ageing 87
see Psychological illness
Quality of life studies, Ukraine 226
see Radiation phobia
Reduction in ability to work 87
Skin grafting 87
see Suicides
Foodstuffs
see Contamination
Contribution to total dose from different radionuclides to 10 May 1986 188
European Commission foodstuff recommended activity limits 192
Evaluation of dose from ingestion of foods 183
see Farming, contamination
see Fish, contamination
Food contamination *versus* food consumption for ^{137}Cs, graph 191
Intervention levels, international recommendations, WHO, FAO 190
see Vegetables and fruit
see Wild animals
Forests
Area of contaminated forests 197, 207
Contamination 18
see Decontamination
Dose rates in pine needles and mosses 198
Doses received by forest areas, effect on tree growth 222
Gomel, forest fire 196
Lethal radiation doses received by coniferous trees 197
the Red forest 196, 197
Forsmark NPP, Sweden, 28
April detection of radioactive cloud from Chernobyl 48, 53
France
Anti-nuclear cartoon 1986 140
Dose from natural background compared to Chernobyl dose 58
First year effective dose from Chernobyl 176
Lifetime dose from natural background 173
Measurements of ^{131}I in milk in May 1986 185
Nuclear power 134, 136
Fraudulent reports, *see* Press reports
Fuel, *see* Reactors
Fusion
and Nuclear weapons 60

G

G7 countries, 1995 Memorandum of Understanding, Chernobyl NPP closure 143
Gale, Robert 92, 288
Gamma dose rate, maximum inside Sarcophagus, June 1998 25
Gaussian distribution 252, 237, 324
Garino, Thomas, fraudulent video of a burning cement factory bought by American TV 49
Gaussian distribution 237, 324
Genetic effects, *see* Stochastic effects
Geology of the Chernobyl NPP region 118

Germany
  Anti-nuclear cartoon 1998 142
  Area (in 1000 km^2) contaminated by ^{137}Cs in the range 10–185 kBq/m^2 213
  Dose from natural background compared to Chernobyl dose 59
  First year effective dose from Chernobyl 176
  Lifetime dose from natural background 173
  Measurements of ^{131}I in milk in May 1986 185
  Nuclear power 136, 139
  Wind and rainfall pattern in the early days post-accident, Bavaria 182
Glasnost and perestroika 33
Goitre, *see* Thyroid diseases
^{198}Gold (^{198}Au) 150
Gomel
  Bryansk–Belarus *hot spot*, ^{137}Cs 211, 215, 216
  Forest fire, June 1986 196
  see Radioactive plume
  Reproductive health patterns 238
Gorbachenko, Nikolai, radiation monitoring technician 29
Granulocyte 324
Gray, as the SI unit of absorbed dose, definition 2
Greece
  Abortions in 1987, radiation phobia 137
  Anti-nuclear demonstration 1986 140
  Area (in 1000 km^2) contaminated by ^{137}Cs in the range 10–185 kBq/m^2 213
  Dose from natural background compared to Chernobyl dose 58
  First year effective dose from Chernobyl 176
  Hospital ward, invalid children 241
  Lifetime dose from natural background 173
  Measurements of ^{131}I in milk in May 1986 185
  Reproductive health patterns 238
Grishenko, Anatoly, test pilot, helicopters 31
Grodno 270, 271, 276, 277
Ground contamination, *see* Environmental contamination
Ground zero, Semipalatinsk 62, 64
Grouse shooting 28
Gubaryev, Vladimir, Science correspondent of Pravda, author of play Sarcophagus 288
Gulf war syndrome
  see Depleted uranium
  Heavy metal poisoning 247
Guskova, Angelina
  Experience of treating over 1000 victims of radiation accidents 92
  Failure of bone marrow transplants 91
  Gale, Robert of little help 92
  Grouping of 56 patients by medical history of radiation burns 83
  Liquidators, radiation sickness, chromosome aberrations 167
  see Moscow Hospital No. 6

Photographs of firemen who died 81, Plate III
Photograph of Guscova 97
Survival data of 28 NPP workers and firemen 81

H

Haematological diseases
Agranulocytosis 242, 243, 244
Applastic anaemia 242, 243, 244
Histiocytosis X 242, 243, 244
IPHECA WHO heamatology study 243, 244
see Leukaemia
Myelodysplastic syndrome 242, 243, 244
Sasakawa Memorial Health Foundation screening programme 242
Half-life
see Individual radioactive isotopes
Hammer, Armand 92
Harrisburg, Pennsylvania, USA, *see* Three Mile Island
Heart disease, *see* Coronary heart disease *and* Cardiovascular problems
Heavy water 15
Heinrich, Oleg, control room operator 30
Helicopter pilots
Afghanistan war, comparison of hazards with Chernobyl 43
Avoiding power lines 46
Deaths due to radiation exposure 43
Fatal accident and memorial 45, 47
see Fires after explosions
Lead covering to helicopter floors 31
Lifting of protective dome over the reactor using MI-26 helicopter, failure 31
Missions 43
Neutron absorbers did not reach the core of the RBMK 45
Overflying speed 45
Radiation doses 27
Radiation safety Colonel, eyewitness account 31
Hereditary effects, *see* Stochastic effects
Herodotus 309
Hiroshima and Nagasaki atomic bombs
A-bomb dome, Hiroshima 86
Acute and late effects 87
Atomic bombs, ^{239}Pu and ^{235}U 12, 14, 57, 58, 324, 328
Bomb blast, thermal rays *and* radiation 57, 58
Buildings destroyed 87
Chromosome aberrations in children 168
Comparisons with Chernobyl 50, 57
Discrimination against survivors 115
Energy release 57, 58
Epilation, Hiroshima 86, 87
Fat Man, Nagasaki ATB 328
Little Boy, Hiroshima ATB 325
Number of deaths by end December 1945 87
Recommended books on medical effects by *Hersey, John* and *Chisholm, Ann* 85, 93
Shadow of man and ladder imprinted on a wooden wall, Nagasaki 86

Hiroshima and Nagasaki survivors
- *A-bomb disease* and *A-bomb neurosis* 232
- *see* Absolute risk
- Aortic aneurysm, relative risk 237
- *see* Attributable fraction
- Brain damage *in utero* 237
- Calculus, kidney, ureter, relative risk 237
- Cardiovascular disease 236
- Case histories 93
- Cataract, relative risk 237
- Cervical polyp, relative risk 237
- Chromosomal aberrations, dose estimation methods 164, 167
- Cirrhosis, liver, relative risk 236
- Colon cancer, excess relative risk 280
- Dementia, relative risk 237
- Dosimetry, T65D and DS86 165
- DS86 and T65D dosimetry 153
- Duodenal ulcer, relative risk 237
- Gastric ulcer, relative risk 237
- Hepatitis, liver 236
- Hypertension, relative risk 237
- Late health effects 281, 282
- *see* Leukaemia
- Liver cancer, excess relative risk 280
- Lung cancer, excess relative risk 280
- Lymphoma, relative and excess risks 280, 283
- Mean annual cancer incidence crude rates, Belarus and Hiroshima 262
- Mean annual cancer incidence crude rates, Nagasaki 263
- Mental retardation, *in utero* study of prenatally exposed survivors 237
- *see* Mortality ratio
- Multiple myeloma, relative and excess risks 280, 283
- Myocardial infarction 236, 237
- Non-cancer diseases 235, 236
- Photographs of survivors 86
- Prostate cancer, excess relative risk 280
- Prostate, hyperplasia, relative risk 237
- Psychological illness 232, 233
- Relative risk as a function of organ dose 283
- *see* Relative risk
- Schizophrenia 232
- Stomach cancer, excess relative risk 280
- Suicides 232
- Uterine absorbed dose, small head circumference, severe mental retardation 237
- Uterine myoma 236, 237
- Year of development of cancers 284

Hospitals and medical centres
- *in* Chernobyl town 94
- Kiev University Hospital 30, 79
- *see* Moscow Hospital No. 6
- *in* Slavutich 94

Hungary
- Dose from natural background compared to Chernobyl dose 58
- First year effective dose from Chernobyl 176
- Measurements of ^{131}I in milk in May 1986 185

^{2}Hydrogen, *see* Deuterium
^{3}Hydrogen, *see* Tritium
Hydrocephalus 238
Hyperthyroidism 325
Hypothyroidism 325

I

IAEA International Chernobyl Project 213, 216, 217, 218, 219, 239, Plate VIII
ICRP recommended dose limits 4, 246
Icons and religious wall paintings 115
Industrial disasters with major loss of life, before the Chernobyl accident
  India, Bhopal, poison gas leak accident, 6954 deaths 225, 226
  China, coal dust explosion, 1572 deaths 227
  Colombia, dynamite truck explosion, 1100 deaths 226
  France, river dam collapse, 421 deaths 226
  Germany, chemical plant explosion, 561 deaths 226
  India, Bhopal, poison gas leak accident, 6954 deaths 225, 226
  India, mine explosion, 431 deaths 226
  Italy, reservoir accident, 2600 deaths 226
  Mexico, natural gas explosion, 452 deaths 226
  USA, fertilizer ship explosion, 562 deaths 226
  USSR, Novosibirsk, biological/chemical warfare plant accident, 300 deaths 226
Ignalina, Lithuania, RBMK reactor 15
Ilyin, Leonid, Academician 29
India
  Bhopal, poison gas leak accident 225, 226
  and Radioactive plume 53
INSAG reports 32, 33, 35, 135, 137, 325
International Atomic Energy Agency (IAEA), Vienna
  August 1986 Post-accident review meeting 33, 39, 79, 305, Plate III, Plate IV
  1987 meeting on Problems of skin burns 83
  see INSAG reports
  International Nuclear Event Scale 69
International Classification of Diseases (ICD) 6
International Commission on Radiological Protection (ICRP)
  Committed dose as recommended by ICRP 3
  ICRP recommended dose limits 3
International Commission on Radiological Units (ICRU)
  Definition of the roentgen unit 2
Intervention levels, international recommendations, WHO, FAO 190
Stable Iodine for thyroid blocking of radioactive iodine to the thyroid
  to Chernobyl NPP workers 79
  in Kiev 79
  in Poland 28
  in Pripyat 79

Russian *sedative mixture* as an alternative 79
in USSR, including lack of necessary quantities in reserve 77, 78, 79
Iodine deficiency 78
Iodine liquid given to children because of lack of tablets, resulting gastric problems 78
Iodine metabolism 77
^{131}Iodine
Deposition in Gomel 77
see Environmental contamination
Half-life 54, 63
Kiev Urology Institute, renal counters used for thyroid measurements 76
see Milk, contamination
Releases from the accident 5, 52, 53, 54
Time evolution of exposure pathways 56
Thyroid ^{131}I uptake measurements 76
Worldwide activity release from atmospheric testing and Chernobyl 63
^{132}Iodine
Decay product of ^{132}Te 53
^{133}Iodine
Releases from the accident 54
^{135}Iodine
Decay produces ^{135}Xe 14
IPHECA, WHO projects
Brain damage *in utero* 238, 239
Dosimetry methods, thyroid gland and whole-body 155
Haematology 241, 243, 244
Improvement in diagnostic facilities, Belarus, Russia, Ukraine 242
Map of regions covered by IPHECA 274
Thyroid study 277
IQ tests
for the General population, Gaussian 237
and Mental retardation 237, 238
Radiation induced downward shift 237
Wechsler Intelligence Scale 238
Ireland
Dose from natural background compared to Chernobyl dose 58
First year effective dose from Chernobyl 176
Lifetime dose from natural background 173
Measurements of ^{131}I in milk in May 1986 185
Irish sea 205
Irmolenko, fireman 29
^{55}Iron (^{55}Fe)
Half-life 63
Worldwide activity release from atmospheric testing 63
Irtysh river 62
Israel
Measurements of ^{131}I in milk in May 1986 185
Italy
Area (in 1000 km^2) contaminated by ^{137}Cs in the range 10–185 kBq/m^2 213
Dose from natural background compared to Chernobyl dose 58
First year effective dose from Chernobyl 176
Lifetime dose from natural background 173

Measurements of ^{131}I in milk in May 1986 185
Soveso chemical accident in 1976 226
Ivankov 95

J

Japan
Chiba Radiation Research Centre 71
see Hiroshima and Nagasaki
Measurements of ^{131}I in milk in May 1986 185
Mox, uranium and plutonium mixed oxide fuel 234
and Radioactive plume detection 53
Sasakawa Memorial Health Foundation screening programme 242, 244, 245
see Tokaimura accident
Tokyo University hospital 71
Joachimsthal 171
Jülich 157
Junke Fenix, Juarez, Mexico, ^{137}Cs accident 151

K

Kalashnikov rifles 130, 131
Kaluga 215, 217, 274
Kassabian, Mihran 84
Kazakhstan, *see* Semipalatinsk
Keloid scars 85, 86
KGB 107, 331
Kharkov 168
Khmelnitzi and Rovno power plants (K2/R4), Ukraine 141, 144
Khodemchuk, Valery, reactor operator, body never recovered 29
Kibenok, fireman lieutenant 74
Kiev
see Chernobyl Museum, Contractoviya Square, city Fire Department
Distribution of stable iodine tablets, delays 79
Electricity grid controller requests continued supply 25 April 34
Founding of Kiev 310
Golden Gates 314
Institute of Haematology and Blood Transfusions 76
Kiev reservoir 16
Kreshchatik 310, 321
see Polissya
Population *and* location from Chernobyl NPP 16
Reservoir 16, 119, 204, 217
St Cyril church 312
St Sophia cathedral 312
Urology Institute 76
Köningsberg 313
Kopach 220, 223
^{85}Krypton
Gaseous release from accident 14, 51, 54
Half-life 54
Kurchatov city 62
Kurchatov Institute of Atomic Energy, Moscow 31, 33, 287
Kushnin, Anatoli, air force Colonel, helicopters 31
Kyrgystan
see Suicides
see Techa river area
Kyshtym accident, USSR
Description of accident 68
East Urals Radioactive trace (EURT) 68
High level nuclear waste tank 68

Lack of information from USSR 48
Long-term committed doses and main radionuclides 175
see Mayak Production Association

L

Lappland, reindeer, grazing on lichen, ^{137}Cs contamination 184
Lava, nuclear fuel masses after the accident
Chernobylite 128, Plate V
Constituents 128, 130
Elephant's foot 118, Plate V
Flow out of the central reactor hall 126
Initially could not be located 20, 126, 128
Melting point 128
Remaining radioactivity, uranium and plutonium 126, 216
Sample analysis, use of Kalashnikov rifles 130, 131
Lead poisoning, Latvian study 247
Lecture on radiation protection, to control room operator 30
Legasov, Valery
the Army 294
Arrival at the NPP 290
Dosimetric control 293
Equipment faults 299
Euphoria and tragedy 301
Evacuation of Pripyat 292
Impossible to find a culprit 300
the Information service 294
Lack of equipment and facilities 293
Like a samovar 299
and Man–machine interface, RBMK 33
Margarita Legasov's reminiscences 303
NPP workers and managers 293
Organization and responsibility 298
Politically incorrect 33
Problems in the development of atomic energy 296
RBMK reactors 297
Suicide 33, 288, 303
Training and education 295
Victory Day 301
Visit of Prime Minister Ryzhkov 294
Leiden 168
Lenin, Vladimir Ilyich
Chernobyl NPP named after Lenin 26
and *Hammer, Armand* 92
the Lenin Room in schools 76
Portrait 241
Quotation on communism and electrification 26
Stained glass window, Central V I Lenin Museum, Moscow 248
Leukaemia
Attributable fraction, background deaths, excess deaths, liquidators 251, 285, 286
Brain damage, mental retardation, children treated for leukaemia and for brain tumours 238
Relative risk at 1 Sv, ATB survivors 280
Incidence 1945–80 252, 280

Induction period, schematic diagram, risk as a function of age at exposure 279
Latent periods, atomic bomb survivors 278
and *Legasov, Valery* 33
Lymphoid leukaemia 262
Mean annual cancer incidence crude rates, Belarus and Hiroshima 262
Mean annual cancer incidence crude rates, Nagasaki 263
Monocytic leukaemia 262
Mortality ratio 252
Myeloid leukaemia 262
Prediction modelling 253, 285, 286
Russian liquidators, cancer, respiratory and digestive organs 257
USA leukaemia statistics, incidence, survival, mortality 280

Liquidators
see Acute radiation syndrome
Attributable fraction, leukaemia 251, 285, 286
Attributable fraction, solid cancers 251, 285
Cancer, respiratory and digestive organs, Russian liquidators 257
Classification and group, e.g. gas protection regiment 90
Classification and type of work 90
Coal miners 119
Collective effective doses 179
Definition 27
Distribution of registered doses, Belarus, Ukraine, Russia 179
see Doses *and* Dosimetry
EPR dosimetry, tooth enamel 159
see Firemen
Military liquidators, doses 153, 154
Population and sub-population numbers of liquidators 27
see Psychological illness
Quality of life studies 226
Time evolution of important exposure pathways 56
Time schedule for shift work 76, 110

Liquidator certificates
Benefits 99, 100, 101
Category I certificate 96
Categories of certificate 98
Legal decree 95
for Medical students working in the 30 km zone, May 1986 76

Lithuania
Area (in 1000 km^2) contaminated by ^{137}Cs in the range 10–185 kBq/m^2 213
Ignalina RBMK reactor 15

Liver cirrhosis
in Atomic bomb survivors 236
one of 14 deaths by end of 1995 82

Los Alamos, USA 138, 148, 151

Luxembourg
Dose from natural background compared to Chernobyl dose 58
Exclusion zone area 1.5 times that of Luxembourg 215
First year effective dose from Chernobyl 176
Lifetime dose from natural background 173

Lymphocytes
  and Leucocyte 166
  Manual counting, initial medical tests 77
  Tokaimura workers, dose estimation, lymphocyte counting 71
Lymphoma, ATB survivors, relative risk 280

M

Makarov 204
^{54}Manganese
  Half-life 63
  Worldwide activity release from atmospheric testing 63
Maps
  Chernobyl NPP in relation to Pripyat, Kiev and the Kiev reservoir 16
  Kiev and Pripyat in relation to Belarus, Poland and other neighbouring countries 17
Markov, Georgi, Poisoned umbrella murder 150
Mayak Production Association
  Circumstances leading to major exposures 64, 65
  Doses 64, 159
  EPR dosimetry, tooth enamel 159
  Karachai lake 67, 68
  Plutonium production 50
  Radiochemical plant 64
  Reactor 64
  Release data for 1945–60 162
Medical centre, Chernobyl town 94
Medical diagnostic x-rays
  Annual effective collective dose 172
  Chest x-rays 170
  CT scans 174
  Dental x-rays 174
  Fluoroscopy 174
  Japan, annual dose equivalent 174
  Radiography 174
Medical examinations immediately after the accident
  Assessment in schools, haematologist, endocrinologist, opthalmologist 76
  see Belokon, Valentin
  Biochemistry tests 77
  Blood counts 77
  Children's health assessment, triage, healthy, probably healthy, sick 76
  of Eyes 77
  see Guscova, Angelina
  Medical students drafted 76
  Mobile dosimetry laboratories 75
  see Moscow Hospital No. 6
  Personnel 75
  Thyroid ^{131}I uptake measurements 76
  Working hours of medical teams 76
Mediterranean sea 205
Mental retardation, *see* Brain damage *in utero* and mental retardation
Metabolism of iodine 77
Meteorology of Chernobyl region 118
Mettler, Fred 29
MI-26 helicopter 31
Milk, contamination
  Contaminated milk, Ukraine, collective and private farms 186
  see ^{131}Iodine

Measurements of ^{131}I in milk in May 1986, by country 185
Minsk 16, 157, 168, 204, 215, 270, 271, 276, 277
Modelling, *see* Prediction modelling
Moderator, reactor 13
Mogilev 212, 215, 238, 239, 270, 271, 274, 276, 277
Moldovia
Area (in 1000 km^2) contaminated by ^{137}Cs in the range 10–185 kBq/m^2 213
^{99}Molybdenum
Core inventory 54
Half-life 54
Release from the accident 54
Morozov, Iourii, holder of a liquidator category 1 certificate 96
Mortality ratio, leukaemia, ATB survivors 252
Moscow Hospital No. 6 30, 75, 79
Mox, uranium and plutonium mixed oxide fuel 234
Mule spinner's cancer 256
Multiple myeloma, ATB survivors, relative risk 280
Munich and Neuherberg 157, 168
Myelocyte 327

N

Nagasaki, *see* Hiroshima and Nagasaki
Narodichi village school 76, 209
National Institute of Standards and Technology, (NIST), USA 160
Nazis, World War II 111, 112, 113
Necrosis as a late effect of irradiation 6, 10
^{239}Neptunium
Core inventory 54
Half-life 54
Release from the accident 54
Netherlands
Dose from natural background compared to Chernobyl dose 58
First year effective dose from Chernobyl 176
Lifetime dose from natural background 173
Measurements of ^{131}I in milk in May 1986 185
Neutrons
Delayed neutrons 12
Fast neutron 13
Neutron absorbers including phenomenon of xenon poisoning 13, 14
Prompt neutrons 12, 13
Quality factor (Q in H=DQN) 3
Slow neutron 13
Thermal neutron 13
^{95}Niobium
Techa river contamination 65
Time evolution of exposure pathways 56
Non-stochastic effects, *see* Somatic effects
Normal distribution, *see* Gaussian distribution
North Anna-1 power station, Virginia, USA, in 28 April 1986 TASS report 48
Norway
Area (in 1000 km^2) contaminated by ^{137}Cs in the range 10–185 kBq/m^2 213

Dose from natural background compared to Chernobyl dose 58
First year effective dose from Chernobyl 176
Lifetime dose from natural background 173
Measurements of ^{131}I in milk in May 1986 185
Nuclear accidents other than Chernobyl
see Kyshtym
TASS report of 28 April 1986, accidents in USA 45
see Three Mile Island
Nuclear facilities, non-nuclear research industry, radiation medicine, fatal accidents
Critical assembly 138, 148, 149
Industrial radiography 138, 148, 149
Reactors 138
Reports to IAEA 138, 148, 149
Nuclear fission, *see* Fission
Nuclear fuel, *see* Reactors
Nuclear fuel masses after the accident, *see* Lava
Nuclear power for electricity generation
Anti-nuclear power demonstrations 137, 138, 139, 140, 142
Expansion 1980–96 135
in France 134, 136
in Germany 136, 139
INSAG recommendations 137
in Japan 136
Khmelnitzki and Rovno (K2/R4), Ukraine 141, 144
Long-term committed doses and main radionuclides 175
Nuclear power reactors, worldwide, at end of 1992 139
Nuclear share of electricity generation, IAEA data by country 136
Obninsk, Russia 134
Outlook to year 2015 141
Shippingport, USA 134
see Three Mile Island
in United Kingdom 134, 136
Nuclear reactors, *see* RBMK reactors, *see* Reactors
Nuclear weapons
Atmospheric testing 1945–80, number of tests and fission yield 61
Chain reaction 12
Dose reconstruction using EPR dosimetry 160
Fission device 60
Fusion device 60
see Hiroshima and Nagasaki atomic bombs
Kurchatov city 62
Long-term committed doses and main radionuclides 175
Missiles, radar early warning screen, near Chernobyl 48
Nagasaki-type atomic bomb tests, Nevada, USA 153
Nuclear testing 60
see Polygon
Radioactive releases, comparison with those from Chernobyl 50
Safety trial type of experiment 60
see Semipalatinsk, test site
Soviet tests, atmospheric *and* underground 63
Totskoye text site, Urals 160

O

Obninsk, Russia 134, 157
Ocular disease 245–248
Oncogenesis due to radiation 258
Orel 215, 217, 274
Orthodox Church, *see* Churches

P

Paracelsus 171
Peripheral blood stem cell transplantation (PBSCT)
  Advantages over autologous bone marrow transplants 71
  Treatment of workers irradiated at Tokaimura 70, 71, 72
Petrovskii, Alexandr, fireman 28
Pikalov, Vladimir, general 154, 198, 291
Plume, *see* Radioactive plume
Plutonium as an envirinmental pollutant, Lithuanian study 247
^{238}Plutonium
  Core inventory 54
  Half-life 54
  Release from the accident 54
  Residual activity in the global environment after 70 years 57
^{239}Plutonium
  Core inventory 54
  Fission product, Chernobyl 14
  Half-life 50, 54, 63
  Nagasaki atomic bomb 12, 58
  and Nuclear weapons 60
  Release from the accident 54
  Residual activity in the global environment after 70 years 57
  Worldwide activity release from atmospheric testing and Chernobyl 63
^{240}Plutonium
  Core inventory 54
  Fission product, Chernobyl 14
  Half-life 50, 54, 63
  Release from the accident 54
  Residual activity in the global environment after 70 years 57
  Worldwide activity release from atmospheric testing and Chernobyl 63
^{241}Plutonium
  see ^{241}Americium
  Core inventory 54
  Fission product, Chernobyl 14
  Half-life 50, 54, 63
  Release from the accident 54
  Residual activity in the global environment after 70 years 57
  Time evolution of exposure pathways 56
  Worldwide activity release from atmospheric testing and Chernobyl 63
^{242}Plutonium
  Release from the accident 54
Poisoned umbrella murder 150
Poland
  Area (in 1000 km^2) contaminated by ^{137}Cs in the range 10–185 kBq/m^2 213
  Distribution of stable iodine, thyroid blocking 28
  Dose from natural background compared to Chernobyl dose 58

First year effective dose from Chernobyl 176
Gdansk 313
Institute of Oncology, Warsaw 28
Measurements of ^{131}I in milk in May 1986 185
Polish–Belorussian border nearest to Chernobyl 28, 290
Polissya region
Animals in Polissya Ecological Reserve 186, 187, Plate VII
Apple Saviour festival 318
Architecture 314
Bee keeping 312
Chervets red dye 313
Chervone Polissya state farm 318, 319
Grey wolves 187, Plate VII
Rushniks 315, 317, 318
Saint Eliah 112, 315, 316
Tolsty Les 317
Wild boars 187, Plate VII
Politics, *see* USSR
Polovinkin, Andrei, fireman 28
the Polygon, Semipalatinsk 62
Populations exposed
see Evacuees
Evolution over time and exposed groups 53, 56
Irradiation stages 56
see Liquidators
see Residents of contaminated territories
Time evolution of important exposure pathways 56
Portugal
Dose from natural background compared to Chernobyl dose 58
First year effective dose from Chernobyl 176
Lifetime dose from natural background 173
Measurements of ^{131}I in milk in May 1986 185
^{40}Potassium (^{40}K)
Content in typical soils 182
in Muscle 171
Pott, Percival 256
Pravik, fireman lieutenant 28, 74
Prediction modelling
see Absolute risk
Additive model 253, 254
Atomic bomb survivors, cancer 253
BEIR V model 254, 255, 256
see Cancer
Chernobyl, cancer 253
see Leukaemia
Linear model, dose-response relationship 259
Multiplicative model 254
see Relative risk
RNMDR modelling 255, 257, 267
Thyroid cancer, Russian liquidators 267
Prefixes for factors of 10 such as μ for 10^{-6} 5
Press reports
American satellite detection of accident not publicly reported 48
Comparisons with Hiroshima and Nagasaki 50
Exaggerated reports of the initial number of deaths 49
see Forsmark power station, Sweden
Fraudulent video shown on American and Italian TV, burning cement factory in Trieste 49

Radioactive cloud over Kiev, fraudulent photograph, Sunday Times *and* Time 49
28 April 1986 TASS 45, 48
Pripyat marshes 17
Pripyat town
Children's playground 109, Plate V
Distribution of stable iodine, thyroid blocking 79
Evacuation 103, 105, 107
Helicopter view in 1991 108
Population and location from Chernobyl NPP 16
Thyroid dose estimates 157
Pripyat river 16, 27, 76, 119, 188, 204, 217
Protons 3
Psychological illness
A-bomb disease and *A-bomb neurosis* 232
in Atomic bomb survivors 232, 233
Comparison study between residents in contaminated and in clean areas 229, 230
Escapism through alcohol and drugs 228
see Follow-up of the liquidators
International Chernobyl Project study, IAEA, perception of illness 229
Major psychological and social problems, documented by UNESCO 228
Perceived risk of premature death, city of Kiev and village of Bogdany 230
see Radiation phobia
Schizophrenia 232
Stress-related illness 229, 231
see Suicides
Teenagers, their opinions on Chernobyl 228, Plate VI

Q

Quality factor for ionizing radiation in formula for the weighted absorbed dose 3
Quality of life
of Techa river population, Chelyabinsk province, Kyrgystan 227
of Ukranian liquidators 226
Quality of a radiation beam 2
Quotation, *Lenin* on communism and electrification 26

R

rad, unit of absorbed dose 2
Radiation burns
from Carrying *Vladimir Sashenok* 30
Case histories 80, 93, 94, Plate II, Plate III
Characteristics of 1st–4th degree radiation burns 6, 30, 80
Degrees of x-ray burn, defined in 1904 85
to Firemen 28, 81
Grouping of 56 patients by medical history of radiation burns 83
Hands of x-ray physician in 1903 85
IAEA 1987 meeting on problems of skin burns 83
Late effects after the accident 6, 7, 30
Moderate to severe reactions 83
Percentage of body surface, burns 83

Photographs of burns 7
Skin grafts 30, 83, 93, 94
THIRD Degree burns, member of the public fishing in the NPP cooling pond 7
Tokaimura workers 71
Transient erythema 83
Ulcers at the site of the burns 6, 81
Widespread erythema 83
Radiation effects
Acute effects, definition 11
see Acute radiation syndrome (ARS)
Blood formation depression 10, 80
Cataract formation 11
Chronic effects, definition 11
Diarrhoea 10
Epilation 10, 11, 81, 93
Fibrosis 11
Keloid scars 85, 86
Leukopenia 80
Lymphocyte and granulocyte count (g/litre) reduction 10, 80
Nephritis 10
Organ atrophy 11
see Radiation burns
see Skin erythema
see Somatic effects
see Stochastic effects
Sterility in males 10, 11
Tissue necrosis 10
Vomiting 6, 74, 80
Radiation induced cancer
see Cancer
see Hiroshima and Nagasaki
Not after use of ^{131}I for treatment of thyrotoxicosis 77
Röntgen hands 84
see Stochastic effects
see Thyroid cancer
after X-ray beauty treatment, skin cancer 77
after X-ray treatment for ringworm, skin cancer 77
Radiation monitoring
see Doses
see Dosimetry
inside Sarcophagus 24, 25
Radiation oncogenesis 258
Radiation oncology, *see* Radiation therapy
Radiation phobia
Abortions in Greece in 1987 137
see China syndrome
and Chukseyev, Dmitri 117
Coffin bonuses, workers at Chernobyl NPP after the accident 88
Evacuees called *Fire flies* and *Glow worms* 114
Wives of liquidators afraid to have more children 87
Radiation sickness, *see* Acute radiation syndrome
Radiation syndrome, *see* Acute radiation syndrome
Radiation therapy for cancer
Accident in Exeter, United Kingdom 147
Accident in Juarez, Mexico 149, 151
Accident in Plymouth, United Kingdom 147, 148
Accidents reported to IAEA 148, 149
Brain damage, mental retardation, children treated for leukaemia and for brain tumours 238
Cataract after radiation therapy 246
Cataract after total body irradiation 246

^{60}Co, ^{137}Cs, ^{192}Ir or ^{198}Au sources 70, 150, 258
Overdoses, medico–legal proceedings 147
Shielding of the eye during radiation therapy 246
Radiation units *see* Units, radiation, definitions
Radioactive plume
Arrival times of detectable activity in air, Europe, Scandinavia 53, 55
Atmospheric transport 53
Direction of travel 50
and Gomel, incidence of thyroid cancer in children and adolescents 50
Height of plume 50
and Sweden 50, 53
Radioactive releases from the Chernobyl accident
Comparison with those from nuclear weapons testing 50
Comparison with those of Three Mile Island 50
Components, two difference types 51
Core inventory 51, 54
Daily pattern of releases 51, 52, 53
Fuel particle elements 54
see also Individual radioactive isotopes, *including those of* caesium, iodine, strontium
Noble gases 54
see Populations exposed
see Radioactive plume
Refractory elements 54
Release phases, four 51
Residual activity in the global environment after 70 years 55, 57
Sequence and composition 51
Total activity of the 800 waste sites within the 30 km zone 5
Total release from the accident, including ^{131}I and ^{137}Cs 5, 50, 54
Volatile elements 54
Radioactive waste
Drum bomb, Semipalatinsk 63, 65
Radionuclide production, long-term committed doses *and* main radionuclides 175
Radiotherapy, *see* Radiation therapy
Radium 1, 2, 28, 171, 258
RBMK reactors
Biological shield, 2000 tonne 14
see Chernobyl RBMK
Condenser 19
Control and protection system basic functions 21
Control rods 13, 19, 35–38
Faults prior to the Chernobyl accident 15
Feedwater system transients 15,37,38
Functional grouping of the 211 absorbing rods 22
Geometrical arrangement of the core 20
Graphite moderator 20, 21
History of development 15
at Ignalina, Lithuania 15
at Kursk 15
at Leningrad/St Petersburg 15
Measurements and subsystems 22
Measures to improve safety of RBMKs post-accident 25, 26

Pitfalls other than defects in design of control and safety rods 40
Pump failure 40
Safety systems 21, 22, 25, 26
see Reactors
at Smolensk 15
Steam drums 19, 20, 22, 36
No containment building 14
Reactors
CANDU reactors 15
Chain reaction 12, 13
Cladding of fuel, zirconium–niobium alloy 13, 20
Containment building 14
Control rods/absorbing rods 13, 19, 21, 22, 26, 35–38
Coolant *and* Cooling system 15, 20, 21, 26, 36, 38
Core 13, 19, 20, 22, 26, 38, 51
Criticality 13, 38
Enriched uranium 13
Fast breeder reactors 15
Fuel 12, 13, 14, 19, 20, 24
Fuel channel rupture 38
Fuel pins, dimensions 20
Gas cooled power reactors 15
Heavy water reactors 15
Light water reactors 15
Magnox reactors 15
Moderator: hydrogen or carbon 13, 19, 20
Operation 12, 13
Power density 14, 15, 21, 22
Pressurized water reactors (PWR) 15
Prompt critical excursion 38
see RBMK reactors
Reaction rate 13, 14
Reactivity, positive *and* negative 13, 26, 36, 40
Refuelling 14, 19
Scram 14, 21, 38, 39
Shutdown 14
Spent fuel 14
Start-up 14, 21
Turbogenerators 20, 36
Types 15
Xenon poisoning 14
Relative biological effectiveness (RBE) 2, 3
Relative risk
Breast cancer, BEIR V data, function of attained age and age at time of exposure 255
Definition 250
and Excess relative risk 250
Incidence of non-cancer diseases, ATB survivors 237
Leukaemia, ATB survivors 251, 280
Liver disease, chronic, ATB survivors 236, 237
Lymphoma 280
Multiple myeloma 280
Myocardial infarction, ATB survivors 236, 237
and Multiplicative model 254
Solid cancers, excess relative risk 280
Uterine myoma, ATB survivors 236, 237
Releases of radioactivity, *see* Radioactive releases
Relocation, *see* Evacuation and Resettlement
rem, unit of RBE 3
Reproductive health patterns
in Belarus, 1982–90 240
Congenital abnormalities 240
Down's syndrome 238
Hospital ward for invalid children, Gomel 241
Hydrocephalus 238
Infant mortality 240

IAEA International Chernobyl Project findings 239
Low birth weight 240
Maternal morbidity 240
Meningocele 238
Neo-natal morbidity 240
Perinatal mortality 239
Post-natal death rates 239
Stillbirths 240
TASS report 238
Thalidomide 238
Twins, incidence, mothers evacuated from contaminated territories, Ukraine 241
Resettlement of evacuees
Building costs 104
Construction statistics 110
Discrimination against evacuees from Pripyat 114
Problems 104, 114, 116
see Radiation phobia
see Returnees
Residents of contaminated territories
Collective effective doses 175, 176, 177
Comparisons between populations in contaminated and in *clean* areas 229, 230
Escapism through alcohol and drugs 228
Fatigue and loss of appetite 227
International Chernobyl Project study, IAEA, perception of illness 229
Major psychological and social problems, documented by UNESCO 228
Perceived risk of premature death, city of Kiev and village of Bogdany 230
Population numbers 207
Special controlled zones (SCZ) 207
Stress related illness 229, 231
Teenagers, their opinions on Chernobyl 228, Plate VI
Time evolution of important exposure pathways 56
Wish to relocate from their homes 228
Returnees
Ex-partisans from World War II 111
Garden vegetable plots 112
Number of persons 112
Swedish newspaper portrait award photograph 114
Richter scale 119
Ringworm 77
Risk
see Absolute risk
see Attributable fraction
from Cancer, specification 249–259
Confidence interval and confidence limits 252
see Leukaemia
Possible cohorts for useful data in the future, airline pilots, Chernobyl, Chelyabinsk 259
see Prediction modelling
Problems in estimating radiation induced cancer risks in man 259
see Relative risk
Standard error 253
Romania
Area (in 1000 km^2) contaminated by ^{137}Cs, 10–185 kBq/m^2 213
Dose from natural background compared to Chernobyl dose 58

First year effective dose from Chernobyl 176
Measurements of ^{131}I in milk in May 1986 185
Rovno 215
Rovno and Khmelnitzi power plants (K2/R4), Ukraine 141, 144
Russia
Area (in 1000 km^2) contaminated by ^{137}Cs in the range 10–185 kBq/m^2 213
see Bryansk region
Klintsy 198
see Kyshtym
Leningrad region 198
see Mayak Production Association
Moscow 30, 75, 79, 157
see Nuclear weapons
Obninsk 134, 157
see Semipalatinsk
St Petersburg 157
see Techa river area
see USSR
^{103}Ruthenium
Core inventory 54
Half-life 54, 63
Release from the accident 54
Techa river contamination 65
Time evolution of exposure pathways 56
Worldwide activity release from atmospheric testing and Chernobyl 63
^{106}Ruthenium
Atmospheric fallout from nuclear weapons testing 61
Core inventory 54
Half-life 54, 63
Release from the accident 54
Techa river contamination 65
Time evolution of exposure pathways 56
Worldwide activity release from atmospheric testing and Chernobyl 63
Ryzhkov, Nikolai, Prime Minister, USSR 31, 294, 299

S

Sarcophagus
Aerial view during construction 127
Architectural drawings 120, 121
Biological shield, *Elena*, 2000 tonne 14, 40, 45, 120, 121, 131
Buttresses wall 120, 121, 122, 129
Cascade wall 120, 121, 122, 124, 125, 129
Chernobylite 128, Plate V
after Completion, photographs 129, 132
Completion date 122
the Complex Expedition to locate the missing fuel 128
Construction 120–125
Dust suppression unit 132
see Elephant's foot
Entry to the Sarcophagus 122
Fire hazard of electrical cables 122, 123
Future options 144, 145
Hazard warning notice inside Sarcophagus 132
Holes in the walls of the Sarcophagus 122
Interior views 122–126
see Lava
Likelihood of partial collapse 40, 112, 144

Sarcophagus (continued)
Mammoth beam 120, 122
Measurement recording laboratory, dosimetric and temperature monitoring 25
Radiation certificate after visit 133
Radiation dose rates inside the building 14, 124, 128, 133
Radiation monitoring 14, 24, 25
Radioactive dust problems 122, 131, 132, 144
Radioactive water problems 131
Roof 124, 128
see Second Sarcophagus
also called Shelter, Ukritiye 120, 144
Step wall, *see* Cascade wall
Wall built between Units No. 3 and 4 120, 122, 125, 130, 131
Sasakawa Memorial Health Foundation screening programme 242, 244, 245
Sashenok, Vladimir, NPP worker, died within 12 hours of accident 29, 30, 72
Satellite pictures 8, 48
Satellite re-entries, long-term committed doses and main radionuclides 175
Scherbina, Boris, Council of Ministers of the USSR 289, 292
Sculpture, which marks the boundary of the NPP on the road to Chernobyl 11
Second Sarcophagus
Problem of expense 145
Tender document 40, 43, 146
Semipalatinsk test site for nuclear weapons
Drum bomb 63, 65
Goose neck 64
Ground zero 62, 64
Lake Balapan, *Atomic lake* 62, 66
the Polygon 62
Tests 1949–62 63
Shippingport, USA 134
Siemens 142
Sievert, as the SI unit of dose equivalent, definition 3
Shavrei, Ivan, fireman 28
Shcherbak, Yuri, author, Ukranian Ambassador to USA 29
Skin erythema
Definition 1
see Radiation burns
Slavutich 94, 110
Slovakia
Area (in 1000 km^2) contaminated by ^{137}Cs in the range 10–185 kBq/m^2 213
Slovenia
Area (in 1000 km^2) contaminated by ^{137}Cs in the range 10–185 kBq/m^2 213
^{24}Sodium (^{24}Na)
Blood activation methods for dose estimation, Tokaimura 71
Somatic effects
Definition 9
see Radiation effects
Radiation exposure thresholds for selected somatic effects 10
Sosnovyi Bor, reduction of ^{137}Cs activity with time 208
Soviet Union, *see* USSR

Spain
  Dose from natural background compared to Chernobyl dose 58
  First year effective dose from Chernobyl 176
  Lifetime dose from natural background 173
  Measurements of ^{131}I in milk in May 1986 185
Spanish–American War in 1898 84, 85
Standard error, definition 253
Stari Sokoly control point at the border of the 30 km zone 220
Stochastic effects
  Definition 10
  No threshold dose 10
^{89}Strontium
  Core inventory 54
  Half-life 54, 63
  Releases from the accident 54
  Techa river contamination 65
  Worldwide activity release from atmospheric testing and Chernobyl 63
^{90}Strontium
  Absorbed in gastrointestinal tract 183
  Atmospheric fallout from nuclear weapons testing 61
  Concentration in tooth enamel 160
  Core inventory 54
  Half-life 54, 63
  Releases from the accident 54
  Techa river contamination 65
  Time evolution of exposure pathways 56
  Worldwide activity release from atmospheric testing and Chernobyl 63
Submarine accident, nuclear, USSR 148, 151
Suicides
  in Atomic bomb survivors 232, 233
  ^{137}Cs, 15 Gy, USSR 148
  ^{137}Cs capsule, Bulgaria 149
  Kyrgystan, population of 2093 liquidators 227
  see Legasov, Valery
  in Liquidator population of 1986–87 87
  Lithuania, population of 546 liquidators 227
  one of 14 deaths by end of 1995, those with ARS 82
  Reported to IAEA 148, 149
Survival
  of Chernobyl victims with ARS 6, 7, 81
  see Follow-up of survivors
  Later deaths, after acute phase, not correlated with original severity of ARS 82
  of Tokaimura heavily irradiated workers 71
  to end of 1995 82
  the 106 who recovered from ARS 87
Sverdlovsk 68
Sweden
  Area (in 1000 km^2) contaminated by ^{137}Cs in the range 10–185 kBq/m^2 213
  Contamination of lake fish in Swedish Lappland 188
  Dose from natural background compared to Chernobyl dose 58
  First year effective dose from Chernobyl 176

Forsmark nuclear power station, 28 April detection of Chernobyl radioactive cloud 48, 53
^{131}I deposition on grass 184
Lifetime dose from natural background 173
Measurements of ^{131}I in milk in May 1986 185
Newspaper award, photograph of returnees 114
Policy of phasing out nuclear power 137
and Radioactive plume 50, 53
Reindeer, grazing on lichen, ^{137}Cs contamination, Lappland 184
Stockholm 48
Switzerland
Anti-nuclear demonstration in 1986 140
Area (in 1000 km^2) contaminated by ^{137}Cs in the range 10–185 kBq/m^2 213
Dose from natural background compared to Chernobyl dose 58
First year effective dose from Chernobyl 176
Lifetime dose from natural background 173
Measurements of ^{131}I in milk in May 1986 185
Perch, contamination in lake Lugano 186

T

Techa river area
Contamination levels 65, 67
Dose estimation using EPR tooth enamel dosimetry 159
Drinking water 65
Metlinsky village pond 65, 67
see Kyshtym
see Mayak Production Association
Quality of life of Techa river area population 227
^{129m}Tellurium
Releases from the accident 54
^{135}Tellurium
Decay to ^{135}Xe 14
Releases from the accident 52
Telyatnikov, Leonid, fire chief at Chernobyl NPP 88, 102
Temperature of nuclear fuel masses
in 1986 25
Maximum in June 1998 25, 51
Ternopolskoye village 111, 319
Thalidomide 238
Three Mile Island accident
Causes of the accident 59
Comparison of radioactive releases with those from Chernobyl 50, 59
Consequences 135
Containment building 14
Cost of dismantling TMI-2 60
Fuel rod damage 59
Legal action 59
Metropolitan Edison 60
Release of ^{95}Kr and ^{133}Xe into the atmosphere 59
in TASS report of 28 April 1986 48
TMI-1 begins generating power again 59
TMI-2 permanently shut down 59
Thyroid cancer
in Adults living in contaminated territories 276
Age as a prognostic factor 263

Thyroid cancer (continued)
Age at time of irradiation and age at time of diagnosis, Belarus children 272
Age at time of irradiation and age at time of diagnosis, Russian children 273
Age-specific incidence rates, Belarus, Estonia, Denmark 264
Age-specific incidence rates, Germany, USA-SEER programme 265
in Atomic bomb survivors, risk factors, adults 276
Before and after the Chernobyl accident, Belarus 270, 277
in Children and adolescents, Belarus including Gomel 268–272
in Children, Ukraine 273, 275
Distant metastases 264
as a Function of ^{137}Cs contamination level, Russia 273
see Gomel
Incidence in Ukraine and Belarus 274
Internal exposure, inhalation and ingestion 56
see ^{131}Iodine
^{131}I therapy 266
IPHECA WHO thyroid study 277
in Lithuanian liquidators 267
Mean annual cancer incidence crude rates, Belarus and Hiroshima 262
Mean annual cancer incidence crude rates, Nagasaki 263
Multicentricity 264
Nodal metastases as a prognostic factor 263
Pathology 263, 266
Prediction modelling 267, 276, 277
Radiation induced 77
see Radioactive plume
in Russian liquidators 264, 267
Sasakawa thyroid screening programme 245, 278
Screening 277, 278
Survival 266
Thyroid diseases, non-malignant
Cystic lesion 245
Goitre 77, 78, 245, 324
Hyperthyroidism 245, 325
Hypothyroidism 245, 325
Iodine deficiency, dietary 243
Possible causes for the increase in thyroid diseases 244
Prevalence, Russian liquidators and all Russian population 242
Sasakawa thyroid screening programme 245
Thyroid hormone, free T_4 244, 329
Thyroid nodules 244, 245
Thyroid stimulating hormone (TSH) 244, 330
Thyroid doses in adults 158
Thyroid doses from ^{131}I received by Ukranian population 209, 210
Thyroid ^{131}I uptake measurements 76
Thyrotoxicosis 77
Timofeyeva, Natasha, schoolgirl eyewitness 27
Tiredness, chronic, as a characteristic of acute radiation syndrome 6
Tokaimura accident
Accident description 69, 70

Bone marrow transplantation (BMT) not used 9
see Criticality accident
Dose estimates for the three severely irradiated workers 9
IAEA fact finding mission 70
see Peripheral blood stem cell transplantation (PBSCT)
Population evacuation 69
Workers exposed 69
Totskoye nuclear weapons test site, Urals 160
Towpik, Edward, cancer surgeon eyewitness 28
Toxic waste dump, Love Canal, USA 135
Toy tank and camera used to search for missing fuel 20, 126
Trial of Chernobyl managers and engineers, jail sentences, current work and health status
Bryukhanov, Victor, Director 300
Djatlov, Anatoly, Deputy chief engineer 300
Fomin, Nikolai, Chief engineer and deputy director 300
Kovalenko, Alexander, Chief of the reactor room 301
Laushkin, Yuri, State inspector of Gosatom and Energonadzor 301
Rogoshkin, Boris, Shift chief 300
Translocation 166
Transplantation of human embryonic liver cells (THELC)
see also Bone marrow transplantation (BMT)
Indications for liquidators 92
Causes of death post-THELC 92
Trieste 49
Tritium
Atmospheric fallout from nuclear weapons testing 61
Half-life 63
Worldwide activity release from atmospheric testing 63
Tula 215, 217, 274
Turbine hall, *see* Chernobyl RBMK
Turkey
Measurements of ^{131}I in milk in May 1986 185
Tyumen 68

U

Ukraine
Area (in 1000 km^2) contaminated by ^{137}Cs in the range 10–185 kBq/m^2 213
see Costs
see Environmental contamination
see Evacuation
see Kiev
National Academy of Sciences 220, 221
Population 16
see Water contamination
Ulcers at the site of radiation burns 6
United Kingdom
British Nuclear Fuels (BNFL), Sellafield, quality control records falsified 234
Dose from natural background compared to Chernobyl dose 59

Wild animals, contamination
Boars and grey wolves in Polissya Ecological Reserve 186, 187, Plate VII
Boar, deer, fox, lynx, moose, within the 30 km zone 184, 187
Rabbit 182
Reindeer, grazing on lichen, ^{137}Cs 184
Windscale
British Nuclear Fuels (BNFL), Sellafield, quality control records falsified 234
Nuclear accidents 69, 175
World War II 111, 112, 113
Wormwood xviii, 308

X

X-rays
Degrees of x-ray burn, defined in 1904 85
Discovery by Wilhelm Röntgen in 1895 1, 3, 85
see Medical diagnostic x-rays
Röntgen hands, x-ray physician in 1903 84
X-ray burn in 1898 84
X-ray induced cancer following treatment, ringworm, *beauty treatments* 77
^{133}Xenon
Gaseous release from accident 14, 51, 54
Half-life 54
^{135}Xenon
Half-life 14
Production and decay 14
Xenon poisoning 14

Y

Yanov railway station 105, 220, 224
^{91}Yttrium
Half-life 63
Worldwide activity release from atmospheric testing 63
Yugoslavia
Dose from natural background compared to Chernobyl dose 58
First year effective dose from Chernobyl 176
Measurements of ^{131}I in milk in May 1986 185

Z

Zeleny Mys 94, 110, 111, 211, 220
Zhitomir 215, 274
Zhlobin 198, 204
Zirconium
Zr alloy fuel channels and welds with steel piping, RBMK reactors 40
^{95}Zirconium
Atmospheric fallout from nuclear weapons testing 61
Core inventory 54
Half-life 54, 63
Release from the accident 54
Techa river contamination 65
Worldwide activity release from atmospheric testing and Chernobyl 63
Zones
10 km zone 103
Definition by contamination level 104
see Exclusion zone, 30 km
Strict control zone (SCZ) 103
in Ukraine 104